Animal Welfare

Volume 16

Series Editor

Professor Clive Phillips
Foundation Chair of Animal Welfare
Centre for Animal Welfare and Ethics
School of Veterinary Science
University of Queensland
Gatton 4343, QLD
Australia

More information about this series at
http://www.springer.com/series/5675

Clive J.C. Phillips
Editor

Nutrition and the Welfare of Farm Animals

Springer

Editor
Clive J.C. Phillips
Centre for Animal Welfare and Ethics
University of Queensland
Gatton, Australia

ISSN 1572-7408
Animal Welfare
ISBN 978-3-319-27354-9 ISBN 978-3-319-27356-3 (eBook)
DOI 10.1007/978-3-319-27356-3

Library of Congress Control Number: 2016933444

Printed on acid-free paper

This Springer imprint is published by Springer Nature
The registered company is Springer International Publishing AG Switzerland

This book is dedicated to the memory of David E. Beever, who for many years toiled to improve the feeding of dairy cows and who sadly passed away on 16 June 2014.

Preface

Adequate nutrition is something that most of us take for granted. Yet, millions of farm animals face welfare and production problems because we, either wittingly or unwittingly, expose them to food shortages. Rangeland livestock have to cope with diminished food supplies during drought and flood; chickens and pigs are deprived of adequate food for periods during their production cycle; and dairy cows almost invariably suffer inadequate food supplies in early lactation to meet their very considerable energy requirements. In some cases, inadequate planning for extreme weather or the provision of unsuitable feeds renders farm animal production both inefficient and unproductive, as well as having major adverse effects on the welfare of the animals concerned. In extensive production systems, livestock owners are often inclined to keep as many animals as possible even if those cannot be fed adequately. Better quality and quantity of feed to enable the animals to grow to their potential would improve both their welfare and the profitability of the enterprise.

This document is divided into 11 chapters, each of which focuses on a different aspect of the interaction between nutrition and welfare, with emphasis on the provision of recommendations on how to alleviate nutritional and welfare problems. Welfare is perceived by animals through emotions, and the emotion most closely associated with nutrition, that of hunger, is given detailed consideration, particularly in relation to pigs and poultry fed restricted diets.

Starvation is commonly perceived to be an unacceptable level of malnutrition, in which the animal's function is severely compromised and hence its welfare. It results from an even longer period of feed restriction than that causing hunger. In this book, it is considered particularly in relation to rangeland cattle and sheep, which are frequently exposed to reduced feed levels as a result of climatic variation. Herbivores in extensive systems face temporal and spatial variation in the availability and quality of food and water, the chemical and morphological defences of plants, thermal stress, disease, predation, and competition. Many of the challenges herbivores face in extensive systems derive from the variability and unpredictability of the environment. Nutrient imbalances and natural toxins are also of greater importance than for livestock under intensive management systems.

Ruminant livestock are most often kept on marginal lands, with significant variation in feed supply and vulnerability to climate change. They show remarkable behavioural adaptability, not only for sward composition and height, but also for hot and cold environments, as well as enduring long periods foraging to find suitable pasture. However, under rangeland conditions, poor nutrition is the biggest welfare issue that livestock are confronted with, according to an Australian survey (Phillips et al. 2009). Undernutrition is therefore the biggest welfare problem for large numbers of stock. Advance preparation and low stocking rates will help farmers manage drought on rangelands, but it is anticipated that starvation in ruminant livestock will increase in future as a result of anthropogenic climate change and a shortage of energy crops to provide feed supplements.

The problems associated with short-term food deprivation are also considered in this book, particularly in the stressful conditions of animal transport, which has profound implications for vulnerable classes of livestock, such as calves. Feeding behaviour can be used as a tool to detect or prevent short-term problems. For example, automatic feeders can be used to detect changes in the feeding patterns of individual animals in a group, and these changes in turn can be used to identify health problems, often before clinical symptoms are present. Appropriate feeder design can improve animal welfare by eliminating competition between animals. Competition leads to uneven distribution of the feed between animals in a group, in which the lower ranking individuals lose out.

Provision of roughage is essential for ruminant livestock, but it is increasingly evident that it is also beneficial for monogastrics. Provision of roughage or fibre may in part alleviate hunger and reduce aggression in gestating sows and broiler breeders, may prevent the development of outbreaks of feather pecking in laying hens, and can provide environmental stimulation for growing pigs. Establishing a natural nutritional behavioural pattern in farm animals may be used to improve welfare as well as productivity. For example, post-weaning stress in piglets may be reduced by utilizing the innate curiosity and sensory modalities of this species, to the animals' advantage.

Diseases associated with malnutrition are given careful attention, arising both from specific nutrient deficiencies and from immunosuppression as a result of restricted macronutrient intake. Good nutrition can minimize tissue damage (from nutrient deficiencies or excesses) and ensure an optimal immune system activity. This helps to ensure freedom from disease (infectious or parasitic), but inflammatory responses should also be kept to a minimum by avoiding energy excesses and digestive disorders and by supplying specific compounds (omega-3 PUFA, CLA PUFA, antioxidants, etc.).

Throughout, it is recognized that early detection of welfare problems is a key to successfully controlling their impact, particularly on animal production, using tools such as regular observation of animal behaviour and condition. However, sometimes this is not possible such as when disasters occur, including earthquakes, floods, and wars. We are all too aware of these calamities in relation to their effects on humans, but we rarely consider that farm animals will be also severely affected, and in particular their feed and water supply. The emergency provision of feed to animals in such situations is considered, and a chapter is devoted to provision of water, a key nutrient, to farm animals, since its deprivation, often in the extreme situations referred to above, results in loss of life faster than from any other nutrient deficiency.

This publication derives from an Expert Meeting co-ordinated by Harinder Makkar on the impact of nutrition on animal welfare, which took place in FAO Headquarters in Rome on 26–30 September 2011. The objectives of the meeting were to assess the current state of knowledge of the influence of nutrition of animals on their welfare and to identify feeding options for different livestock production systems that improve animal welfare, while increasing profitability of the livestock producers and ensuring safety and quality through the food chains. The meeting was attended by 11 leading scientists in the field, and a report from the meeting is available at http://www.fao.org/docrep/017/i3148e/i3148e00.pdf. A document containing case studies on enhancing animal welfare and farmer income through strategic animal feeding also emerged from this meeting and is available at http://www.fao.org/docrep/017/i3164e/i3164e00.pdf.

I hope that this book will stimulate researchers, teachers, government officials, international organization officers, donors, extension workers, veterinarians, and farmers to consider the important role that nutrition has to play in the welfare and productivity of farm animals.

Gatton, Australia Clive J.C. Phillips

Reference

Phillips CJC, Wojciechowska J, Meng J, Cross N (2009) Perceptions of the importance of different welfare issues in livestock production in Australia. Animal 3:1152–1166

Animal Welfare Series Preface

Animal welfare is attracting increasing interest worldwide, especially in developed countries where the knowledge and resources are available to, at least potentially, provide better management systems for farm animals, as well as companion, zoo, laboratory, and performance animals. The key requirements for adequate food, water, a suitable environment, companionship, and health are important for animals kept for all of these purposes.

There has been increased attention given to animal welfare in the West in recent years. This derives largely from the fact that the relentless pursuit of financial reward and efficiency, to satisfy market demands, has led to the development of intensive animal management systems that challenge the conscience of many consumers in this part of the world, particularly in the farm and laboratory animal sectors. Livestock are the world's biggest land users (FAO 2002), and the farmed animal population is increasing rapidly to meet the needs of an expanding human population. This results in a tendency to allocate fewer resources to each animal and to value individual animals less, for example in the case of farmed poultry where flocks of over 20,000 birds are not uncommon. In these circumstances, the importance of each individual's welfare is diminished.

In developing countries, human survival is still a daily uncertainty, so that provision for animal welfare has to be balanced against human welfare. Animal welfare is usually a priority only if it supports the output of the animal, be it food, work, clothing, sport, or companionship. However, in many situations, the welfare of animals is synonymous with the welfare of the humans who look after them, because happy, healthy animals will be able to assist humans best in their struggle for survival. In principle, the welfare needs of both humans and animals can be provided for, in both developing and developed countries, if resources are properly husbanded. In reality, the inequitable division of the world's riches creates physical and psychological poverty for humans and animals alike in many parts of the world.

Increased attention to welfare issues is also evident for zoo, companion, laboratory, sport, and wild animals. Of growing importance is the ethical management of breeding programmes, since genetic manipulation is now technically advanced, but there is less public tolerance of the breeding of extreme animals if it comes at the expense of animal welfare. The quest for producing novel genotypes has fascinated breeders for centuries. Dog and cat breeders have produced a variety of deformities that have adverse effects on their welfare, but nowadays the breeders are just as active in the laboratory, where the mouse is genetically manipulated with equally profound effects.

The intimate connection between animals and humans that was once so essential for good animal welfare is rare nowadays, having been superseded by technologically efficient production systems where animals on farms and in laboratories are tended by increasingly few humans in the drive to enhance labour efficiency. With today's busy lifestyles, companion animals too may suffer from reduced contact with humans, although their value in providing companionship, particularly for certain groups such as the elderly, is beginning to be recognized. Animal consumers also rarely have any contact with the animals that are kept for their benefit.

In this estranged, efficiency-driven world, people struggle to find the moral imperatives to determine the level of welfare that they should afford to animals within their charge. A few people, and in particular many companion animal owners, strive for what they believe to be the highest levels of welfare provision, while others, deliberately or through ignorance, keep animals in impoverished conditions in which their health and well-being can be extremely poor. Today's multiple moral codes for animal care and use attempt to satisfy people with a broad range of cultural identities, who are variously influenced by media reports of animal abuse, guidelines on ethical consumption, and campaigning and lobbying groups.

This series has been designed to contribute towards a culture of respect for animals and their welfare by producing learned treatises about the provision for the welfare of the animal species that are managed and cared for by humans. The early species-focused books were not detailed management blueprints; rather they described and considered the major welfare concerns, often with reference to the behaviour of the wild progenitors of the managed animals. Welfare was specifically focused on animals' needs, concentrating on nutrition, behaviour, reproduction, and the physical and social environment. Economic effects of animal welfare provision were also considered where relevant, as were key areas where further research is required.

In this volume, the series again departs from the species focus to address the relationship between animal welfare and nutrition, with a particular focus on farm animals. Good nutrition might seem to be a necessary input for any economic animal production system, as without it animals will not grow, lactate, or reproduce to generate the products we require from them. However, for grazing animals the vagaries of the weather produce variation in the growth of the plants that grazing animals consume, and during disasters such as floods and fire food supply can be seriously disrupted. Intensively managed farm animals are often fed an excessively nutrient-rich diet causing a number of metabolic disorders. This is frequently provided in a manner that minimizes the need for animals to search for food, as they would do naturally, resulting in abnormal behaviours indicative of an unfulfilled need. At other times, such as during transport or in an attempt to prevent animals becoming excessively fat, feed availability is restricted and the ensuing hunger reduces welfare. This book collects together a range of experts in farm animal nutrition and welfare to make a unique focus on a topic that is acknowledged to be important but rarely evaluated.

Gatton, Australia Clive J.C. Phillips

Reference

Food and Agriculture Organization (2002) http://www.fao.org/ag/aga/index_en.htm

Acknowledgements

We are grateful to the participants of the Expert Meeting on 'Impact of animal nutrition on animal welfare', held at FAO headquarters in Rome in September 2011, for their assistance in providing the chapters contained in this document.

Thanks are also due to Daniela Battaglia and Harinder P.S. Makkar who organized and managed the Expert Meeting.

List of Abbreviations

ADG	Average daily gain
AGRP	Agouti-related protein
ANSA	Armed non-state actors
BC	Body condition
BCS	Body condition score
BW	body weight
C_3	Carbon-3 fixation
C_4	Carbon-4 fixation
Ca	Calcium
Cl	Chlorine
CLA	Conjugated linoleic acid
CLW	Critical Live Weight
Co	Cobalt
CPA	Conditioned Place Aversion
CPP	Conditioned Place Preference
CRED	Center for Research on the Epidemiology of Disasters
CT	Condensed tannin(s)
DHA	Docosahexaenoic acid
DM	Dry matter
DMI	Dry-matter intake
DOM:CP	Digestible organic matter to crude protein [ratio]
EFSA	European Food Safety Authority
EPA	Eicosapentaenoic acid
ERC	Emergency Relief Coordinator
EU	European Union
FAO	Food and Agriculture Organization of the United Nations
FAWC	Farm Animal Welfare Council
FCE	Feed conversion efficiency
Fe	Iron
FFA	Free fatty acid(s)

FNS	Food and Nutrition Service [USDA]
FPT	Failure of passive transfer
FSNAU	Food Security and Nutrition Analysis Unit
g	Gram
GIS	Geographical information system
HLR	Heterophil-to-lymphocyte ratio
Hp	Haptoglobin
HPA	Hypothalamopituitary axis
I	Iodine
IASC	Inter-Agency Standing Committee
IDP	Internally displaced person
Ig	Immunoglobulin
IL-1	Interleukin 1
INSARAG	International Search and Rescue Advisory Group
ISPA	International Society for the Protection of Animals
L	Litre
LEGS	Livestock Emergency Guidelines and Standards
LPS	Lipopolysaccharides
ME	Metabolizable energy
Mg	Magnesium
MJ	Megajoule
MOU	Memorandum of Understanding
MP	Metabolizable proteins
mRNA	Messenger ribonucleic acid
Na	Sodium
NDF	Neutral detergent fibre
NEB	Negative energy balance
NE_L	Net energy for lactation
NF	Nuclear factor
NF-kB	Nuclear Factor kappa B
NGO	Non-governmental organization
NH_3	Ammonia
NSP	Non-starch polysaccharide(s)
OIE	World Organisation for Animal Health (formerly Office International des Epizooties)
OM	Organic matter
P	Phosphorus
PCC	Plasma cortisol or corticosterone concentration
PeNDF	Physically effective fibre
PeNDF	Physically effective fibre
PPARα	Peroxisome proliferator-activated receptor α
PSC	Plant secondary compound(s)
PSM	Plant secondary metabolite(s)
PUFA	Polyunsaturated fatty acid

ROM	Reactive oxygen metabolites
SAR	Search and rescue
SARA	Subacute ruminal acidosis
Se	Selenium
SRW	Standard Reference Weight
T_h1	Type 1 helper T cell(s)
T_h2	Type 2 helper T cell(s)
THI	Temperature-humidity index
TMR	Total mixed ration
TNF-α	Tumour necrosis factor alpha
UDAW	Universal Declaration on Animal Welfare
UN	United Nations
UNHCR	UN High Commissioner for Refugees
UNICEF	United Nations Children's Fund
USAID	US Agency for International Development
USDA	United States Department of Agriculture
VFA	Volatile fatty acid(s)
WHO	World Health Organization
WSPA	World Society for the Protection of Animals
Zn	Zinc

Contents

About the Authors

Alex Bach is an ICREA (Institució Catalana de Recerca Estudis Avancats) Research Professor and Director of the Department of Ruminant Production of IRTA (Institut de Recerca i Tecnologia Agroalimentàries) devoted to the study of ruminant production systems (nutrition, management, and development). He conducts research on ruminant nutrition and metabolism and dairy cow and replacement management. He has received several awards in recognition of his research activity, has spoken at more than 90 international congresses, and is author or co-author of more than 100 peer-review publications, more than 90 extension articles, and 10 books or book chapters. He is section editor and sits in the editorial boards of several scientific journals, is member of scientific committees, and has served as a scientific expert for the European Food Safety Authority.

David Beever was Emeritus Professor of Animal Production at The University of Reading, UK, and part-time Nutrition Director and Scientific Advisor to Richard Keenan & Co since 2004. He had previously spent 22 years at the Grassland Research Institute, Hurley, before becoming Professor of Animal Production at The University of Reading and Director of the Centre for Dairy Research (CEDAR) in 1992. His research focused on the physiology and nutrition of ruminant livestock, principally dairy cows and beef cattle, and with colleagues he published over 500 peer-reviewed papers, conference presentations, and scientific abstracts. More recent research interests included the importance of physical nutrition to optimize rumen function and the efficiency of conversion of feed into milk or meat.

Giuseppe Bertoni is a former Director of Istituto di Zootecnica, Faculty of Agriculture, and teacher of Animal Physiology. The main topics of his research have been the effects of ruminant nutrition on physiological response, health (inflammatory response), and milk quality. He was also involved in the development of a model for the objective evaluation of cow welfare in dairy farms. He is now the coordinator of a project for secure, safe, and sustainable food production in countries at different development levels, namely Italy, India, and Democratic Republic of Congo.

Dominique Blache holds a PhD in Neuroscience from the University of Bordeaux (France) and has been working at the University of Western Australia since 1993. For over a decade, Associate Professor Blache has investigated the mechanisms linking nutrition to reproductive status in sheep, goat, emus, and alpacas. In 2000, he developed the original Animal Ethics and Welfare unit at UWA. His research in the field of animal welfare was initially concentrated on the biological mechanisms that underpin temperament and the consequences of selection for temperament on production, physiology, and behaviour. He has expanded his research interests to other stress-affected aspects of the production process: reproduction, growth rate, and immune function. He has contributed to over 150 peer-reviewed publications and several books in the field of farm animal reproduction, including *Animal Welfare*. Ed. M. Appleby, J. Mench, A. Olsson and B. Hughes. 2nd ed. CABI, 2011.

Luigi Calamari is a researcher at the Istituto di Zootecnica, Faculty of Agriculture, Catholic University of the Sacred Heart, Milan, and teacher of Animal Husbandry. The main topics of his research are on the effects of ruminant nutrition, heat stress, and breeding conditions, on metabolic response, milk yield, and composition, in particular on cheesemaking properties. He is also involved in the development of a model for the objective evaluation of cow welfare in dairy herds. He developed and has continuously updated software for diet formulation for cattle, pigs, sheep, and goats, used by researchers as well as by many technicians in the field.

Rick D'Eath As a behavioural scientist, Rick D'Eath studies behaviour as a way of measuring animal welfare and to understand and solve welfare problems that involve behaviour in captive animals, particularly farmed livestock. He studied Biological Sciences at the University of Oxford and stayed on there to do his DPhil on Social recognition of flockmates in laying hens, supervised by Dr. Marian Dawkins. He then worked as a postdoctoral research scientist at SLU in Sweden with Dr. Linda Keeling, before taking up his present position at SRUC. He teaches scientific methodology and molecular genetics of behaviour and supervises dissertation projects for the Edinburgh University MSc Applied Animal Behaviour and Welfare. His research interests include animal personality and temperament, in particular social behaviour, measurement of welfare (developing and validating measurement and sampling methodologies that work on a farm scale), and measurement of hunger and satiety in poultry, pigs, rodents, and humans.

Trevor Devries is associate professor and Canada Research Chair in Dairy Cattle Behaviour and Welfare at the Department of Animal Biosciences, University of Guelph, in Canada. He has a PhD in dairy cattle behaviour and welfare and works on the links between dairy cow behaviour, nutrition, and welfare. His research explores the development and learning of feeding behaviour in dairy cattle, as well as the physiological consequences of that behaviour. He leads a number of research projects on feeding management and behaviour in ruminants, aimed to address practical approaches to improving the housing, management, and feeding of dairy cattle in efforts to improve production, health, and welfare.

Derek Haley is an Assistant Professor of Applied Ethology and Animal Welfare in the Ontario Veterinary College and Associated Faculty with the Campbell Centre for the Study of Animal Welfare, University of Guelph. His responsibilities include teaching veterinary students and graduate students about the behaviour and welfare of farmed animals. He directs a research programme specializing in the behaviour and welfare of beef and dairy cattle. Prior to his academic career, Derek was the Manager of Livestock Welfare for Agriculture and Rural Development, Government of Alberta, Canada.

James Hogan graduated with BSc Agriculture (Sydney) and PhD (Aberdeen). He worked as a Research Scientist in the Commonwealth Scientific and Industrial Research Organization, Australia, on quantitative aspects of the digestion of temperate and tropical pastures in the stomach and intestines of sheep and cattle. The concepts developed have been incorporated into publications regarding feeding standards for ruminants in Britain and Australia. He has applied this knowledge in training courses and study groups with the Animal Health and Production Section of the International Atomic Energy Agency. He has contributed to the estimation of methane emissions by Australian livestock and in recent years has reviewed several aspects of the nutritional welfare of ruminants in collaboration with Professor Clive Phillips of the University of Queensland.

Jos Houdijk graduated in Animal Sciences (Wageningen University, The Netherlands) where he also did his PhD in Animal Nutrition (1998). Since then, he has worked at SRUC (Scotland's Rural College), Edinburgh, where he is now a Senior Research Scientist and Reader in Animal Nutrition and Health, in its Disease Systems Team. Dr. Houdijk has a research interest in animal nutrition per se, as well as nutritional sensitivity of health, disease, and environment, and works with both farm animals and laboratory rodents to address both strategic and applied research questions.

Ingrid De Jong is a senior scientific researcher in behaviour and welfare of poultry at the Animal Welfare group of Wageningen UR Livestock Research, The Netherlands. She has a PhD in animal physiology and behaviour, and the focus of her research is behaviour and welfare of broilers, broiler breeders, and laying hens. She has carried out several studies on the welfare aspects of feed restriction in broiler breeders. She coordinates several research projects on topics such as ontogeny of behaviour, transgenerational effects of stress, welfare and health of broilers in commercial systems, and welfare of broiler breeders.

Marina (Nina) A. G. von Keyserlingk (BSc, MSc, PhD, Professor) is a Natural Sciences and Engineering Research Council Industrial Research Chair holder in Animal Welfare at The University of British Columbia (UBC) and is recognized internationally for her research on care and housing for cattle. She completed her undergraduate in Agricultural Sciences at UBC, her MSc at the University of Alberta, and PhD in Animal Sciences at the University of British Columbia. Following completion of her PhD, she worked in agribusiness for 6 years before returning to academia as a postdoctoral fellow in animal behaviour and welfare. She was appointed as an Assistant Professor in the Animal Welfare Program in 2002, promoted to Associate Professor in 2006, and achieved the rank of Professor in 2010. Dr. von Keyserlingk is an astonishingly productive scientist having published over 120 refereed research publications during her career as a scientist. She also recently co-authored the definitive scholarly text—The Welfare of Cattle—in her core area of research (Springer Verlag, 2008). Dr. von Keyserlingk is the 2012 recipient of the Canadian Animal Industries Award in Extension and Public Service in recognition of her outstanding service to the animal industries of Canada in technology transfer, leadership, and education in animal production. She is also the 2013 Recipient of the Elanco Award for Excellence in Dairy Science in recognition of her outstanding research in dairy science and 2013 recipient of The Metacam 20 Award for her outstanding service in the area of cattle welfare.

Shane Maloney is a comparative physiologist in the School of Anatomy, Physiology and Human Biology at the University of Western Australia. His research focuses mainly on the thermal physiology of mammals and birds, with a particular interest in the responses of production mammals to extremes of temperature. He has contributed to more than 100 peer-reviewed publications and several book chapters. He contributed to M. Appleby, J. Mench, A. Olsson and B. Hughes (editors). *Animal Welfare*. 2nd ed. CABI, 2011.

Xavier Manteca is Professor of Animal Behaviour and Animal Welfare at the School of Veterinary Science in the Autonomous University of Barcelona, Spain. His main areas of interest are the productive implications of social and feeding behaviour and the assessment of animal welfare, particularly in pigs and ruminants. Xavier has an MSc degree in Applied Animal Behaviour and Animal Welfare from the University of Edinburgh and a PhD from the Autonomous University of Barcelona.

Birte L. Nielsen is director of research in animal behaviour at the National Institute of Agricultural Research (INRA) in France. She comes from a previous position as head of the research unit for Behaviour and Stress Biology at Aarhus University, Denmark. Over the last 20 years, Dr. Nielsen has worked mainly on research in basic and applied ethology. She has in-depth experience in experimental behaviour science on growing pigs, dairy cows, broiler chickens, and broiler breeders and has been actively involved in work on horses, sows, laying hens, and ostriches. She has authored and co-authored a number of papers on feeding behaviour. From 2005 to 2010, she was chairperson of the Danish Animal Welfare Council and a member of the National Committee for Agricultural Animals under the Danish Ministry of Justice, advising on the implementation of international animal welfare legislation and recommendations into Danish law. She forms part of the editorial board for the journal '*Animal*' in the section for Behaviour, Health and Welfare. Dr. Nielsen has published extensively in international scientific journals and has contributed to book chapters and numerous conference abstracts, as well as publications in the popular and agricultural press.

John F. Patience is a professor in the Department of Animal Sciences at Iowa State University, having joined the faculty in 2008. Prior to this, he was President and CEO of the Prairie Swine Centre, a non-profit research and technology transfer company located in Saskatoon, Canada. He received his PhD from Cornell University and BSc (Agr) and MS degrees at the University of Guelph in animal science and animal nutrition, respectively. John is responsible for research in applied swine nutrition and, as such, undertakes investigations on energy metabolism, ingredient evaluation, and general wean-to-finish feeding and management. He also has responsibilities in extension, providing service to pork producers across Iowa as well as the USA generally. John has published over 80 refereed journal articles, 160 refereed abstracts, 140 conference proceedings, and almost 300 other miscellaneous articles. He has presented more than 300 invited talks at industry and academic meetings in North and South America, Europe, and Asia. In 2010, he was appointed a Fellow of the CSAS. He has held formal office in animal science societies in North America.

Clive J.C. Phillips studied agriculture at Reading University and obtained a PhD in dairy cow nutrition and behaviour from the University of Glasgow. After lecturing at the Universities of Wales and Cambridge in farm animal production and medicine, he was appointed in 2003 to the Foundation Chair in Animal Welfare at the University of Queensland, where he is now Director of the Centre for Animal Welfare and Ethics. His research interests include the welfare of farm, zoo, and companion animals, with a particular focus on behaviour and nutrition, the ethics of animal use, and attitudes to animals. Recent books include *Principles of Cattle Production* (2nd ed., CABI), *The Welfare of Animals – the Silent Majority* (Springer), and *The Animal Trade* (CABI).

Larry W. Roeder Jr. has an MS in Information Science. He held several positions in government and the private sector, including UN Affairs Director for the World Society for the Protection of Animals, Executive Director for the Global Disaster Information Network, Editor in Chief for Climate Caucus, and Policy Adviser on Disaster Management in the US Department of State. He has also made the protection of all animals an avocation from childhood, when he cared for camels, water buffalo, and horses in Egypt. He subsequently worked with Bedouin in the Sinai and Navajo and Pueblo Indians in the American South West to protect livestock from disasters, and he now volunteers as a wildlife rehabilitator. He also recently advised the President of Somaliland on developing a national veterinary service. In addition, he has published articles and textbooks on animal welfare and humanitarian issues.

Athanasiadou Spiridoula is a veterinary surgeon qualified from Thessaloniki, Greece; she completed her PhD at the University of Edinburgh on secondary plant compounds and their role in controlling parasites in livestock. She has since developed an interest in the molecular interactions between nutrition and immunity, which she explores in rodent models and livestock. Her areas of research include host × parasite interactions and how these are affected by the immune-modulatory effects of the host's nutrition; the use of molecular markers for detection of disease susceptibility and nutritional imbalance; the use of non-forage crops in animal nutrition; to minimize drug input into the food chain and identify novel antimicrobial and antiparasitic agents for sustainable disease control; and investigation and implementation of sustainable control of parasitic and microbial endemic diseases in livestock.

Bert J. Tolkamp studied Animal Science at Wageningen University, The Netherlands, and spent a year as special trainee (ruminant nutrition and extensive beef production systems) at the International Centre for Tropical Agriculture (CIAT, Cali, Colombia). After obtaining an MSc in Tropical Animal Production (1977), he was employed for some years as a lecturer at Wageningen University and subsequently as an FAO associate expert at a dairy training centre in Namaacha, Mozambique. Upon return to Wageningen University, he had various teaching and research contracts until 1994 and completed during this period a PhD thesis (with Jan Ketelaars) 'Towards a new theory of feed intake regulation in ruminants' on the basis of research with goats and sheep. From 1995, he has worked on a range of research projects (including rodents, poultry, pigs, sheep, and cattle) and teaching activities at the Scottish Agricultural College (now: SRUC) in Edinburgh, Scotland. Before retirement, he was a Reader in Animal Science with special interests in (modelling of) food intake, diet selection, feeding behaviour, and nutrition in relation to animal health and welfare.

Erminio Trevisi is Associate Professor at the Istituto di Zootecnica, Faculty of Agriculture, Food and Environmental Sciences. His first degree was in Agricultural Science and his PhD was in Molecular Biotechnology. Currently, he teaches Animal physiology and physiopathology of nutrition and Valorization of Animal Origin Food. Research activities have mainly addressed the role of nutrition and the environment on physiopathological responses of the periparturient dairy cow, the assessment of animal welfare on dairy farms, the interaction between nutrition and gut (gene expression, inflammatory response, and microbiota), and milk quality. He has been involved in several scientific projects funded by Italian Ministries and the European Community, including RuminOmics and Fecund.

Phil Vercoe is Programme Leader of Animal Production Systems in the Institute of Agriculture at the University of Western Australia, where there is currently an emphasis on developing strategies to mitigate methane production from ruminants.

He has a PhD in the molecular biology of ruminal microbes and over 20 years' experience in animal production systems. His research interest is in the role that ruminal microbial ecology plays in developing 'clean, green, and ethical' ruminant production systems, with a specific focus on plant secondary compounds and their potential to manipulate fermentation in the rumen to reduce methane emissions. The overarching objective of his research programme is to encourage greater diversity in grazing systems by changing how we value native plants to develop grazing systems that are productive and profitable with less environmental impact. He has published a number of international scientific publications and book chapters and edited three books in this field.

Juan J. Villalba is an Associate Professor in the S.J. & Jessie E. Quinney College of Natural Resources at Utah State University, Logan, Utah, USA. He also serves as Research Coordinator of the programme BEHAVE, a worldwide network of scientists, producers, land managers, and extension personnel committed to integrating behavioral principles and processes with local knowledge to enhance ecological, economic, and social values of rural and urban communities and landscapes. He has a PhD in Range Science and his research focuses on understanding the mechanisms that influence foraging behaviour in herbivores, with the aim of creating efficient alternatives to manage animals and their environment. He has written more than 80 peer-reviewed papers on these subjects and has been an Associate Editor of leading journals in the field of animal and range sciences. He has also been asked to write several synthesis papers for peer-reviewed journals and books and has been invited to speak at national and international symposia.

Tina M. Widowski earned Masters and PhD degrees in Animal Sciences at the University of Illinois-Urbana, where she specialized in the behaviour and physiology of farm animals. She is Professor of Applied Animal Behaviour & Welfare in the Department of Animal Biosciences, Director of the Campbell Centre for the Study of Animal Welfare, and Egg Farmers of Canada Research Chair in Poultry Welfare at the University of Guelph, Guelph, Ontario, Canada. Her research currently focuses on assessing and improving the welfare of poultry and pig.

Chapter 1
Introduction to Welfare and Nutrition

C.J.C. Phillips

Abstract Animal welfare can be defined in a number of ways, and the use of each definition should be related to the context. In relation to nutrition, cognitive responses are mostly poorly understood; therefore, a concept of animal welfare revolving just around an animal's sentience is not of much practical benefit. Linking animal welfare to responses to the environment is more useful, but this ignores the impact of congenital diseases and other internal problems that the animal may face. This chapter recommends that an animal's nutritional experiences in life, and the benefit or harm that they bring to the animal, offer the best concept of welfare that is both practically relevant and meaningful to the animal. Assessment of welfare can range from simple environmental measurements, which may be questionable in their relation to the animal's state, to complex models that attempt to organize individual influences into a coherent, broad-ranging scheme for practical use. People's perception of animal welfare is driven by a number of different factors, which may lead to views that are at variance with the scientific literature. Gender is one of the most important, with women displaying greater empathy than men unless they are in a subservient role to a partner's views. Pet ownership in childhood is another, which increases empathy towards animals in a fundamental way. Cultural drivers have also been identified. It is concluded that animal welfare is a complex concept with many influences, which is supported by a variety of perspectives on definition and assessment. A thorough investigation of key concepts is necessary before this term can be reliably used in animal science, humanitarian, legal, cultural, or public contexts.

1.1 Introduction

Animal welfare is under greater scrutiny than ever before. Nutrition is acknowledged to be one of the most important drivers of the welfare of livestock, particularly in a rangeland context (Phillips et al. 2009; Phillips and Phillips 2010, 2012).

C.J.C. Phillips (✉)
Centre for Animal Welfare and Ethics, School of Veterinary Science, University of Queensland, Gatton, Australia
e-mail: c.phillips@uq.edu.au

C.J.C. Phillips (ed.), *Nutrition and the Welfare of Farm Animals*, Animal Welfare 16, DOI 10.1007/978-3-319-27356-3_1

This chapter introduces the major concepts of animal welfare definition and assessment, with particular reference to nutrition, as well as outlining the key drivers for the public attitude towards animal welfare.

1.2 Definition and Assessment of Animal Welfare in a Nutritional Context

The definition of animal welfare that we choose must relate to the purpose for which we use the term. Lawyers may need a more precise definition than the general public, who need a definition in terms that can be easily understood. Scientists require a definition in terms that can be measured, and animal managers or keepers require one that is practically relevant and can be easily applied to their animals.

The word 'welfare' comes from an old Norse word, *velferth*, meaning 'good travel'. Across the world, two main concepts arise in describing animal welfare: that of wellness of animals at the current moment in time, often referred to as well-being (or being in a good state: *bienestar* in Spanish, *bien-être* in French, and *bemestar* in Portuguese), and wellness over a more prolonged period (*wohlfahrt* in German, for example).

In relation to nutrition, it is evident that nutrients are one of the most fundamental inputs in life processes. Unlike when animals are in an uncomfortable environment or experiencing a negative emotion, animals can simply not survive without appropriate nutrients. Broom (1986) provided one of the first and most widely used definitions, namely, that animal welfare is 'the state of an animal with regard to its ability to cope with its environment'. Whilst we may prefer to aim to provide an environment that the animals can take pleasure in, rather than just 'cope', and there may also be elements of the animal's internal state that affect their welfare that are unrelated to their environment, such as congenital diseases (Phillips 2009). Nevertheless, this definition has been adopted by many scientists, philosophers, and governments worldwide.

The main alternative definition considers animal welfare to relate primarily to animals' feelings (Duncan and Fraser 1997). Although feelings are undoubtedly one measure, if indeed they can be measured, of animal welfare, Broom (1992) proposes that there are other important indicators such as longevity, production, and reproduction rates, as well as physiological, immunological, and behavioural measures. The 'feelings' approach denies that non-sentient animals have welfare status, whereas the public at least believe that these animals are worthy of moral consideration (McGrath et al. 2013). In all likelihood, sentience is not something that is or is not possessed by animals; there are gradations of sentient ability, but these are poorly understood. Feelings are transient and relate primarily to short-term considerations of animal well-being and not long-term welfare assessment. For example, an animal may select food items that are desirable in the short term, especially if they are carbohydrate rich, because they provide a pleasurable feeling and adaptive benefit in the short term (Matsuno and Thibault 1995), but in the long term, exclusive selection of these items may reduce welfare by causing disease, such as dental decay or obesity. Some events that damage an animal's welfare in the long term, such as removal of a limb under anaesthetics, will not result in any adverse feelings at the time. However, it is evident that a long-term summation of feelings could ultimately give, in an ideal world, a useful measure of just one component of welfare.

We can overcome the problem of uncertainty about animals' feelings by relating welfare to animals' experiences, where these are defined as events participated in or lived through (Phillips 2009). For example, when an animal approaches a food item, it senses the presence of the food through sight and touch, and we can understand these processes and even measure them, but we have little comprehension of what higher level cognitive processes, such as pleasure, excitement, etc., occur in response to the food. Whether such experiences contribute positively or negatively to animal welfare, i.e. contribute benefit or harm, can be determined experimentally by offering animals choices, by monitoring their responses to specific imposed situations, or by soliciting expert or even public opinion. The approach will vary with the situation. Thus, it appears preferable to consider welfare, at least for any legal or scientific purpose, using only our knowledge of situations or processes that are known or can be demonstrated to affect welfare. If an animal's processing of the events must be considered, it is prudent to only consider the awareness that occurs at a simple level, principally phenomenal awareness that uses seeing, hearing, touch, etc., rather than attempting to estimate more complex forms of awareness, such as executive awareness that recognizes goals and intentions (Young 1994).

Evaluation of animals' welfare over a long period is preferable and accords with how we often evaluate life's processes on our own welfare, or should do. Short-term evaluations can be misleading. For example, giving livestock a choice between concentrate feeds and forage may lead them to exclusively choose the former, which might suggest that this is good for their welfare, but their welfare in the long term will be adversely affected by malfunction of the digestive processes.

Because of the dangers of a short-term, narrow assessment, animal welfare assessment is often broad ranging. A commonly used framework for such an assessment is the Five Freedoms, developed in the United Kingdom by the Brambell Committee and the Farm Animal Welfare Council (FAWC 2009) in response to public concern about animal welfare. All of these, not just Freedom from hunger or thirst, can be related to the impacts of nutrition on animal welfare (Table 1.1). A key Freedom, which addresses the public concern about unnaturalness in farm animal production systems, is the Freedom to perform normal behaviour (both appetitive and consummatory).

Animal welfare assessment is, in many people's view, related to an animal's relationship to the natural world. People have a strong affinity to nature (Kellert and Wilson 1993), perhaps because our own separation from nature can have profoundly negative consequences on our own welfare. A strong relationship to nature would have been of adaptive benefit over the course of our evolution and is therefore seen as an essential component in animal welfare. Scientific evidence

Table 1.1 Five Freedoms for assessment of animal welfare and their relation to nutrition

Freedom	Relation to nutrition
Freedom from hunger or thirst	At least short-term malnutrition can result in the emotions of hunger or thirst in animals. Long-term consequences of malnutrition are uncertain.
Freedom from discomfort	Many consequences of malnutrition are likely to lead to discomfort: bloating, acidosis, ketosis, deficiencies, toxicities, etc. Some diseases causing discomfort have many predisposing factors, of which malnutrition can be one, e.g. lameness.
Freedom from pain, injury, or disease	If left unchecked, malnutrition often ends in a diseased state, usually accompanied by pain. Malnutrition, particularly quantitative restriction, can also cause aggression between animals, leading to injury.
Freedom to express normal behaviour	Diets are often too concentrated in energy and other essential nutrients to firstly ease distribution to the animals and secondly to accelerate production or reproduction rates. Because of the inverse relationship between energy and fibre concentration in forages, concentrated feeds are often deficient in fibre. Livestock that evolved to consume fibrous herbage or forage for a substantial portion of the day display abnormal oral behaviours in response to highly concentrated diets that can be rapidly eaten and digested.
Freedom from fear and distress	Aggression associated with food restriction is likely to produce fear and distress in some animals. Responses to repeated nutritional stress may invoke fear responses.

is, however, lacking for animal benefit, for example, of natural enrichment over artificial enrichment.

A recent theme of animal welfare conceptualization is that the Five Freedoms are unnecessarily focused on avoiding negative consequences of our actions on animals. Promoting positive welfare should be at least as important. Others may speculate that welfare cannot be dichotomized into just positive and negative components, and some behaviour, such as play, may have elements of both.

There have been other useful concepts relating to the definition of animal welfare. One is that we should attend first to the animal's 'needs' (resources that are essential for life to be sustained in the long term) and only when these are met should we attend to their 'desires' (resources that will improve the quality of life, but are not essential). For life preservation, needs such as an adequate supply of food and water are most likely to be limiting, whereas factors such as temperature, spatial, or social stress may limit reproduction or production, but are not necessarily life threatening, at least in the short term, and may be considered to be desires. Another useful concept recognises that animals' choices can indicate their welfare needs, which when associated with good health can provide a framework for welfare evaluation (Dawkins 2008). Introducing an ethical component into welfare conceptualization has produced the notion that animals' lives should be 'worth living' (FAWC 2009). Whilst laudable in its objectivity, the inevitable questions arise of 'who decides' and 'how do you measure it'. A utilitarian type calculation is one obvious method of measurement, but the units and parameters of such measurements are unclear.

As welfare assessment has been discussed in many forums, it has become clear that there is no universal method that will suit all purposes. Scientists are generally agreed that animal assessments are more directly related to welfare than environmental assessments, as indicated by the foregoing discussion on definition of welfare as animal (feeling) based or environment based. However, as with feelings vs. environment, measurements on animals are more time-consuming and problematic, as well as often being less exact. Animals are unpredictable and respond differently to welfare challenges, and a risk-based assessment has been advocated by the European Food Safety Authority (EFSA 2012). When faced with an unsuitable diet, some animals will continue to thrive; others will develop ill health or abnormal behaviour. The risk management approach aims to minimize the chance of animals developing these undesirable outcomes.

1.3 Public Opinion on Animal Welfare

Animal welfare is a concept that can be defined scientifically, according to the parameters mentioned in the previous section; however, people vary in the emphasis that they place on different components. At its most fundamental, this reflects individual philosophies of life, with those with a more utilitarian philosophy having less benign attitudes (Bjerke et al. 1998), that focus on concern for the practical and material value of animals (Kellert 1985). Those who believe in animals' cognitive capabilities are more likely to be more empathetic (Hills 1995). This in part will depend on people's own experiences; for example, those who have experienced grief themselves are more likely to acknowledge that animals can feel grief (McGrath et al. 2013). Empathy for animals can be diminished by animal workers' need to become inured to the animals' suffering for their own mental health or other benefit, as in the case of those who work in intensive farming enterprises (Hills 1995). Similarly, those who grow up in rural areas or on farms generally have less concern for animal welfare than those who grow up in urban environments (Kendell et al. 2006). At the other end of the spectrum, those who are involved in animal protection organizations have greatest concern for animals, and they believe that animals' sentience capacity is closer to humans, compared with those who are not involved (Phillips and Izmirli 2012). They also have more concern for unrelated social issues, which suggests that there is a personality trait that covers caring attitudes generally. Similarly, those who do not eat meat because of a concern for animal suffering have a greater ability to recognize sentience in the different animal species (Izmirli and Phillips 2011).

One of the most important drivers of attitudes to animals is pet ownership in childhood (Ellingsen et al. 2010), which fosters a more empathetic attitude (Paul and Serpell 1993; Bjerke et al. 1998, 2001). Pet ownership also encourages humanistic and moralistic attitudes and a less utilitarian perspective towards animals (Bjerke et al. 1998). Such attitudes towards animals appear to be established early in life, and make people more likely to be concerned about animal welfare (Driscoll 1992) and oppose animal testing (Kruse 1999).

Gender is another important driver of attitudes towards animals. Females generally have greater concern for animal welfare and rights than males, but they are only likely to express this view in households (Peek et al. 1997) and countries (Phillips et al. 2011) in which there is a low level of dependence of women on men. If women are highly dependent on a male head of the household, they are likely to adopt his attitudes (Peek et al. 1997). However, there is other evidence of female dominance in animal welfare concerns. Members of animal rights activist organizations are predominantly female, at least in North America (Einwohner 1999) and Eurasia (Phillips and Izmirli 2012), and more women than men are vegetarians (Neumark-Sztainer et al. 1997; Phillips et al. 2011). Females tend to characterize meat and meat-eating experiences negatively, and red meat-eating in particular is more common among males than females (Beardsworth and Bryman 1999; Kubberød et al. 2002). This may be due to the relational role of women in society, having primary responsibility for nurturing, empathy, and care towards others (Peek et al. 1996). Patriarchal domination, and a tendency for women to try to escape from this by having a stronger relationship to animals in the household, is an alternative reason (Peek et al. 1997). Cultural feminist theory suggests that women make moral judgements based on relations rather than universal standards of right and wrong. Kruse (1999) suggests that males support a primarily Darwinian view of animals, in which the natural world is exploited and controlled, whereas females have primarily a romantic view of animals with greater affection for them and concern for their ethical treatment.

The cultural differences in attitudes to animal welfare appear to relate primarily to regions of the world, rather than nationhood or ethnic group (Phillips et al. 2012). Eastern Mediterranean countries appear to have particularly high levels of concern compared with other Eurasian regions, perhaps because of the favourable climate for crop production and classical culture. Northern European countries historically have greater reliance on meat for food in the colder climate, although there have for several decades been high levels of concern in the United Kingdom, which may relate to the rapid development of intensive animal farming in that country. People in Asian countries typically have less concern than those in European countries, but this is at least partly explained by their lower income level (Phillips et al. 2012).

A large multinational survey found little or no effect of religion across a sample of Eurasian countries, when other factors, such as nationality, had been accounted for (Izmirli and Phillips 2011; Phillips et al. 2012). However, another study found that regular church attendees were less supportive of animal welfare than those who did not attend church (Deemer and Lobao 2011), although within churchgoers those reflecting deeply on religion or belief systems had more concern than those who did not. In North America, Protestants had less concern than those who had no faith or were Roman Catholics (Driscoll 1992), and those in Australia who supported more liberal churches had greater concern than those supporting a conservative church (Bowd and Bowd 1989).

Attitudes towards animal welfare can sometimes influence personal behaviour, such as purchasing habits, although there is sometimes a disconnect between ideas about animals and actions towards them (Braithwaite and Braithwaite 1982). Students who support animal protection organizations are more likely to avoid meat products (Phillips and Izmirli 2012). Vegans have greater concern for animals than vegetarians, who in turn have greater concerns than omnivores. However, veganism and vegetarianism are likely to have little effect by themselves in reducing the overall number of animals raised for food, and reductions in the amount of meat being eaten by omnivores are likely to have a much greater effect. A reduction in meat consumption is most strongly related to concerns about killing animals (Izmirli and Phillips 2011), and many people express great concern about inhumane abattoir practices that are nowadays broadcast widely (Tiplady et al. 2013).

1.4 Conclusions

Identifying a precise understanding of what is meant by the concept of animal welfare is important for effective improvement in practical situations. Any definition should be appropriate to the situation in which the concept is used. The extent of good and bad events that animals experience provides a practical, meaningful theoretical concept of animal welfare. The Five Freedoms can provide a useful framework to assess the long-term influence of nutrition on animal welfare, which includes the impact on hunger and thirst, disease, pain, and behaviour. Finally, public attitude towards animal welfare is important to determine the action taken and is affected by a variety of factors, in particular gender, pet ownership in childhood, and culture.

Acknowledgement The author is grateful to Harinder Makar (FAO) for helpful contributions to this chapter.

References

Beardsworth A, Bryman A (1999) Meat consumption and vegetarianism among young adults in the UK: an empirical study. Br Food J 101(4):289–300

Bjerke TB, Odegardstuen TS, Kaltenborn BP (1998) Attitudes toward animals among Norwegian adolescents. Anthrozoös 11(2):79–86

Bjerke T, Kaltenborn BP, Odegardstuen TS (2001) Animal-related activities and appreciation of animals among children and adolescents. Anthrozoös 14(2):86–94

Bowd AD, Bowd AC (1989) Attitudes toward the treatment of animals: a study of Christian groups in Australia. Anthrozoös 3(1):20–24

Braithwaite J, Braithwaite V (1982) Attitudes toward animal suffering: an exploratory study. Int J Study Anim Probl 3(1):42–49

Broom DM (1986) Indicators of poor welfare. Br Vet J 142(6):524–526

Broom DM (1992) Animal welfare: its scientific measurement and current relevance to animal husbandry in Europe. In: Phillips CJC, Piggins D (eds) Farm animals and the environment. CABI, Wallingford, pp 245–253, 430

Dawkins MS (2008) The science of animal suffering. Ethology 114(10):937–945

Deemer DR, Lobao LM (2011) Public concern with farm-animal welfare: religion, politics, and human disadvantage in the food sector. Rural Sociol 76(2):167–196

Driscoll JW (1992) Attitudes towards animal use. Anthrozoös 5(1):32–39

Duncan IJH, Fraser D (1997) Understanding animal welfare. In: Appleby MC, Hughes BO (eds) Animal welfare. CABI, Wallingford, pp 19–31

EFSA (European Food Safety Authority) (2012) Scientific opinion. Guidance on risk assessment for animal welfare EFSA panel on animal health and welfare (AHAW). EFSA J 10(1):2513

Einwohner RL (1999) Gender, class and social movement outcomes: identity and effectiveness in two animal rights campaigns. Gend Soc 13(1):56–76

Ellingsen K, Zanella AJ, Bjerkås E, Indrebø A (2010) The relationship between empathy, perception of pain and attitudes towards pets among Norwegian dog owners. Anthrozoös 23 (3):231–243

FAWC (Farm Animal Welfare Council) (2009) Farm animal welfare in Great Britain: past, present and future. Farm Animal Welfare Council, London

Hills AM (1995) Empathy and belief in the mental experience of animals. Anthrozoös 8 (3):132–142

Izmirli S, Phillips CJC (2011) The relationship between student consumption of animal products and attitudes to animals in Europe and Asia. Br Food J 113(2–3):436–450

Kellert SR (1985) Public perceptions of predators, particularly the wolf and coyote. Biol Conserv 31(2):167–189

Kellert S, Wilson EO (1993) The biophilia hypothesis. Island Press, Washington, DC

Kendell HA, Lobao LM, Sharp JS (2006) Public concern with animal well-being: place, social structural location, and individual experience. Rural Sociol 71(3):399–428

Kruse CR (1999) Gender, views of nature and support for animal rights. Soc Anim 7(3):179–199

Kubberød E, Ueland Ø, Tronstad Å, Risvik E (2002) Attitudes towards meat and meat eating among adolescents in Norway: a qualitative study. Appetite 38(1):53–62

Matsuno AY, Thibault L (1995) Effect of sucrose and fructose macronutrient diets on feeding behavior of rats. Physiol Behav 58(6):1277–1280

McGrath N, Walker J, Nilsson D, Phillips CJC (2013) Public attitudes towards grief in animals. Anim Welf 22(1):33–47

Neumark-Sztainer D, Story M, Resnick MD, Blum RW (1997) Adolescent vegetarians: a behavioral profile of a school-based population in Minnesota. Arch Pediatr Adolesc Med 151 (8):833–838

Paul ES, Serpell JA (1993) Childhood pet keeping and humane attitudes in young adulthood. Anim Welf 2(4):321–337

Peek CW, Bell NJ, Dunham CC (1996) Gender, gender ideology, and animal rights advocacy. Gend Soc 10(4):464–478

Peek CW, Dunham CC, Dietz BE (1997) Gender, relational role orientation and affinity for animal rights. Sex Roles 37(11–12):905–920

Phillips CJC (2009) The welfare of animals: the silent majority. Springer, Dordrecht, p 220

Phillips CJC, Izmirli S (2012) The effects of levels of support for animal protection organisations on attitudes to the use of animals and other social issues. Anim Welf 21:583–592

Phillips CJC, Phillips AP (2010) Attitudes of Australian sheep farmers to animal welfare. J Int Farm Manage 5(Ed.2):17–43

Phillips CJC, Phillips AP (2012) Animal welfare in Australia: a tour of cattle and sheep farms. Amazon Kindle e-books. http://www.amazon.com/Animal-Welfare-Australia-Cattle-Sheep-ebook/dp/B00AHJNIJO/ref=sr_1_25?s=digital-text&ie=UTF8&qid=1386814321&sr=1-25&keywords=animal+welfare

Phillips CJC, Wojciechowska J, Meng J, Cross N (2009) Perceptions of the importance of different welfare issues in livestock production in Australia. Animal 3:1152–1166

Phillips CJC, Izmirli S, Aldavood SJ, Alonso M, Choe BI, Hanlon A, Handziska A, Illman G, Keeling L, Kennedy M, Lee GH, Lund V, Mejdell C, Pelagic VR, Rehn T (2011) An international comparison of female and male students' attitudes to the use of animals. Animals 1:7–26

Phillips CJC, Izmirli S, Aldavood SJ, Alonso M, Choe BI, Hanlon A, Handziska A, Illman G, Keeling L, Kennedy M, Lee GH, Lund V, Mejdell C, Pelagic VR, Rehn T (2012) Students' attitudes to animal welfare and rights in Europe and Asia. Anim Welf 21(1):87–100

Tiplady CM, Walsh D-AB, Phillips CJC (2013) Public response to media coverage of animal cruelty. J Agric Environ Ethics 26(4):869–885

Young AW (1994) Neuropsychology of awareness. Chapter 8. In: Revonsuo A, Kamppinnen M (eds) Consciousness in philosophy and cognitive neuroscience. Erlbaum, Hillsdale, NJ

Chapter 2
Hunger Associated with Restricted Feeding Systems

Bert J. Tolkamp and Rick B. D'Eath

Abstract Hunger and satiety are normally alternating animal states that determine feeding patterns in animals with free access to food. However, chronic hunger can result if access to food is restricted deliberately over long periods. Such systems of quantitative food restriction are widely applied during rearing and/or the dry period in some animals, such as broiler breeders and sows, to avoid the negative health and welfare consequences of ad libitum feeding in animals destined for breeding. Quantitative restriction is associated with signs of hunger such as increased feeding motivation, activity, and stereotypic behaviour. An alternative approach to restrict energy intake is to provide food of a reduced quality. The benefits of these alternative diets for animal welfares are controversial: some researchers conclude that they result in more normal feeding behaviour, promote satiety, and thus improve animal welfare; others argue that 'metabolic hunger' remains irrespective of how the restriction of energy intake is achieved. These disagreements between researchers are a result of differences in (1) assumptions about what determines feeding behaviour and controls food intake, (2) the assessment of the value of the different methodologies that have been used to measure animal hunger, and (3) the weighting placed on 'naturalness' of behaviour as a determinant of welfare. Some problems associated with commonly-used behavioural and physiological measures of hunger are discussed. The development of welfare-friendly feeding systems for broiler breeders and sows would benefit greatly from a better understanding of the controls of food intake and from better methodologies to measure animal hunger.

2.1 Introduction

In many farming systems (fast-growing pigs or beef bulls, and lactating dairy cows), high intakes of food (i.e. energy and nutrients) are beneficial because these allow the achievement of high performance levels. However, in other farm animals (such as dry cows, sows or ewes, and broiler breeders during rearing and

---

B.J. Tolkamp (✉) • R.B. D'Eath
Animal and Veterinary Sciences Group SRUC, The Roslin Institute Building, Easter Bush, Midlothian EH25 9RG, UK
e-mail: Bert.Tolkamp@sruc.ac.uk; Rick.DEath@sruc.ac.uk

C.J.C. Phillips (ed.), *Nutrition and the Welfare of Farm Animals*, Animal Welfare 16, DOI 10.1007/978-3-319-27356-3_2

lay), as well as in many pets, zoo animals, and even humans, high-energy intakes associated with the ad libitum intake of high-quality diets can have very detrimental effects on health and welfare (Schwitzer and Kaumanns 2001; Gough 2005; Stubbs and Tolkamp 2006; D'Eath et al. 2009). High-energy intake in such animals is associated with extensive lipid deposition and (associated) lowered fertility, ascites, lameness, heart problems, diabetes, metabolic diseases, etc. (Hocking and Whitehead 1990; Aviagen 2001; Mench 2002). The only way to avoid these health and welfare problems is to apply feeding systems that deliberately prevent the sustained daily consumption of large amounts of energy. Animals in several of these systems are assumed to be chronically hungry, which is a considerable welfare concern (D'Eath et al. 2009). In addition, restricted access to food (sometimes complete withdrawal of food for up to 1 or 2 weeks) is still practised in some countries to induce moulting in laying hens (Bell 2003). The practice is considered particularly detrimental for animal welfare (Webster 2003) and is therefore not applied at all in some countries, such as the UK ('In no circumstances may birds be induced to moult by withholding feed and water'; Defra 2002: 11). Alternative methods, frequently relying on the supply of low-quality poultry food (such as alfalfa; McReynolds et al. 2009) or with specific nutrient imbalances (such as excessive zinc; Park et al. 2004), may achieve the same goals and help combat disease (such as *Salmonella* infections), as well as addressing the welfare concern caused by hunger associated with food withdrawal (Aygun and Yetisir 2009; Howard et al. 2012). For that reason, the emphasis here will be on hunger associated with feeding systems of sows and poultry breeders. Questions are discussed relating to normal feeding behaviour, the nature of hunger, possible ways of measuring it, the type of feeding systems that are thought to be associated with hunger, and possible ways of alleviating the welfare problems associated with quantitative food restriction.

2.2 What Is Hunger?

Feeding is amongst the most natural of all animal behaviours as it has a profound effect on the survival of the individual and its reproductive success, i.e. its evolutionary fitness. Normal feeding behaviour of an animal with continuous access to food shows a typical temporal structure of feeding bouts (frequently called meals) separated by nonfeeding intervals (Tolkamp et al. 2011). It is generally assumed that animals end a meal when they are satiated (Metz 1975; Le Magnen 1985; Forbes 1995; Zorrilla et al. 2005). As a result, the probability is very low of animals starting a meal shortly after finishing one. With the passing of time, however, satiety wears off and the gradually increasing feeding motivation or hunger will systematically increase the probability of animals starting a meal. Similarly, an animal starting a meal is assumed to be hungry, whilst the increasing amount of food consumed during the meal will gradually decrease hunger and increase satiation, thereby increasing the probability of animals ending the meal (Tolkamp

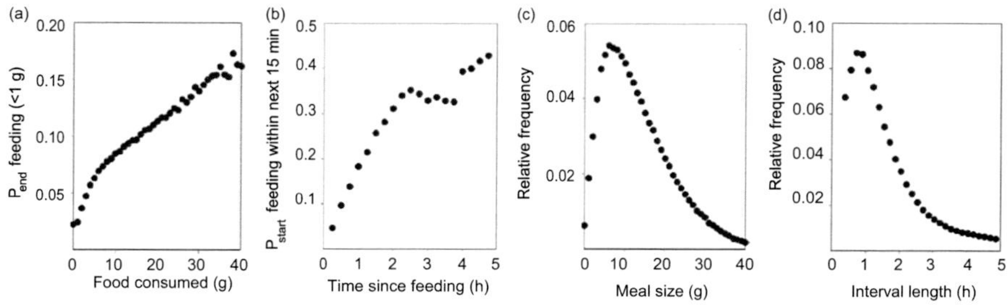

Fig. 2.1 (**a**) Probability of broilers ending a meal before consuming another gram of food (P_{end}) in relation to the amount of food already consumed; (**b**) the probability of broilers starting a meal (P_{start}) in relation to the time since the last meal; (**c**) the frequency distribution of meal size; (**d**) the frequency distribution of the length of between-meal intervals. Re-drawn after data presented by Howie et al. (2009a) and Tolkamp et al. (2011)

et al. 2011). Figure 2.1 illustrates these principles and their consequences for the distribution of meal size and between-meal interval length for broiler chickens. Similar relationships can be found for other species, including pigs, cows, ducks, turkeys, rats, and dolphin calves (Tolkamp et al. 2011). This illustrates the fact that 'hunger' can be considered as a natural motivational state that leads to feeding behaviour, and hunger has, therefore, been seen as one of the 'most basic, primitive, and unremitting of all motivating forces' (Webster 1995). Patterns of normal feeding behaviour as characterized in Fig. 2.1 occur in semi-wild (dolphin calves; Tolkamp et al. 2011) as well as domesticated animals and seem to be virtually unaffected by genetic selection for productive traits in intensively selected farm animals, such as broiler chickens (Howie et al. 2009b; Tolkamp et al. 2011, 2012). These findings contradict the view that animals from genetic lines that have been intensively selected for productive traits are in a constant state of hunger, even when fed ad libitum (Burkhart et al. 1983; Bokkers et al. 2004).

For this reason, it may appear odd that 'freedom from hunger and thirst' features high in the 'Five Freedoms' that form part of most codes of recommendation for animal welfare (FAWC 1992; AWAC 1994). It is obvious that the term 'hunger' can have very different connotations in different contexts (for a brief discussion of thirst, see Kyriazakis and Tolkamp 2011). On the one hand, feeding might never occur unless an animal experiences hunger as a (short-term) motivational state that will result in a search for, and consumption of, food. On the other hand, hunger can also be defined (as in this chapter) as a 'negative subjective state of an animal that is chronically undernourished' (D'Eath et al. 2009). In this definition, the longer term is emphasized because (1) it differentiates clearly from the (appetitive) hunger associated with control of short-term feeding behaviour, and (2) chronic restriction has a larger effect on feeding motivation than acute deprivation (Savory et al. 1993; Ferguson and Paule 1995, 1997; Bokkers et al. 2004). The term 'undernourished' can refer to energy as well as nutrient supply; restricted feeding regimes are, however, almost invariably aimed at restricting energy intake and the term will be interpreted here in that sense. The definition emphasizes animals' subjective states because these are a critical component of animal welfare (Dawkins 1990; Duncan 1993; Fraser and Duncan 1998; Fraser et al. 1997), even though these may not always be easily measured.

2.2.1 Measurement of Hunger in Relation to Quantitative Food Restriction

There are three broad types of indicators of hunger (including the stress associated with it) that have been used to evaluate hunger in relation to quantitative restriction of access to food: (1) (changes in) observed behaviour, (2) physiological indicators, and (3) specific behavioural tests. These have been reviewed extensively by D'Eath et al. (2009) and are briefly summarized and updated here.

Behavioural Indicators of Stress Associated with Hunger The feeding motivation of animals subjected to quantitative food restriction is usually very high. Any food made available is consumed immediately and various behaviours occur after feeding ends that are thought to be a result of unsatisfied feeding motivation (D'Eath et al. 2009). Quantitative restriction generally results in an increase in behaviours relating to foraging (Epling and Pierce 1988; Dewasmes et al. 1989; Koubi et al. 1991; Weed et al. 1997) and in overall activity (indicated by reduced resting; Savory and Lariviere 2000; Hocking 2004). These are generally redirected oral behaviours, such as spot-pecking by boiler breeders (Sandilands et al. 2005, 2006; Nielsen et al. 2011), and chain or bar chewing and rooting in substrate by sows (Appleby and Lawrence 1987). The drinker can also be targeted and excessive drinking or water spillage occurs frequently in feed-restricted animals (Rushen 1984; Sandilands et al. 2005). These redirected oral behaviours can become stereotypic (i.e. repetitive and unvarying whilst not serving an immediately obvious function; Mason and Latham 2004). Stereotypic behaviour generally increases with the level of food restriction in chickens (Savory et al. 1996) and pigs (Brouns et al. 1994; Bergeron et al. 2000). Such stereotypies represent the expression of foraging behaviour, modified by the physical constraints of the environment (Lawrence and Terlouw 1993). In summary, the behavioural effects of quantitative restriction are largely agreed upon by different researchers and there is almost universal agreement that these behavioural effects indicate that quantitative restriction results in hunger (D'Eath et al. 2009).

2.2.2 Physiological Indicators of Stress Associated with Hunger

Some indicators of feeding and stress physiology have been used extensively by researchers that investigated hunger in broiler breeders and sows and the most relevant are briefly reviewed here.

Plasma Cortisol or Corticosterone Concentration Measures of stress physiology such as glucocorticoids [Plasma Cortisol or Corticosterone Concentration (PCC)] and other measures of hypothalamic–pituitary–adrenal axis (HPA) functioning are widely used in animal welfare [reviewed in Mormède et al. (2007)]. PCC is probably the most widely used indicator of stress associated with chronic hunger in broilers and sows (D'Eath et al. 2009). Although PCC may seem an ideal indicator of stress at first sight, there are a number of interpretational problems (D'Eath et al. 2009; see also discussion below). There is, in addition, an issue related to the use of PCC as an indicator of hunger that was raised by D'Eath et al. (2009) but has been analysed only superficially before (Kyriazakis and Tolkamp 2011). In some studies (Hocking et al. 1996; Kubíková et al. 2001; de Jong et al. 2003), PCC is recorded in birds of a given age subjected to different nutritional treatments (frequently ranging from severe restriction to ad libitum). Such studies generally show that PCC at a given age is higher in birds with more restricted access to food than birds with more liberal (or ad libitum) access to food. This has been interpreted as evidence for higher stress levels associated with restricted feeding regimes, at least at severe restriction levels (de Jong et al. 2003). The differential feeding regimes in these studies had, however, also strong effects on the daily amount of energy and nutrients available to the birds for body weight gain and, therefore, on the weight that was attained at the age when PCC was measured (with higher weights always associated with more liberal feed supply). The consequences are that there seems to be a strong relationship between bird body weight and PCC in these studies (see Fig. 2.2). That raises the question whether the variation in PCC in these studies was caused by variation in levels of stress associated with feeding level or (partly at least) by variation in body weight per se. Doubt about the validity of PCC as an indicator of hunger is reinforced by data obtained in studies in which PCC has been recorded repeatedly (i.e. at different ages) in birds subjected to the same feeding regime. Such studies suggest that body

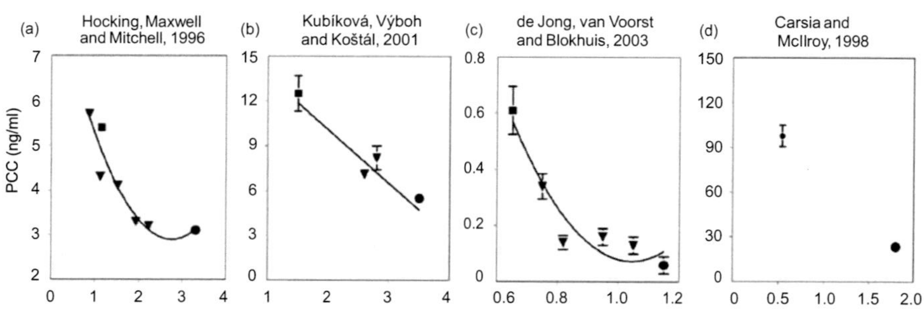

Fig. 2.2 Plasma corticosterone concentration (PCC) against group mean body weight during rearing in broiler breeders aged 7 weeks (**c**), 12 weeks (**a**), or 13 weeks (**b**), and turkeys aged 5 weeks (**d**). Feeding treatments were high-quality food, fed either ad libitum (*large filled circle*) or in restricted amounts, i.e. close to commercial recommendations (*square*), other levels of restriction (*triangles*), or low-protein food fed ad libitum (small *filled circle*). *Drawn lines* are the best fitting linear plus quadratic [as suggested by Hocking et al. (1996)] or linear regression lines of group mean PCC on group mean body weight

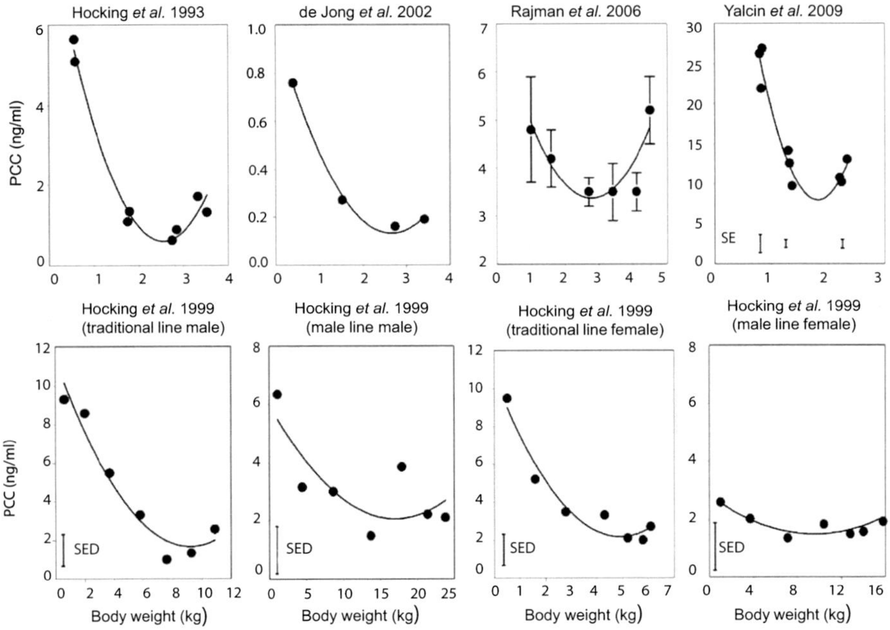

Fig. 2.3 Relationship between body weight and plasma corticosterone concentration (PCC) in broiler breeders (*top panels*) and turkeys (*bottom panels*) with continuous ad libitum access to high-quality food. *Drawn lines* are the best fitting linear plus quadratic regression lines of group mean PCC on group mean body weight [as suggested by Hocking et al. (1996)]

weight in itself has a considerable effect on PCC in ad libitum-fed as well as restricted animals (Kyriazakis and Tolkamp 2011). Figure 2.3 shows examples of the observed relationship between PCC and body weight in five studies in which all birds were fed ad libitum. Most graphs show that, in ad libitum-fed birds, PCC initially declined fast with increasing age and body weight, but in some studies subsequently stabilized (or even slightly increased again) at higher body weights. This decline in PCC is unlikely to be associated with a decrease in hunger with age or body weight because in these studies all birds were fed ad libitum throughout.

Other Physiological Indicators of Hunger-Related Stress Ratios of white blood cells [the frequently analysed heterophil-to-lymphocyte ratio (HLR)] are a common method of animal welfare assessment, including stress related to hunger, especially in broiler breeders. Some of the criticism applied to PCC also applies to this indicator and it is therefore hardly surprising that the results are confusing (D'Eath et al. 2009). Different studies find that HLR increases, stays the same, or decreases in response to different levels of quantitative restriction (D'Eath et al. 2009). A similar confused picture emerged for both poultry and pigs when other immune markers were evaluated, and we concluded that these are not reliable indicators of hunger (D'Eath et al. 2009).

The concentrations of several plasma nutrients, metabolites, and hormones (such as glucose, nonesterified fatty acids, beta-hydroxybutyrate, insulin, etc.) or their ratios have been measured as alternative indicators of chronic hunger in feed-restricted animals (de Jong et al. 2003, 2005; Nielsen et al. 2011). As discussed by D'Eath et al. (2009), the concentrations of such substances usually vary throughout the day (also in response to variation in meal patterns) and are intimately linked with energy supply and/or energy status of the animal, which make them less suitable as indicators of hunger as it is experienced by the animal. More recent attempts to quantify hunger is the measurement of gene expression for peptides such as neuropeptide Y or Agouti-related protein (AGRP), the latter sometimes being referred to as the 'hunger peptide'. Although there is strong evidence that expression of such genes is related to levels of feed restriction (Boswell et al. 1999; Dunn et al. 2012), the evidence that this is related to animals' subjective state is at present not entirely convincing. There are, therefore, in our view, at present no clear physiological indicators of hunger as a result of quantitative feed restriction. This is a serious limitation for those interested in quantifying the seriousness of hunger states associated with various forms of energy intake restrictions as currently practised in animal farming.

2.2.3 Specific Behavioural Tests of Feeding Motivation to Evaluate Hunger

In the absence of unequivocal clear physiological indicators, tests of feeding motivation seem a valuable tool in the assessment of hunger because feeding motivation is expected to reflect directly the animal's own subjective state of hunger (Dawkins 1990). There are a number of types of test that have been used to measure feeding motivation as an indicator of hunger, and below is an updated summary of the detailed review by D'Eath et al. (2009).

2.2.3.1 Feeding Rate and Compensatory Feed Intake

Although increasing levels of food deprivation usually increase short-term feeding rate, there are considerable problems associated with the interpretation of such tests (D'Eath et al. 2009). There is the question of when to test (relative to normal feeding time in the restricted treatments) and the problem that animals subjected to feed restriction are accustomed to feeding very fast, especially when there is competition with conspecifics (Nielsen 1999). Decisions that are made to overcome these problems may make treatment comparisons rather arbitrary (Sandilands et al. 2005), especially when different food qualities are tested (D'Eath et al. 2009).

2.2.3.2 Operant Responding

Operant conditioning for a food reward is an effective way of measuring feeding motivation, as the ratio of cost (lever pressing) to reward (food) can be altered by the experimenter. The method has been validated for quantitative restriction, and increasing restriction generally results in increasing feeding motivation (D'Eath et al. 2009). The method is, however, problematic when comparing treatments with access to different food qualities (see below).

2.2.3.3 Choice Testing and Conditioned Place Preference or Aversion Testing

Recently, attempts have been made to quantify hunger associated with restricted feeding practices with the aid of choice and Conditioned Place Preference (CPP) or Conditioned Place Aversion (CPA) testing. Buckley et al. (2011) used a Y-maze paradigm in which food-restricted broilers could choose between differently coloured arms of a Y-maze to obtain a small or large food reward. The original hypothesis (i.e. the more severe the food restriction and resulting hunger, the sooner the birds would learn the task) was proven false, however, as the most severely restricted birds were least likely to complete the test successfully. The authors concluded that severe food restriction may limit the birds' ability to learn and that, as a result, the test could not quantify the seriousness of the hunger associated with different restriction levels. Other methods such as CPP and CPA can be useful tools in animal welfare assessment because they allow measurement of the reinforcing properties of a stimulus in extinction. Dixon et al. (2011) used CPP/CPA techniques in a series of experiments with broiler breeders to determine the best methodology for establishing CPP/CPA. The experiments showed, however, a number of methodological issues (especially strong preferences for environments birds were not housed in immediately before the test) that restricted the suitability of these tests for the quantitative welfare assessment of chronically food-restricted birds (Dixon et al. 2013).

2.2.4 Possible Ways of Alleviating Hunger Associated with Quantitative Food Restriction

In ruminant production systems, avoidance of excessive gain in condition (or moderate weight loss in over-fat animals) is generally achieved by giving animals ad libitum access to lower quality foods (usually a forage) by which the voluntary energy intake is reduced. A typical example is the provision of dry cows with no or limited amounts of concentrate and ad libitum access to forage such as grazing, silage, or straw. Such systems allow animals to express their normal patterns of feeding behaviour and are not associated with (stress caused by) hunger (Tolkamp et al. 2002).

In response to the concerns over the negative animal welfare consequences of hunger associated with quantitative food restriction, dietary manipulation has also been proposed as an alternative to quantitative restriction for monogastrics like broiler breeders and dry sows (Meunier-Salaün et al. 2001; Mench 2002; Hocking 2004; Ru and Bao 2004; D'Eath et al. 2009). If food is reduced in quality, but offered ad libitum, this is called qualitative restriction (Sandilands et al. 2006). Such diets can restrict intake because ad libitum-fed animals consume less energy from low-quality foods (Brouns et al. 1995; Savory et al. 1996; West and York 1998; Whittemore et al. 2002; Tolkamp et al. 2005; Johnston et al. 2006). Diets where food quality is reduced but food is not offered ad libitum (Zuidhof et al. 1995; Danielsen and Vestergaard 2001; de Jong et al. 2005) can be called 'rationed alternative diets'. Here, we refer to qualitative restriction and rationed alternative diets as 'alternative diets'. Such alternative diets usually involve using foods of lowered quality through the addition of bulky ingredients containing dietary fibre, such as sugar beet pulp (Whittaker et al. 2000; Danielsen and Vestergaard 2001), wheat bran and cobs (Robert et al. 1997), or oat hulls (Sandilands et al. 2005), although other preparations known to suppress appetite have also been used in poultry (phenylpropanolamine, Oyawoye and Krueger 1990; monensin sodium, Savory et al. 1996; calcium proprionate, Savory and Lariviere 2000), sometimes in combination with fibrous ingredients (Sandilands et al. 2005, 2006).

Whether such alternative diets reduce hunger and so enhance animal welfare in comparison with quantitative restriction is controversial, and the reasons behind this controversy are briefly discussed below. Most research involves broiler breeders (the parent stock of meat chickens) that are severely restricted during rearing (25–50 % of ad libitum intake, Savory et al. 1993; Renema et al. 2007) or non-lactating sows, which are restricted during pregnancy (50–60 % of ad libitum intake, Lawrence et al. 1989). Although the problem also clearly applies to companion animals (Butterwick and Hawthorne 1998; Umeda et al. 2006), we focus on these groups for the evidence that is briefly reviewed below.

2.3 Alternative Diets

2.3.1 Effects on Behaviour

The provision of alternative diets, in comparison with quantitative restriction of food intake, generally leads to a number of behavioural changes and there is broad agreement between studies about the type of changes that can be observed (D'Eath et al. 2009). Feeding and foraging behaviour appear more 'natural', i.e. the animal is in control of its feeding behaviour, meals are ended whilst food is still available, and more normal meal patterns can be observed over the day (Savory et al. 1996; Meunier-Salaün et al. 2001; Hocking 2004). However, 'naturalness of behaviour' is considered more important for the assessment of animal welfare by some (Kiley-Worthington 1989; FAWC 1998: 'Freedom to express normal behaviour') than by others (Dawkins 1990; Broom and Johnson 1993; Duncan 1993). In addition, redirected oral behaviour and stereotypies generally decrease considerably (D'Eath et al. 2009). It is generally assumed that (1) quantitative food restriction plays a central role in the development of post-feeding oral stereotypies (Dantzer 1986; Appleby and Lawrence 1987; Lawrence and Terlouw 1993; Mason and Latham 2004), (2) these appear to reflect frustrated motivation (Mason et al. 2007), and (3) are a sign of reduced welfare as they develop under suboptimal environmental conditions (Dantzer 1986; Mason and Latham 2004). A reduction in redirected oral behaviours and stereotypies can then be interpreted as signalling improved animal welfare (Bergeron et al. 2000; Danielsen and Vestergaard 2001; Zonderland et al. 2004; Sandilands et al. 2005). At the same time, however, the duration of feeding activity is usually much longer under qualitative compared with quantitative restriction, and the total duration of oral behaviours is frequently similar under quantitative and qualitative restriction regimes (D'Eath et al. 2009). If oral behaviours are substitutable and functionally equivalent outlets for unsatisfied feeding motivation, a decrease of redirected oral behaviours and stereotypies under qualitative restriction is then not necessarily a sign of improved welfare (Savory and Maros 1993; Dailey and McGlone 1997; McGlone and Fullwood 2001). A reduction in stereotypic oral behaviour in animals subjected to qualitative compared with quantitative restriction is, therefore, considered a clear indication of reduced hunger by some—but certainly not by all—investigators (D'Eath et al. 2009; Kyriazakis and Tolkamp 2011).

2.3.2 Effects on Physiological Indicators

At present, there are no physiological indicators that can be used with confidence to assess the effects of various degrees of quantitative restriction of food intake as discussed above. It is even more problematic to use indicators such as PCC to assess the effects on (the stress caused by) hunger associated with alternative diets in comparison with quantitative food restriction. As discussed by D'Eath et al. (2009), these problems include (1) the considerable diurnal variation in PCC in relation to variation in feeding patterns associated with different feeding regimes; (2) PCC also rises in response to pleasurable or exciting stimuli, reflecting emotional arousal; (3) PCC is not usually a good measure of a chronic stress (such as might result from chronic food restriction), since the HPA axis adapts and PCC levels tend to return to baseline; and (4) the important metabolic role of corticosterone makes it particularly difficult to interpret PCC as a welfare measure to determine levels of hunger in different feeding treatments. Birds subjected to qualitative food restriction that results in similar reductions in growth rate (compared with ad libitum feeding) as quantitative food restriction generally have comparable PCC (Sandilands et al. 2005). But in view of the arguments mentioned earlier, it is not clear at all whether this means that there is no difference in stress between qualitative and quantitative food restriction or that (in our view more likely) PCC is not suitable to measure stress caused by chronic hunger associated with different feeding systems. Similar critical comments can be made for the other proposed physiological indicators (D'Eath et al. 2009; Kyriazakis and Tolkamp 2011).

2.3.3 *Effects as Measured via Specific Tests*

Although operant testing has been successfully applied to quantify hunger associated with different levels of quantitative restriction and to compare effects of different fibre types (Souza da Silva et al. 2012), the application and interpretation is much more problematic when alternative diets are considered in comparison with quantitative restriction. Because animals in these regimes are accustomed to different foods, it is not immediately clear which food should be offered during the test (the same problem applies for the tests of feeding rate and compensatory feeding). For instance, when the reward consists of the usual (treatment) food (Robert et al. 1997; Savory and Lariviere 2000), it is difficult to decide what constitutes 'equivalent responding' (equal weights of food, equal energy per reward, equal access time?) when comparing across treatments (D'Eath et al. 2009). When all treatments are tested with the same food (Lawrence et al. 1989; Ramonet et al. 2000), the reward food is often of high quality. The food reward then represents a greater contrast with their usual food for animals fed alternative diets than for animals subjected to quantitative restriction, which in itself may result in larger responses in such animals (Ramonet et al. 2000; Savory and Lariviere 2000). As discussed by D'Eath et al. (2009), these differences in reward type can explain much of the variation in the findings of different studies using sows and their interpretation. Similar problems apply to the question of the timing of the tests. With alternative diets, operant responding for food after a meal is typically low and stays low for longer before beginning to rise and is high again before the next meal (Robert et al. 1997; Day et al. 1996). In contrast, under quantitative restriction, operant responding for food is similarly high at different tested time points (Lawrence et al. 1989; Ramonet et al. 2000; Savory and Lariviere 2000). This is usually interpreted as indicating that alternative diets result in satiety for at least part of the day (Robert et al. 1997; Savory and Lariviere 2000), although others have interpreted this result more in terms of a constraint, where, for example, gut distension inhibits further feeding, despite a continuing demand for energy (metabolic hunger; Day et al. 1996).

2.3.4 *Perspectives*

At present, the use of qualitative food restriction as an alternative to quantitative restriction to alleviate hunger in (mainly) broiler breeders and sows is controversial. The contrast in policy and practice between these commercially important monogastric species is also confused. EU recommendations ensure that sows are given qualitative restriction or more usually rationed alternative diets in which they have access to fibre in addition to concentrate. This is done on the basis that it will alleviate hunger. In contrast, simple quantitative restriction of broiler breeders is still routine practice throughout the world.

To decide whether or not qualitative restriction can alleviate the welfare problem of hunger in such animals, two important questions need to be addressed. The first relates to the characterization or quantification of levels of animal hunger. Good decisions about the suitability of different feeding systems in relation to animal welfare can be taken only if the consequences of these different systems for the animals' experience in terms of hunger can be measured. For this purpose, modified behavioural tests (such as choice, operant responding, and CPP/CPA) and Qualitative Behavioural Assessment (Wemelsfelder et al. 2001, 2009) and perhaps also (changes in) the expression of specific genes may prove effective. The second question relates to the structure of the animals' foraging programmes, notably their apparent goals. If animals are pure intake rate maximizers subject to constraints, there is perhaps little hope that alternative feeding strategies can be developed that achieve both welfare goals (i.e. avoiding the negative welfare consequences associated with ad libitum feeding of high-quality food and the hunger associated with restrictive feeding practices). However, if animals are optimizers (i.e. trying to achieve the best possible combination of costs and benefits with their behaviour; Illius et al. 2002; Tolkamp et al. 2002; Forbes 2007), efforts can be made to identify what it is exactly that animals respond to and perhaps more appropriate diets can subsequently be developed. Until these issues have been resolved, the search for feeding systems that alleviate hunger and avoid the negative consequences of excessive energy intake at the same time may prove to be elusive.

2.4 Conclusions

Some researchers have concluded that welfare of animals such as dry sows and broiler breeders is improved by alternative diets because foraging behaviour is more normal and characteristics of the food enhance satiety (Zuidhof et al. 1995; Robert et al. 1997; Zonderland et al. 2004; Souza da Silva et al. 2012). This view is reflected in policy: in the EU, there has been a requirement that sows receive 'a sufficient quantity of bulky or high-fibre food as well as high-energy food' since 2001 (EC Council 2001), and this is implemented in the member states through legislation (Netherlands: de Leeuw et al. 2004) or codes of practice (UK: Defra

2003). Other researchers, however, emphasize the conflicting evidence and interpretations of the welfare benefits of alternative diets (Day et al. 1996; Meunier-Salaün et al. 2001; Mench 2002; Hocking 2004; Ru and Bao 2004; de Jong et al. 2005; D'Eath et al. 2009). A third group has argued that there is no convincing evidence of any welfare benefits of such alternative diets because (1) the hunger indicators of total oral behaviours and activity remain similar between quantitative restriction and alternative diets, and (2) that if nutrient requirements and energy needs remain unsatisfied, 'metabolic hunger' will occur (Lawrence et al. 1989; Owen 1992; Savory et al. 1996; Dailey and McGlone 1997; Savory and Lariviere 2000; McGlone and Fullwood 2001). Hence the development of welfare-friendly feeding systems for broiler breeders and sows would benefit greatly from a better understanding of the controls of food intake and from better methodologies to measure animal hunger.

Acknowledgements SRUC receives funding from the Scottish Government.

References

Appleby MC, Lawrence AB (1987) Food restriction as a cause of stereotypic behavior in tethered gilts. Anim Prod 45(1):103–110

Aviagen (2001) Parent stock management manual: ross 308. Aviagen, Midlothian

AWAC [Animal Welfare Advisory Committee] (1994) Annual report. Ministry of Agriculture and Fisheries, Wellington

Aygun A, Yetisir R (2009) Researches on the response of different hybrid layers with respect to egg production performances of forced molting programs with and without feed withdrawal. J Anim Vet Adv 8(12):2680–2686

Bell DD (2003) Historical and current molting practices in the US table egg industry. Poult Sci 82:965–970

Bergeron R, Bolduc J, Ramonet Y, Meunier-Salaun MC, Robert S (2000) Feeding motivation and stereotypies in pregnant sows fed increasing levels of fibre and/or food. Appl Anim Behav Sci 70(1):27–40

Bokkers EAM, Koene P, Rodenburg TB, Zimmerman PH, Spruijt BM (2004) Working for food under conditions of varying motivation in broilers. Anim Behav 68:105–113

Boswell T, Dunn IC, Corr SA (1999) Hypothalamic neuropeptide Y mRNA is increased after feed restriction in growing broilers. Poult Sci 78(8):1203–1207

Broom DM, Johnson KG (1993) Stress and animal welfare. Chapman and Hall, London

Brouns F, Edwards SA, English PR (1994) Effect of dietary fiber and feeding system on activity and oral behavior of group-housed gilts. Appl Anim Behav Sci 39(3–4):215–223

Brouns F, Edwards SA, English PR (1995) Influence of fibrous feed ingredients on voluntary food intake of dry sows. Anim Feed Sci Technol 54(1–4):301–313

Buckley LA, McMillan LM, Sandilands V, Tolkamp BJ, Hocking PM, D'Eath RB (2011) Too hungry to learn? Hungry broiler breeders fail to learn a Y-maze food quantity discrimination task. Anim Welf 20(4):469–481

Burkhart CA, Cherry JA, Vankrey HP, Siegel PB (1983) Genetic selection for growth-rate alters hypothalamic satiety mechanisms in chickens. Behav Genet 13(3):295–300

Butterwick RF, Hawthorne AJ (1998) Advances in dietary management of obesity in dogs and cats. J Nutr 128(12; Suppl):2771S–2775S

Carsia RJ, McIlroy PV (1998) Dietary protein restriction stress in the domestic Turkey (*Meleagris gallopavo*) induces hypofunction and remodeling of adrenal steroidogenic tissue. Gen Comp Endocrinol 109(1):140–153

D'Eath RB, Tolkamp BJ, Kyriazakis I, Lawrence AB (2009) 'Freedom from hunger' and preventing obesity: the animal welfare implications of reducing food quantity or quality. Anim Behav 77(2):275–288

Dailey JW, McGlone JJ (1997) Oral/nasal/facial and other behaviors of sows kept individually outdoors on pasture, soil or indoors in gestation crates. Appl Anim Behav Sci 52(1–2):25–43

Danielsen V, Vestergaard EM (2001) Dietary fibre for pregnant sows: effect on performance and behaviour. Anim Feed Sci Technol 90(1–2; Special Issue):71–80

Dantzer R (1986) Behavioral, physiological and functional aspects of stereotyped behavior: a review and a re-interpretation. J Anim Sci 62(6):1776–1786

Dawkins MS (1990) From an animal's point of view: motivation, fitness and animal welfare. Behav Brain Sci 13(1):1–61

Day JEL, Kyriazakis I, Lawrence AB (1996) The use of a second-order schedule to assess the effect of food bulk on the feeding motivation of growing pigs. Anim Sci 63(3):447–455

de Jong IC, van Voorst S, Ehlhardt DA, Blokhuis HJ (2002) Effects of restricted feeding on physiological stress parameters in growing broiler breeders. Br Poult Sci 43(2):157–168

de Jong IC, van Voorst AS, Blokhuis HJ (2003) Parameters for quantification of hunger in broiler breeders. Physiol Behav 78(4–5):773–783

de Jong IC, Enting H, van Voorst A, Blokhuis HJ (2005) Do low-density diets improve broiler breeder welfare during rearing and laying? Poult Sci 84(2):194–203

de Leeuw JA, Jongbloed AW, Verstegen MWA (2004) Dietary fiber stabilizes blood glucose and insulin levels and reduces physical activity in sows (*Sus scrofa*). J Nutr 134(6):1481–1486

Defra [Department for Environment, Food and Rural Affairs] (2002) Code of recommendations for the welfare of livestock: laying hens. Defra Publications, London

Defra (2003) Code of recommendations for the welfare of livestock: pigs. Defra Publications, London

Dewasmes G, Duchamp C, Minaire Y (1989) Sleep changes in fasting rats. Physiol Behav 46 (2):179–184

Dixon LM, Sandilands V, Bateson M, Brocklehurst S, Tolkamp BJ, D'Eath RB (2011) Using conditioned place aversion techniques as a novel tool for measuring affective states: broiler breeders find unpredictable wind aversive. Br Poult Abstr 7(1):40–41

Dixon LM, Sandilands V, Bateson M, Brocklehurst S, Tolkamp BJ, D'Eath RB (2013) Conditioned place preference or aversion as animal welfare assessment tools: limitations in their application. Appl Anim Behav Sci 148(1–2):164–176

Duncan IJH (1993) Welfare is to do with what animals feel. J Agric Environ Ethics 6(Special Suppl 2):8–14

Dunn IC, Wilson PW, Smulders TV, D'Eath RB, Boswell T (2012) Hypothalamic agouti-related peptide mRNA levels as a potential integrated measure of hunger state in birds. Br Poult Abstr 8(1):20–21

EC Council (2001) Council Directive 2001/88/EC of 23rd October 2001 amending Directive 91/630/EEC laying down minimum standards for the protection of pigs. Off J Eur Communities 316:1–4

Epling WF, Pierce WD (1988) Activity-based anorexia: a biobehavioral perspective. Int J Eat Disord 7(4):475–485

FAWC [Farm Animal Welfare Council] (1992) FAWC updates on the five freedoms. Vet Rec 131:357

FAWC [Farm Animal Welfare Council] (1998) Report on the welfare of broiler breeders. FAWC, London

Ferguson SA, Paule MG (1995) Lack of effect of prefeeding on food-reinforced temporal response differentiation and progressive ratio responding. Behav Process 34(2):153–160

Ferguson SA, Paule MG (1997) Progressive ratio performance varies with body weight in rats. Behav Process 40(2):177–182
Forbes JM (1995) Voluntary food intake and diet selection in farm animals. CABI, Wallingford
Forbes JM (2007) A personal view of how animals control their intake and choice of food: minimal total discomfort. Nutr Res Rev 20(2):132–146
Fraser D, Duncan IJ (1998) 'Pleasures', 'pains' and animal welfare: toward a natural history of affect. Anim Welf 7(4):383–396
Fraser D, Weary DM, Pajor EA, Milligan BN (1997) A scientific conception of animal welfare that reflects ethical concerns. Anim Welf 6(3):187–205
Gough A (2005) Obesity in small animals and longevity in dogs. Vet Times 35:10–11
Hocking PM (2004) Measuring and auditing the welfare of broiler breeders. In: Weeks CA, Butterworth A (eds) Measuring and auditing broiler welfare. CABI, Wallingford, pp 19–35
Hocking PM, Whitehead CC (1990) Relationship between body fatness, ovarian structure and reproduction in mature females from lines of genetically lean or fat broilers given different food allowances. Br Poult Sci 31(2):319–330
Hocking PM, Maxwell MH, Mitchell MA (1993) Welfare assessment of broiler breeder and layer females subjected to food restriction and limited access to water during rearing. Br Poult Sci 34:443–458
Hocking PM, Maxwell MH, Mitchell MA (1996) Relationships between the degree of food restriction and welfare indices in broiler breeder females. Br Poult Sci 37(2):263–278
Hocking PM, Maxwell MH, Mitchell MA (1999) Welfare of food restricted male and female turkeys. Br Poult Sci 40(1):19–29
Howard ZR, O'Bryan CA, Crandall PG, Ricke SC (2012) Salmonella Enteriditis in shell eggs: current issues and prospects for control. Food Res Int 45(2; Special Issue):755–764
Howie JA, Tolkamp BJ, Avendaño S, Kyriazakis I (2009a) A novel flexible method to split feeding behaviour into bouts. Appl Anim Behav Sci 116(2-4):101–109
Howie JA, Tolkamp BJ, Avendaño S, Kyriazakis I (2009b) The structure of feeding behavior in commercial broiler lines selected for different growth rates. Poult Sci 88(6):1143–1150
Illius AW, Tolkamp BJ, Yearsley J (2002) The evolution of the control of food intake. Proc Nutr Soc 61(4):465–472
Johnston SL, Grune T, Bell LM, Murray SJ, Souter DM, Erwin SS, Yearsley JM, Gordon IJ, Illius AW, Kyriazakis I, Speakman JR (2006) Having it all: historical energy intakes do not generate the anticipated trade-offs in fecundity. Proc R Soc Lond B Biol Sci 273(1592):1369–1374
Kiley-Worthington M (1989) Ecological, ethological, and ethically sound environments for animals: toward symbiosis. J Agric Environ Ethics 2(4):323–347
Koubi HE, Robin JP, Dewasmes G, Lemaho Y, Frutoso J, Minaire Y (1991) Fasting-induced rise in locomotor activity in rats coincides with increased protein utilization. Physiol Behav 50 (2):337–343
Kubíková L, Výboh P, Košťál L (2001) Behavioural, endocrine and metabolic effects of food restriction in broiler breeder hens. Acta Vet Brno 70(3):247–257
Kyriazakis I, Tolkamp BJ (2011) Hunger and thirst. In: Appleby MC, Mench JA, Olsson IAS, Hughes BO (eds) Animal welfare. CABI, Wallingford, pp 44–63
Lawrence AB, Terlouw EMC (1993) A review of the behavioural factors involved in the development and continued performance of stereotypic behaviors in pigs. J Anim Sci 71 (10):2815–2825
Lawrence AB, Appleby MC, Illius AW, Macleod HA (1989) Measuring hunger in the pig using operant conditioning: the effect of dietary bulk. Anim Prod 48(1):213–220
Le Magnen J (1985) Hunger. Cambridge University Press, Cambridge
Mason G, Latham N (2004) Can't stop, won't stop: is stereotypy a reliable animal welfare indicator? Anim Welf 13(Suppl):S57–S69
Mason G, Clubb R, Latham N, Vickery S (2007) Why and how should we use environmental enrichment to tackle stereotypic behaviour? Appl Anim Behav Sci 102(3–4):163–188

McGlone JJ, Fullwood SD (2001) Behavior, reproduction, and immunity of crated pregnant gilts: effects of high dietary fiber and rearing environment. J Anim Sci 79(6):1466–1474

McReynolds JL, Genovese KJ, He H, Swaggerty CL, Byrd JA, Ricke SC, Nisbet DJ, Kogut MH (2009) Alfalfa as a nutritive modulator in maintaining the innate immune response during the molting process. J Appl Poult Res 18(3):410–417

Mench JA (2002) Broiler breeders: feed restriction and welfare. Worlds Poult Sci J 58(1):23–29

Metz JHM (1975) Time patterns of feeding and rumination in domestic cattle. Communications of the Agricultural University, No. 75-12. Wageningen, The Netherlands

Meunier-Salaün MC, Edwards SA, Robert S (2001) Effect of dietary fibre on the behaviour and health of the restricted fed sow. Anim Feed Sci Technol 90(1–2; Special Issue):53–69

Mormède P, Andanson S, Auperin B, Beerda B, Guemene D, Malnikvist J, Manteca X, Manteuffel G, Prunet P, Van Reenen CG, Richard S, Veissier I (2007) Exploration of the hypothalamic-pituitary-adrenal function as a tool to evaluate animal welfare. Physiol Behav 92 (3):317–339

Nielsen BL (1999) On the interpretation of feeding behaviour measures and the use of feeding rate as an indicator of social constraint. Appl Anim Behav Sci 63(1):79–91

Nielsen BL, Thodberg K, Malmkvist J, Steenfeldt S (2011) Proportion of insoluble fibre in the diet affects behaviour and hunger in broiler breeders growing at similar rates. Animal 5 (8):1247–1258

Owen JB (1992) Genetic aspects of appetite and feed choice in animals. J Agric Sci 119 (2):151–155

Oyawoye EO, Krueger WF (1990) Potential of chemical regulation of food intake and body weight of broiler breeder chicks. Br Poult Sci 31(4):735–742

Park S, Birkhold SG, Kubena LF, Nisbet DJ, Ricke SC (2004) Effects of high zinc diets using zinc propionate on molt induction, organs, and postmolt egg production and quality in laying hens. Poult Sci 83(1):24–33

Rajman M, Juráni M, Lamošová D, Máčajová M, Sedlačková M, Košťál L, Ježová D, Výboh P (2006) The effects of feed restriction on plasma biochemistry in growing meat type chickens (*Gallus gallus*). Comp Biochem Physiol A Mol Integr Physiol 145(3):363–371

Ramonet Y, Bolduc J, Bergeron R, Robert S, Meunier-Salaun MC (2000) Feeding motivation in pregnant sows: effects of fibrous diets in an operant conditioning procedure. Appl Anim Behav Sci 66(1-2):21–29

Renema RA, Rustad ME, Robinson FE (2007) Implications of changes to commercial broiler and broiler breeder body weight targets over the past 30 years. Worlds Poult Sci J 63(3):457–472

Robert S, Rushen J, Farmer C (1997) Both energy content and bulk of food affect stereotypic behaviour, heart rate and feeding motivation of female pigs. Appl Anim Behav Sci 54 (2–3):161–171

Ru YJ, Bao YM (2004) Feeding dry sows ad libitum with high fibre diets. Asian Aust J Anim Sci 17(2):283–300

Rushen J (1984) Stereotyped behaviour, adjunctive drinking and the feeding periods of tethered sows. Anim Behav 32(4):1059–1067

Sandilands V, Tolkamp BJ, Kyriazakis I (2005) Behaviour of food restricted broilers during rearing and lay—effects of an alternative feeding method. Physiol Behav 85(2):115–123

Sandilands V, Tolkamp BJ, Savory CJ, Kyriazakis I (2006) Behaviour and welfare of broiler breeders fed qualitatively restrictive diets during rearing: are there viable alternatives to quantitative restriction? Appl Anim Behav Sci 96(1–2):53–67

Savory CJ, Lariviere JM (2000) Effects of qualitative and quantitative food restriction treatments on feeding motivational state and general activity level of growing broiler breeders. Appl Anim Behav Sci 69(2):135–147

Savory CJ, Maros K (1993) Influence of degree of food restriction, age and time of day on behavior of broiler breeder chickens. Behav Process 29(3):179–190

Savory CJ, Carlisle A, Maxwell MH, Mitchell MA, Robertson GW (1993) Stress, arousal and opioid peptide-like immunoreactivity in restricted-fed and ad-lib-fed broiler breeder fowls. Comp Biochem Physiol A Physiol 106(3):587–594
Savory CJ, Hocking PM, Mann JS, Maxwell MH (1996) Is broiler breeder welfare improved by using qualitative rather than quantitative food restriction to limit growth rate? Anim Welf 5(2):105–127
Schwitzer C, Kaumanns W (2001) Body weights of ruffed lemurs (*Varecia variegata*) in European zoos with reference to the problem of obesity. Zoo Biol 20(4):261–269
Souza da Silva C, van den Borne JJGC, Gerrits WJJ, Kemp B, Bolhuis JE (2012) Effects of dietary fibers with different physicochemical properties on feeding motivation in adult female pigs. Physiol Behav 107(2):218–230
Stubbs RJ, Tolkamp BJ (2006) Control of energy balance in relation to energy intake and energy expenditure in animals and humans: an ecological perspective. Br J Nutr 95(4):657–675
Tolkamp BJ, Friggens NC, Emmans GC, Kyriazakis I, Oldham JD (2002) Meal patterns of dairy cows consuming diets with a high or a low ratio of concentrate to grass silage. Anim Sci 74 (2):369–382
Tolkamp BJ, Sandilands V, Kyriazakis I (2005) Effects of qualitative food restriction during rearing on the performance of broiler breeders during rearing and lay. Poult Sci 84 (8):1286–1293
Tolkamp BJ, Allcroft DJ, Barrio JP, Bley TAG, Howie JA, Jacobsen TB, Morgan CA, Schweitzer DPN, Wilkinson S, Yeates MP, Kyriazakis I (2011) The temporal structure of feeding behavior. Am J Physiol Regul Integr Comp Physiol 301(2):R378–R393
Tolkamp BJ, Howie JA, Bley TAG, Kyriazakis I (2012) Prandial correlations and the structure of feeding behaviour. Appl Anim Behav Sci 137(1–2):53–65
Umeda TY, Bauer JE, Otsuji K (2006) Weight loss effect of dietary diacylglycerol in obese dogs. J Anim Physiol Anim Nutr 90(5–6):208–215
Webster AJF (1995) Animal welfare—a cool eye towards Eden. Blackwell Science, Oxford
Webster AJF (2003) Physiology and behavior of the hen during induced molt. Poult Sci 82 (6):992–1002
Weed JL, Lane MA, Roth GS, Speer DL, Ingram DK (1997) Activity measures in Rhesus monkeys on long-term calorie restriction. Physiol Behav 62(1):97–103
Wemelsfelder F, Hunter TEA, Mendl MT, Lawrence AB (2001) Assessing the 'whole animal': a free choice profiling approach. Anim Behav 62(2):209–220
Wemelsfelder F, Knierim U, Schulze Westerath H, Lentfer T, Staack M, Sandilands V (2009) Qualitative behaviour assessment. In: Forkman B, Keeling L (eds) Assessment of animal welfare measures for layers and broilers. Welfare Quality Reports, No. 9 (Sixth Framework Programme). Cardiff University, Cardiff, pp 113–119
West DB, York B (1998) Dietary fat, genetic predisposition, and obesity: lessons from animal models. Am J Clin Nutr 67(3; Suppl):505S–512S
Whittaker X, Edwards SA, Spoolder HAM, Corning S, Lawrence AB (2000) The performance of group-housed sows offered a high fibre diet ad libitum. Anim Sci 70(1):85–93
Whittemore EC, Kyriazakis I, Tolkamp BJ, Emmans GC (2002) The short-term feeding behavior of growing pigs fed foods differing in bulk content. Physiol Behav 76(1):131–141
Yalçın S, Bruggeman V, Buyse J, Decuypere E, Cabuk M, Siegel PB (2009) Acclimation to heat during incubation: 4. Blood hormones and metabolites in broilers exposed to daily high temperatures. Poult Sci 88(9):2006–2013
Zonderland JJ, De Leeuw JA, Nolten C, Spoolder HAM (2004) Assessing long-term behavioural effects of feeding motivation in group-housed pregnant sows; what, when and how to observe. Appl Anim Behav Sci 87(1–2):15–30
Zorrilla EP, Inoue K, Fekete EM, Tabarin A, Valdez GR, Koob GF (2005) Measuring meals: structure of prandial food and water intake of rats. Am J Physiol Regul Integr Comp Physiol 288(6):R1450–R1467
Zuidhof MJ, Robinson FE, Feddes JJR, Hardin RT, Wilson JL, McKay RI, Newcombe M (1995) The effects of nutrient dilution on the well-being and performance of female broiler breeders. Poult Sci 74(3):441–456

Chapter 3
Starvation of Ruminant Livestock

J.P. Hogan and C.J.C. Phillips

Abstract Beef cattle and sheep are often raised on marginal lands that provide a highly variable feed supply and are prone to drought and climate change. They evolved a capability of utilizing poor quality feeds and have the potential to survive variation in feed supply through the use of body lipid reserves. Despite this, in regions with marginal land, undernutrition is commonly regarded as the greatest welfare problem for extensively reared cattle and sheep, which can extend to starvation in drought periods. Dairy cows that have been bred for high milk output may also experience starvation in the form of severe macro- or micronutrient deficiencies. Livestock farmers will often only rectify known feed deficiencies if it is economic to do so; hence there can be a risk of starvation when feed costs are high and the value of the animal is low. It is anticipated that the prevalence of starvation, with consequent adverse effects on the welfare of cattle and sheep, is likely to increase with the onset of global human food shortages and increased competition for land resources.

3.1 Introduction

Cattle and sheep were first domesticated approximately 9000 years ago and were unique amongst domesticated livestock in their ability to be kept primarily for the production of both milk and meat for human consumption. They actually produce meat, offal, milk (and associated products), leather, wool, transport, and sporting opportunities. The systems of animal husbandry have been developed over several thousand years and are intended to be sustainable.

The ruminant digestive system includes a large modified forestomach, the rumen, for the processing of coarse fibre and other feeds by microorganisms. It fulfils the purpose of both commencement of digestion and storage of feed. The prolonged digestion process enables ruminants to occupy an ecological niche of consuming fibrous plants, grasses, forbs, and browse material. However, the growth

J.P. Hogan • C.J.C. Phillips (✉)
Centre for Animal Welfare and Ethics, School of Veterinary Science, University of Queensland, Gatton Campus, Gatton 4343, Australia
e-mail: c.phillips@uq.edu.au

C.J.C. Phillips (ed.), *Nutrition and the Welfare of Farm Animals*, Animal Welfare 16, DOI 10.1007/978-3-319-27356-3_3

and hence availability of these food sources are subject to circannual variation. Droughts, floods, and extreme temperatures exacerbate the normal variation in plant growth. Ruminants have developed methods of storing energy reserves as adipose tissue, which can be utilized in times of low feed availability. The storage in hot climates is in concentrated fat depots, in *Bos indicus* cattle and camels there exists one, or in the case of Bactrian camels, two dorsal humps, and in some breeds of sheep fat is deposited in the tail. This enables them to minimize subcutaneous fat and maximize heat loss capacity. In ruminants kept in more temperate regions, bison and *Bos taurus* cattle for example, a greater proportion of fat is stored subcutaneously, which aids in heat conservation. Wherever they are located on the body these fat stores enable ruminant livestock to survive for several months with limited feed supplies.

Most ruminants derive nutrients for survival, growth, and production from the environment in which they live or to which they are moved as part of an animal husbandry system. Experience of the reliability of feed production over many years has led to the optimization of the nature of livestock enterprises in a particular region. Hence, some areas of land are regarded as suitable for dairying, others for breeding lambs or calves to be grown on the breeding property or elsewhere, others for growing or fattening young animals that may or may not be destined for final growth in feedlots, and still others for wool production. Land use for intensive milk production requires a regular and plentiful supply of nutrients; hence, this developed mainly in regions able to produce good quality feed. However, beef cattle, sheep, and dual purpose cattle and sheep are kept in some of the world's least hospitable regions to produce meat and milk, including drought-prone regions and areas with extreme high and low temperatures. The managers, in making decisions on the nature of the enterprise, take into account the risk of failure of the quantity or nutritional quality of the feed supply to match the demands of animals within the enterprise. They attempt to match the time of maximum nutrient demand in late pregnancy and lactation to the anticipated period of maximum nutrient production by adjusting grazing pressure, defined as the rate of dry matter removal from a pasture divided by rate of dry matter accumulation. Thus, the first option to maintain feed supply is to vary stocking rate. However, many of the environmental forces influencing nutrient supply, in particular those concerning climate, are largely uncontrollable, and situations frequently arise in which the supply of feed is not adequate for the animals dependent on it. Pasture production is principally controlled by the adequacy of soil moisture, soil temperature, and sunlight. When animals experience a severe reduction in plane of nutrition the managers must decide whether to sell animals, remove them, or otherwise try to manage the consequences of undernutrition leading, ultimately, if not relieved, to starvation. Starvation may be of rapid or gradual onset, but decisions are usually made to keep the maximum number of animals alive at a minimum cost during a period of feed shortage of unknown duration, which is not usually in the interests of maintaining the welfare of the livestock.

Starvation can occur in numerous situations both on and off commercial properties for the raising of livestock. Apart from plant production failure, two additional reasons are the tethering of animals by the roadside and the failure of livestock to eat under situations of extreme stress. The former mainly involves goats and cattle in developing countries. Animals tethered by the side of the road often have inadequate feed, which encourages them to eat waste material jettisoned from vehicles. Their nutritional challenges may be compounded by stress of isolation and regular proximity of vehicles. The second scenario mainly afflicts animals transported long distances, for example, by ship. In these conditions, sheep especially are prone to cease eating.

Good welfare requires adequate nutrition, which may be defined as the provision to the animal of sources of energy, amino acids, minerals, and trace metabolites appropriate for the physiological requirements of the animal and to support normal growth and development. Inadequate nutrition is scientifically termed malnutrition.

3.1.1 Malnutrition

Malnutrition is a deficit, imbalance, or excess of nutrients with consequential adverse effects on the normal functioning of the animal, including behaviour, physiology, reproduction, health, and growth potential (Hogan and Phillips 2008). It has major potential to affect the welfare of cattle and sheep. Much is known about the physiological and production effects of nutrient deficiencies and toxicities (Underwood and Suttle 1999), and where there are economic benefits in rectifying these problems most farmers will do so. Nutrient deficits can be either general or specific to individual nutrients. Breeding for greater productivity in cattle and sheep has increased their requirements for both micro- and macronutrients, leading to a greater risk of deficiency disorders. For example, the rapid rise to a high milk output at the start of lactation in modern dairy cows leads to a serious risk of calcium deficiency, hypocalcaemia. The associated risk of acidosis, fatty liver, lameness, and mastitis suggests that the limit to physiological homeostasis is being reached in cows bred for very high milk yields. Little is known about the impact of clinical disorders on welfare, although it can be speculated that there are differences from a simple inadequate feed supply, or undernutrition, in that the likelihood of associated subclinical or clinical disease is much greater. The impacts of subclinical disorders on the animal's physiology and production are usually not understood, with little knowledge of their welfare impact.

Apart from mineral deficits, breeding dairy cows to produce in excess of 10,000 L per lactation has exacerbated the energy deficit in early lactation, as nutrient supply has not increased commensurately to satisfy requirements (Kamphues 1998). Increasing the energy concentration in offered diets has associated health problems, such as acidosis, laminitis, acetonaemia, and fatty liver disease (Gearhart et al. 1990). It is currently unclear whether dairy cows in early lactation have negative emotional responses to dietary inadequacies in the form of a persistent unsatisfied hunger (Phillips and Kitwood 2003; Cooper et al. 2010).

Feeding behaviour problems are known to arise from inappropriate feed composition. These include tongue rolling in cattle (and bar biting if they are tethered) (Lindström and Redbo 2000) and wool biting in sheep (Vasseur et al. 2006), which result from feeding a highly concentrated or restricted roughage diet that deprives the animals of adequate oral stimulation. Feed tossing, intersucking, and selfsucking are less common deleterious behaviours associated with inadequate oral stimulation during feeding.

A particular type of malnutrition is undernutrition, 'a prolonged, inadequate supply of nutrients to sustain good health and in the case of immature and underweight animals, growth potential, where prolonged implies that a steady state has been reached' (Agenäs et al. 2006).

3.1.2 Undernutrition

When both existing vegetation and body stores are inadequate, animals experience undernutrition, which is regularly cited as one of the most important welfare problems of rangeland cattle and sheep in Australia (Phillips et al. 2009; Phillips and Phillips 2010). Undernourished animals are presumed to suffer poor welfare, although physiological evidence for this is scarce. In terms of emotional responses, it is likely that undernutrition leads to frustration and exhaustion at least, as well as possible problems arising from associated disease (Hogan and Phillips 2008).

Attempts have been made to characterize the extent of undernutrition using a Standard Reference Weight (SRW: the weight of an animal with the skeleton fully developed and the empty carcass containing 25 % fat) and a Critical Live Weight (CLW: the weight below which an animal cannot walk, graze, or safely obtain drinking water) (CSIRO 2007). SRW corresponds to a body condition score of 3 on a 0–5 scale ranging from emaciated to obese. A score of zero corresponds to the CLW (McGregor 2005), which according to McGregor occurs at about 60–65 % of SRW. However, obtaining accurate animal weights under remote field conditions is difficult and this concept is more likely to be of value in the rare occurrences of undernutrition in intensive farming practices, than in rangeland conditions.

Reducing the welfare impact of climatic variables on ruminant nutrition requires action to be taken sufficiently early, usually in the form of purchasing additional feed (including strategic supplements such as urea or phosphorus), moving stock to areas with better food supplies, or selling stock for slaughter. Much can be done to prepare cattle and sheep stations to withstand drought, such as adopting moderate stocking rates and planting crops such as legumes that are deeper rooting. While increasingly accurate predictions are available about the likelihood of rain, many farmers delay action, e.g. purchasing supplements, because of the high cost, hoping that weather conditions will improve. Legal action by advocacy groups or government to protect undernourished stock is often seen as a last resort, but is logical if farmers are compliant in producing the similar welfare problems for their stock that companion-animal owners would be prosecuted for.

Identification of long-term undernutrition or starvation by body condition scoring is too subjective for legislative use. Ultrasound fat measurements are more objective but are difficult to use under field conditions. Biochemical indicators, such as blood albumin or fructosamine-to-albumin ratio, have potential to be used for this purpose (Agenäs et al. 2006; Strydom et al. 2008). In one study, a blood albumin concentration of <31.5 g/L had a 90 % sensitivity and 98 % specificity in distinguishing beef cattle in low (≤1) body condition score from those in high (>2.5) body condition score (Strydom et al. 2008). Further work is needed to relate biochemically-defined undernutrition to emotions and to define the critical levels of blood albumin and other biochemical indicators. These should take into account breed and concomitant diseases, such as intestinal parasitism, chronic liver disease, and renal disease, which may induce hypoalbuminaemia.

Undernutrition may occur even when animals have food supplied. Sheep that are transported by ship have access to pelleted feed, but a proportion still experience a feed intake restriction or failure to eat, known as inappetence. The aetiology is unclear, but may relate to motion sickness, ammonia accumulation, stress or failure to adapt to novel feeds, or a combination of these. This is the single biggest cause of mortality in sheep transported by ship. Being in fat body condition on entry to the ship approximately doubles the risk of developing inappetence, compared with being thin, suggesting that the fat sheep that develop this condition have insufficient motivation to eat and become totally reliant on body reserves (Higgs et al. 1991). There is a need for studies on the rate of catabolism of lipids and on the rate of production and fate of long chain fatty acids and ketones derived from lipid sources.

The human condition termed 'starvation' may be applied to animals if it is considered to be 'a shortage of nutrients or energy such that the animal starts to metabolize functional tissues rather than food reserves' (Broom and Fraser 2007). However, this definition presents problems. It is difficult to determine, for instance, when the withdrawal of lipids from adipose tissues relates to food reserves or to functional tissues: in addition, in starvation, other tissues are depleted concurrently with adipose tissues, thus rendering a precise point of transition from storage to functional tissue unlikely. It has been proposed (Hogan and Phillips 2008) that starvation, if it refers to an unacceptable level of malnutrition, is restricted to situations in which the functionality of animals is restricted by nutrient supply. An example of this is impaired reproductive function (Buckley et al. 2003), but long-term malnutrition affects non-lactating females and has complex effects on aspects of essential behaviour, such as the length of time devoted to feed gathering. Thus, even with this definition there may be difficulties in determining precise points of transition from adequate nutrition to starvation.

This chapter considers two nutritionally challenging situations commonly found in the husbandry of sheep and cattle. The first is the annual or biennial situation of seasonal changes in the quantity and nutritional value of available pasture or browse, ranging from an excess of nutrients to prolonged undernutrition. The second is the nutritional situation that follows an abrupt loss of access to feed in consequence of disasters such as fire, flood, ice, or snow. The first reflects gradual changes in the nutritional status of the animal, with some nutrients being derived from both the feed and the animal's tissues, and the second the consequences to the animal of suddenly being forced to derive all nutrients from its body reserves.

3.2 Livestock and the Environment

The animal environment, or 'collection of forces that has an influence on the animal' (Purser and Hogan 1993), can be grouped under three headings encompassing Climate, Soil, and Feed Resources (Table 3.1), which interact to influence the supply of nutrients for the animal. The sources of nutrients for domesticated ruminants are principally pasture grasses and legumes, with some contribution from shrubs and trees (Gutteridge and Shelton 1998; Cheema et al. 2011). The growth of different pasture species varies within the different climatic zones, but depends on the climate-driven variables such as light, temperature, and effective rainfall. The last variable is often described as the surplus of precipitation over evaporation but also involves the topography of an area and the capacity of a soil type to absorb and retain moisture. In drought-prone regions, such as the Pilbara region of north-west Australia, this can be limited. Within a macro-environment of semi-arid grassland, for instance, there may be micro-environments that support the growth of plants usually found in wetter regions. A macro-environment often covers a large area of land within which can be seen harsh, endowed, and intermediate regions with wide variations in productivity. The climatic and soil variables determine the varieties of pasture plants suited to a given area. Some are native to a region, but there has been worldwide interest in the introduction from other countries of plant species with superior agronomic and

Table 3.1 The animal's environment (from Purser and Hogan 1993)

Climate	Soil	Feed resource
Light	Topography	Pasture
Temperature	Macro	Type
Humidity	Micro	Quality
Rain	Chemical/physical Structure	Height/density
Wind	Fertility	Maturity
Frost	Hydrology	Browse
Snow		Plant parts
		Nutrient balance
		Acceptability
		Chemical
		Physical
		Supplements
		Water
		Supply
		Quality
		Accessibility

nutritional characteristics. In Australia alone there have been more than 8000 pasture plant introductions (Hogan and Phillips 2011). These include, from Europe, clovers such as Subterranean Clover (*Trifolium subterraneum*), medics such as *Medicago sativa* (lucerne), and grasses like Ryegrass (*Lolium perenne*) in widespread use in the Mediterranean and temperate regions. Plants introduced into tropical areas include grasses such as Rhodes grass (*Chloris gayana*) from Africa and legumes from Central and South America such as Lablab (*Dolichos lablab*) and Stylo (*Stylosanthes* spp.). However, many areas, especially those in pastoral regions, have been degraded by excessive grazing followed by wind and water erosion and invasion by nutritionally inferior introduced species, such as carpet grass (*Axonopus affinis*), couch grass (*Cynodon dactylon*), or Indian couch grass (*Bothriochloa inermis*). As examples of feed sources to ameliorate hunger, the importance of the flowers and leaves of a variety of shrubs and trees as sources of nutrients in both temperate and tropical regions has also become apparent (Wilson 1977; Lowry et al. 1994). In the 1950s and 1960s, FAO had significant involvement in monitoring and controlling plant exploration in remote areas and the dissemination of the plants to other parts of the globe (Whyte 1958).

On a relatively small proportion of the world's surface, the climate-dependent variables permit plant growth at all times of year, but for most areas plant growth ceases temporarily each year. In such situations, the existing plants mature and decline in nutritional value. The length of time during which pasture growth ceases imposes constraints on the management of animal production enterprises. In regions where some plant growth is possible during most of the year, the manager aims to keep a suitable balance between grasses and legumes and adjusts grazing pressure to keep pasture plants in the vegetative state. By contrast, in other regions, it is essential to have sufficient mature or dead pasture available to provide nutrients for livestock when pasture production ceases. This requires a much lower grazing pressure and a proportion of plants reach maturity and set seed before death.

The cessation of pasture growth can be rapid or gradual. Rapid change, which can occur for clover–grass pastures during hot weather, is regularly seen in regions with a Mediterranean climate. However, in other cases growth usually declines more slowly as the reserves of water or nutrients such as nitrate in the soil are exhausted. Many grasses are more shallow-rooted than associated legumes, so that when soil moisture is being depleted grasses tend to be affected by moisture stress before legumes. However, if soil moisture deficit is too great, all pasture material available to the animal dies. The value of the residue to the animal depends on plant's nutritional quality and ease of harvesting, the latter a reflection of plant height and density. Acceptability, often associated with the palatability of plants, can be defined as 'the property of plants providing stimuli to the special senses (taste, touch, sight, smell) of animals thereby affecting feeding behaviour' (Doyle and Egan 1987). Even though animals select the more nutritious components of available feed (Table 3.2), which can be improved by fertilization, this offers little opportunity to improve the quality of feed selected during the dry season. Inevitably, the quality of feed consumed declines as the season progresses (Tables 3.3, 3.4,

Table 3.2 Digestibility of diet on offer and selected by cattle grazing a legume-native grass tropical pasture either unfertilized or fertilized with phosphorus (Gardener et al. 1988)

	Unfertilized		Fertilized	
Season	On offer	Selected	On offer	Selected
Wet	58.6	63.9	61.5	65.6
Wet/dry	53.1	56.8	57.2	59.4
Dry	45.4	53.3	46.5	55.3

Table 3.3 Changes in diet available to cattle grazing a pasture of native Speargrass, *Heteropogon contortus*, in tropical Australia (see Hogan 1996)

Month	Brown leaf (% in sample)	Green leaf (% in sample)	Stem (% in sample)
February	10	73	17
March	13	50	37
August	32	8	60

Table 3.4 Changes in diet available to cattle grazing the shrub legume *Stylosanthes hamata* cv. Verano in tropical Australia (see Hogan 1996)

Month	Green leaf (% in sample)	Flower (% in sample)	Stem (% in sample)
March	31	11	58
May	5	19	76
August	0	0	100

and 3.5). In some plants, the content of nitrogen or sulphur declines asynchronously, and the consequent nutrient imbalance reduces feed intake. The data in Tables 3.3 and 3.4 indicate the combined effects of grazing and the onset of the dry season on the parts of grasses and legumes available to cattle. In both data sets, cattle growth had ceased by July and, with no effective rain expected in the next 100 days, the cattle faced a period of declining nutrition. During that time, effective rainfall might encourage pasture regrowth, but small ineffective showers of rain could, by permitting the growth of fungi, greatly depress the palatability and nutritive value of the roughage. The consequences for annual weight gain and loss, and benefit of phosphorus supplementation in the wet season, are apparent from Table 3.6, a reflection of the length of wet and dry seasons in harsh and endowed regions.

As water from surface dams or ponds becomes depleted, access to the remaining water is often impeded by muddy verges, and bogging in mud is a serious problem for weak animals and may additionally make them vulnerable to predation. Alternative sources of water may be provided from bores or wells, but care must be taken to ensure that the mineral content is sufficiently low (Schlink et al. 2010). Animals accustomed to drinking from surface water may also require training to drink water from bores or wells, especially when offered in troughs. Water supply and animal welfare are considered in Chap. 9 of this book.

Table 3.5 Crude protein (CP) and neutral detergent fibre (NDF) composition (g/kg DM) of diet selected by steers, with periods of zero cattle growth indicated (Hunter et al. 1976)

	Native pasture		Legume–grass pasture	
	CP	NDF	CP	NDF
Late February	144	691	150	700
Late March	156	721	169	590
Mid-April	103	754	131	715
Mid-May	83	783	96	753
	Zero cattle growth			
Early June	64	790	58	769
			Zero cattle growth	
Mid-June	46	814	58	780

Table 3.6 Pattern of weight change (g/day) in Brahman-cross cattle in different periods of the year in harsh and endowed regions of northern Australia, receiving no (−P) or supplementary (+P) phosphorous (Winter et al. 1990)

	Period				
	Early wet season	Late wet season	Early dry season	Late dry season	Annual (kg/year)
Harsh region					
+P	874	652	−149	−425	81
−P	431	266	−153	−315	18
Endowed region					
+P	943	591	29	−54	146
−P	858	449	80	−208	120

The cycle of pasture growth in the wet season followed by maturity and senescence should be taken into account when developing a system of livestock production for a particular region. In such systems, some undernutrition leading to a modest rate of weight loss is commonly accepted during the latter part of the dry season. The management strategy accepts a lower lifetime calf output per breeding cow and some months of additional age before steers reach market weight. The skill of the manager is called on to regulate the grazing pressure in a system in which dry matter is being removed but not added in the form of plant growth. Overstocking of the feed resource or the extension of the dry season can see the reserve of dry matter become so depleted that the animals pass from undernutrition into starvation. The manager then has to consider the options of supplementary feeding, removal of stock to other pastures, or, if that is not physically or financially feasible, the humane slaughter of stock.

3.3 Drought, the Animal and the Ecosystem

The fate of animals during a drought depends on such diverse factors as the ability of the animal to ingest and digest feed, to cope with the many potential health issues, and to meet the demand for nutrients determined by the animal's physiological state (Table 3.7).

In a drought, animals continue to try to satisfy their feed need within an ecosystem in which the reserve of feed is declining in both quantity and nutritional value. Their ability to do so depends first on their capacity to find and harvest feed. Animals must walk to find feed. Under drought conditions, animals congregate at permanent sources of water and graze in a circle outwards from these, usually returning for water daily but sometimes not until after 36 h. When saltbush forms a major part of the diet they must return twice daily to dilute the high salt intake. The radius of the grazing circle, that is, the distance the animal is willing to walk, determines the size of the area of pasture to which the animal has access. With Brahman-Shorthorn cattle in northern Australia, Yeates and Schmidt (1974) observed that some animals grazed only within 4 km of water, whereas others walked about 7 km before commencing grazing and foraged for another 1 km or so. Grazing pressure is heaviest near water and as the feed supply dwindles, the animal is forced to expend more energy in walking. Soundness of legs and feet is important even with high-quality pastures, where animals with infections associated with foot rot, foot abscesses, or chronic forms of foot and mouth disease are so occupied with keeping contact with the flock that they lose grazing time and may become undernourished. Malnutrition of the young, especially with a calcium: phosphorus imbalance, can result in 'pegleg', a permanent form of lameness that reduces the animal's mobility and hence its foraging ability. Phosphorus deficiency can have substantial effects on weight gain (Table 3.6), the difference between harsh and endowed climatic zones being clear.

Table 3.7 Animal–environment interactions (from Purser and Hogan 1992)

Behaviour in the environment	Animal health	Physiological state
Capacity to walk	Pathogens	Pregnancy
Capacity to chew	Toxins	Lactation
Competition for feed	Secondary chemicals	Young growth
Other animals	Paint, rope, wire	Full fleece
Competing insect herbivores	Plastic bag, sheet	Newly shorn
Soil dwelling herbivores	Chemicals	Losing weight
Psychological stresses	Helminths	Barely surviving
Impact of the manager	Ectoparasites	Regaining weight
Dogs and other mammals on the farm		
Flock/herd behaviour		

3.4 Feed Harvesting

A detailed understanding of the different methods of feed harvesting used by ruminants helps to utilize the most appropriate animals in times of limited feed resources. Sheep bite off feed held between the incisor teeth on the lower jaw and a dental pad on the upper jaw. Cattle have a similar dental arrangement but do not graze as close to the soil as sheep and tear off feed wedged between incisors and dental pad by jerking movements of the head. Animals with undershot or overshot lower jaws are disadvantaged in harvesting tough plant material because the incisors and dental pad are not opposed. Attempts to tear off material impose a strain on the teeth that loosens their point of insertion into the jawbone and sets up conditions conducive to periodontal disease. When the feed has been harvested it passes back to be chewed between the cheek teeth of both jaws. Each cheek tooth, or molar, on the upper jaw comes in contact with two lower teeth, and the grooves and ridges of the surfaces of both sets of teeth interdigitate to permit thorough grinding of plant material when upper and lower molars are slid over each other. The permanent teeth of the adult begin to form at 3–6 months of age. An imbalance between calcium and phosphorus in the diet by feeding, for instance, on wheat grain without a calcium supplement causes the development of deformed and misaligned teeth (Franklin 1950) and imposes a permanent penalty on the animal, impairing feed intake. This occurs through the reduction of its ability not only to harvest feed but also to move feed particles to the rumen, following reduction in the particle size of fibrous digesta during rumination. Exposure to fluorosis associated with a concentration of fluoride above 1 mg/L (Schlink et al. 2010) in drinking water also causes problems with the strength and spacing of teeth.

Damage to the teeth, especially the incisors, is frequent during a drought when animals chew timber or bones. Many animals emerge from a drought 'broken mouthed' and their inability to chew properly imposes a permanent nutritional penalty thereafter. Wear and damage to teeth are also caused by the ingestion of sand and gravel (Healey and Ludwig 1965) when sheep, and to a lesser extent cattle, attempt to consume short herbage, fallen plant material, and even the roots of plants. Geophagy, or earth eating, is regularly observed in cattle, especially when held in earth-floored yards (Ryley et al. 1960) in which case 30 % of abomasal contents can be soil. In such circumstances, ruminants could derive some nutritional benefit by recycling fibre and minerals, including nitrogen, deposited in faeces and urine, but the possibilities of parasite transfer must be carefully considered. They may also derive some minerals by licking the coats of companions, although this practice results in the appearance of hair balls and wool balls in the rumen (Ryley et al. 1960).

The nutrition of domestic ruminants may be adversely affected by competitors during drought. More mobile herbivores such as kangaroos or feral camels, or goats in some situations, compete with sheep and cattle and hence have first use of the grazing area. Others such as deer and rabbits, even in small numbers, can not only deprive ruminants of pasture but, by chewing the bark, kill trees and deprive animals of browse. Rats and mice at times appear in plague proportions and eat or foul stored feed. Pastures can be denuded by plague locusts (*Chortoicetes terminifera*), while termites regularly remove an appreciable amount of dead pasture material. Dead pasture material is also rendered inaccessible to livestock by trampling or through wind erosion.

Interruption of grazing time by the imposition of psychological stressors such as people or dogs may exacerbate the effects of drought on individual animals or even the entire flock. Any reduction in feed intake will probably not be compensated for by an extension of grazing time because an isolated grazing animal is vulnerable to predation. Grazing is a highly cohesive activity, and intake is adversely affected by mustering for normal husbandry practices like shearing. Apart from any stress caused by the procedure, animals are brought into the company of unfamiliar companions, some of whom may be aggressive. It may be several days before the animals are returned to their usual pastures and with animals of low calcium status the combination of stress and undernutrition can induce the appearance of hypocalcaemia. When animals are mustered for transport, they are regularly held in yards for 12 h or more before loading (to reduce excretion in the transport vehicle), then generally not fed until they reach their destination, perhaps 48 h later. If transport precedes shipment overseas, after the road or rail transport the animals are held in yards for shearing, health checks, and observation to detect animals reluctant to eat the feed to be offered during the voyage. Reluctant eaters should be removed, but despite this, inanition is the main cause of death of sheep aboard ship (Richards et al. 1989).

3.5 Effect of Drought on Animal Health

Drought-stricken livestock change their eating habits and will attempt to eat many substances that they would normally avoid. Under extensive grazing conditions, sheep and cattle with depraved attitudes (pica) are attracted to old carcasses which may harbour *Clostridium botulinum* and cause death through botulism. Hungry animals may eat toxic plants that they would normally avoid and may venture into waterways and low lying areas which expose them to liver fluke (*Fasciola hepatica*). Care must be taken to prevent access to farm chemicals, including paint, machine oil, and chemicals used as timber preservatives. Plastic bags, sheet, and cord which are used in baling hay and other foreign bodies can, if swallowed, cause blockages in the rumen or abomasum (Hayder et al. 2006). Discarded lead batteries, similarly, are attractive to hungry livestock and have caused fatalities.

The crowding of animals into yards and trucks exposes them to the stresses of confinement and bullying from aggressive companions. The extent to which animals respond by the release of stress hormones, such as adrenalin or cortisol, with concomitant consequences for feed intake and utilization, is largely unknown. Crowding is also conducive to the transfer of ectoparasites such as lice and endoparasites such as helminths. Undernourished animals may also have a reduced capacity to control endopathogenic bacteria species such as *Clostridia*, *Salmonella*, and *Escherichia coli* and are at risk unless vaccinated appropriately.

3.6 Effects of Starvation

3.6.1 Basic Research at Cellular Level

Depletion of nutrients from the digestive tract causes cells to begin recycling proteins and various cell structures by a process called autophagy. In this, a double membrane vesicle, the autophagosome, derived from the outer membrane of mitochondria (Rambold and Lippincott-Schwartz 2010) engulfs the cytosolic or organellar components and then fuses with a lysosome. By doing so, the components are exposed to acid hydrolases and the released amino acids and other metabolites pass into the circulation (Hailey et al. 2010). At the same time, mitochondria begin to elongate in a process that protects them from autophagy and concentrates their capacity to produce ATP (Blackstone and Chang 2011; Gomes et al. 2011). When re-alimentation occurs, the elongation process can be reversed (Hailey et al. 2010) as part of the process of resumption of normal cell function. If starvation continues, the mitochondria undergo autophagy as part of the process of apoptosis or cell death.

3.6.2 Physiological State and Undernutrition

Demand for nutrients is most pressing in females during late pregnancy and lactation. This is particularly so in ewes carrying twin or very large single foetuses in which an interruption to feed supply leads to the catabolism of adipose tissue. The level of circulating growth hormone increases (Goldstein et al. 2011), probably in response to the discharge of the peptide hormone ghrelin (Kojima and Kangawa 2005), with the appearance of increased levels of free fatty acids (FFA) in plasma. The concentration of FFA has been proposed as an indicator of the extent of undernutrition in the ewe (Reid and Hinks 1962), with an inverse correlation between levels of FFA and glucose in plasma. However, Agenäs et al. (2006) did not find it to be useful in the case of starving beef cattle.

Extended periods of undernutrition may lead to the accumulation of fat in the liver with impairment of function. Incomplete metabolism of FFA results in elevated levels of the ketones acetoacetate and beta hydroxy butyrate (Reid and Hinks 1962). Ketones are important metabolic fuels for some tissues, especially the brain in humans. However, in sheep there appears to be little or no oxidation of ketones in the brain (Lindsay and Setchell 1976; Pell and Bergman 1983), and hence the brain competes for glucose with the foetus or the mammary gland. In the undernourished ruminant, glucose supply from conversion of any propionate from the digestive tract, glycerol released during fat catabolism, and glucogenic amino acids is critical (Leng 1970).

Nutritional and psychological stresses lead to cortisol release in pregnant ewes, resulting in the neurological condition 'pregnancy toxaemia' or 'twin lamb disease' with potentially fatal consequences (Reid 1960). This syndrome is most common in fecund breeds in regions providing high-quality pastures in early pregnancy but also subject to climatic conditions such as snow that interrupt feed supply in late pregnancy. Lactating dairy cows most commonly experience deficits of energy during early lactation, resulting in acetonaemia. Sodium deficiencies are also common, particularly in hot climates in which sweating is common, with adverse effects on rumen function (Chiy et al. 1993).

3.6.3 Changes in Body Composition

The effects of prolonged undernutrition on cellular processes vary with factors such as the age and maturity of the animal, its body composition at the start of undernutrition, the proportion of initial body weight lost, and the rate of weight loss (Butler-Hogg 1984). The first author, in measuring weight loss of the fleece-free sheep, separated the animal into three parts: the carcass, without kidneys or associated fat; the viscera, including all internal organs and associated fat; and the remainder, including the head, hide, feet, and blood. His data indicated that in Corriedale sheep reduced in weight from 37 to 30 kg, the carcass was 800 g heavier, the viscera 1800 g lighter and the remainder 500 g heavier than normally fed 30-kg control animals. The fat content of the carcass in weight loss animals was greater than controls, which has been confirmed in adult sheep by Thornton et al. (1979). The loss in weight of the viscera is rapid, evident after even a 72-h fast (Warriss et al. 1987).

When sheep and cattle are re-alimented following a period of undernutrition, their regrowth will be affected by the extent to which the components of cells have been damaged, and, on a macro-scale, by the changes that have occurred in carcass and viscera. Hence, Butler-Hogg (1984) observed relatively slow regrowth of carcasses which had been little affected by undernutrition, but a much faster regrowth of viscera. The composition of gain during re-alimentation is, therefore, likely to be affected by those factors affecting weight loss described above.

3.7 Consequences of Undernutrition and Starvation

Prolonged undernutrition may induce abnormal skeletal and dental development, even though the ability of ruminants to compensate for weight loss during re-alimentation leads many to believe that the consequences of starvation are relatively minor. However, although lambs can survive weight loss following undernutrition post-weaning without long-term restriction of growth (Allden 1968), undernutrition of the dam during pregnancy may have more serious consequences. Recent studies (Wu et al. 2006) have indicated that intrauterine growth retardation can have substantial lifelong effects on metabolism, including stunted growth. Further, undernutrition at the time of conception can affect the glucose insulin and hypothalamic–pituitary–adrenal axes (Jaquiery et al. 2011; Harding et al. 2011), with long-term consequences that include increased susceptibility to disease (Barker and Clark 1997). Severe undernutrition of the dam in late pregnancy may also restrict development of mammary glands and reproduction in the offspring, placing them and even their offspring at risk (Blair et al. 2010). Such prolonged effects reduce the energetic efficiency of the production system.

Undernutrition particularly affects newly shorn sheep, increasing their susceptibility to cold conditions. Losses can be considerable, up to 30 % of lambs born. Feed intake may be interrupted during mustering, reducing heat of fermentation, and increasing susceptibility to inclement weather. Adult non-pregnant females and castrate males are the least vulnerable animals during feed restriction. Provided that they obtain some feed they can survive for long periods, drawing gradually on their energy reserves. As a result, they have lowest priority for supplementary feeding.

3.8 Form and Provision of Nutrients for Survival Feeding

The grasses and legumes available to livestock are those that have adaptive features that permit them to survive in a particular region. Most grasses in the tropics utilize the C_4 metabolic pathway with corresponding anatomical differences from temperate grasses using the C_3 pathway. Fragmentation of plant leaves, an essential feature of the digestive process, is constrained in tropical grasses (Wilson and Kennedy 1996) by epidermal and vascular structures, thick cuticle, and high levels of lignification that are essential for the survival of the plant but reduce nutritive value. Thus, tropical grasses are appreciably less digestible than temperate grasses at the same stages of growth (Minson 1987). There are no such distinctions between temperate and tropical legumes. The distinctions between temperate and tropical grasses become important in considering the nutrition of livestock during drought and will probably become even more important if tropical grasses are required to meet any consequences of global warming.

The nutritional value of a feed depends on intake and digestibility. Feed intake in ruminants is usually driven by the demand for energy, but constrained by the rates at which the residual effects of previous meals have been dissipated (Weston 1996). Energy demands, and hence feed intake, vary with the physiological state of the animal, modulated by factors such as stress associated with heat or cold. Constraints to feed intake, especially with mature roughage, mainly relate to the rate at which fibrous digesta particles can be reduced in size and attain a suitable specific gravity to permit their passage from the rumen (Wilson and Kennedy 1996). The ability of the grazing animal to satisfy the demand for energy depends on those animal–environmental factors described in Table 3.1. When faced with a shortage of feed, the grazing animal can extend the time spent in seeking and eating feed, but grazing will be limited by flock and herd behaviour and by the need for time to rest and to ruminate. As pasture availability declines, grazing livestock experience reduced bite size, but they increase biting rate and grazing time to compensate, if possible. At the commencement of a period of feed shortage, an animal will probably have access to mature forage from the current season, residues from previous seasons, and browse from shrubs and trees.

Availability varies with climatic zone. Thus, in the regions of Mediterranean and temperate climate the feed resource will be composed of temperate grasses, clovers, and medics, while browse will be provided by shrubs and trees suited for those environments. Their nutritional value will depend to some extent on the vigour of growth of the component plants during the preceding one or two seasons and the grazing pressure subsequently applied in the dry season. In one temperate area, Brownlee (1973) found that seed pods of Barrel Medic (*Medicago truncatula*) comprised 70 % of the diet. However, subsequently Denney et al. (1979) found that the nutritional value of the pods resided almost entirely in the seeds, the numbers per pod depending on the growing season. Hence, in a drought following a productive winter and spring, the animals will have access to large amounts of seed, whereas the drought that follows a failed winter and spring commences with a much reduced seed supply. This, too, is the climatic region in which clover-based pastures die suddenly during hot weather, and the nutritive value of the dead residue is low (Doyle and Egan 1987).

Pasture in tropical regions often matures and senesces before it is used by livestock. The physical and chemical composition reflects the preceding growing season. Growing plants take up most of their minerals early in development and then distribute them through the plant organic matter as it is synthesized. A plant that produces much organic matter will have low mineral concentrations. In these situations, the plants may have inadequate concentrations of Na, Cl, I, Co, and Se to maintain livestock, even though for the plants themselves concentrations are adequate. Adequacy of macro-elements Ca, Mg, and P for livestock may depend on the relative concentrations in the soils. Mineral deficiencies may directly affect individual components of the rumen microflora. Co deficiency affects those bacteria producing Vitamin B_{12}, while sulphur deficiency, by impeding the growth of anaerobic rumen fungi, reduces both cellulose fermentation and feed intake. The main mineral deficiency in mature tropical grasses, though, is nitrogen, the lack of which affects the capacity of the rumen microbial population not only to release energy by fermenting the fibre components of the forage but also to meet the needs of the tissues of the animal for essential amino acids.

3.9 Energy and Protein Transactions During Starvation

When an animal gains weight, the proportions of fat, protein, and water in weight gain vary with the stage of development of the animal. Thus, 1 kg gain in a 25 kg lamb might involve 13.8 MJ energy, contained in 130 g protein and 270 g fat. By contrast, at 50 kg the same lamb, as an adult, would store almost twice as much energy, 26 MJ/kg gain, in the form of 90 g protein and 600 g fat (Weston and Hogan 1986). In livestock, weight loss can be regarded as a highly concentrated form of energy contributing to the ME pool. However, whereas ME is usually obtained by the animal as short chain fatty acids, acetate, propionate, and butyrate from the rumen, weight loss provides mainly long-chain fatty acids derived from the catabolism of adipose tissue. The efficiency of use of ME derived from weight loss is assumed to be similar to the highest values for maintenance observed for ME derived from normal diets.

The ME system assumes that all other nutrients are present in adequate amounts. This applies particularly to nitrogen as a source not only of ammonia, amino acids, and peptides for rumen microbes but also for amino acids for the tissues following the digestion of protein in the small intestine. Ruminants depend for protein partly on unfermented dietary protein that bypasses the rumen, but mainly on protein synthesized in the rumen as part of microbial development. The generally low protein content of mature roughage available during a drought limits the level of ammonia in the rumen and thus the supply of protein, derived both by synthesis and from the diet, that reaches the small intestine. In forages, the level of ammonia in the rumen is indicated by the ratio of digestible organic matter to crude protein (DOM:CP) in the diet, with a ratio above about 7:1 indicating a possible ammonia deficit (Weston and Hogan 1968). A similar relationship applies for tropical grasses, but for a given DOM:CP ratio, the concentration of ammonia is lower (Hogan et al. 1989; P.M. Kennedy, pers. comm.). With low-protein feeds, the animal benefits from ammonia derived from urea recycled from the blood, either across the rumen wall or via saliva, a process assisted by mechanisms in the kidney to reduce urea excretion in urine. Despite this, the intake of low-quality, mature tropical grass can be increased by provision of urea and sulphur to cattle (Table 3.8), particularly cows with high demands for energy in late pregnancy. Roughage intake in such cows may be further increased by supplying additional amino acids in the form of a protein meal, such as cottonseed, that is partially protected against deamination in the rumen. Addition of phosphorus is common. It is noticeable (Table 3.8) that a concomitant increased intake of roughage not only reverses cow weight loss but restores calf birth weight from a critical to normal level. Practical difficulties in providing the supplements cannot be ignored, for example, access to land during the wet season may be limited.

Protein released during weight loss becomes available as amino acids for protein metabolism in the liver, kidney, and peripheral tissues. Some amino acids become a critical source of glucose. The nitrogen in amino acids, after deamination to urea, is

Table 3.8 Mean live weight change and roughage intake of cows fed native pasture only (*NP*), native pasture with a urea/sulphur (*US*) supplement, or native pasture with a *US* supplement plus protein (*USP*) for 60 days pre-calving

	Treatment		
	NP	US	USP
Live weight change (g/day)	−815	−308	+750
Roughage intake (kg DM/day)	4.2	6.2	8.1
Calf birth weight (kg)	22.0	30.9	32.1

Data include birth weight of calves (Lindsay et al. 1982)

available for recycling to the rumen and becomes a vital source of ammonia for the rumen microbes.

3.10 Amelioration of Starvation

Starvation can be ameliorated only by supplying a source of energy to the animal. Many producers in temperate and Mediterranean climatic zones conserve roughages for winter feed as hay or silage, but few carry sufficient reserves against prolonged drought, which most commonly occurs in summer. In tropical areas, haymaking occurs when permitted following ending of the wet season, by which time the forage is mature and of relatively low nutritional value. In areas within about 200 km of sugar producing areas, cattle may be supplemented in the dry season by the provision of a mixture of molasses and urea (Ernst et al. 1984). Both ingredients are potentially toxic to cattle: molasses by overconsumption and fermentation of the contained sucrose to lactic acid; and urea by the production of toxic amounts of ammonia in the rumen. In theory, animals need about 3 % urea to meet the demands for ammonia of rumen microbes fermenting the sucrose in molasses, but in practice the incorporation of an additional 5 % urea controls molasses intake and prevents fermentation of the mixture by yeasts. On a smaller scale, mineral blocks fabricated with molasses and urea provide safe access to supplements for animals that have mature roughage available (Leng 1984). However, if feed has to be transported for long distances, the preferred option is usually grain. In tropical conditions with high rainfall, such as northern Australia, roads to cattle during summer rains may be impassable, rendering supplementation impossible. As with molasses, the feeding of grain also presents the risk of accumulation of dangerous levels of lactic acid in the rumen. This can have fatal consequences arising from lactic acidaemia, rumen stasis, and the multiplication in the intestines of pathogenic bacteria such as *Clostridia* and *Salmonella* on a substrate of undigested starch. There is also anecdotal evidence that some animals develop an aversion to grain as the result of overeating and subsequently refuse to eat it, to the extent of dying with empty rumens. These difficulties can be overcome by introducing animals gradually to grain in a roughage–grain mixture in which the proportion of roughage is gradually decreased over 2 or 3 weeks.

The recommendations for survival feeding (Briggs 1968) are to hold the animals in small yards to conserve energy and to offer the available feed such as hay or grain plus ground limestone twice weekly to ensure that even the weakest animals get some feed. An example of this is seen in an experiment by Ryley et al. (1960). Hereford heifers weighing 248 kg were offered, as sole feed, crushed sorghum grain to which they were introduced over 3 weeks in a hay–grain mixture with gradually declining proportions of lucerne hay. Sorghum was fed at the rate of 1.35 kg/day, supplied daily, once weekly, or twice weekly. Twice weekly feeding was the most successful option; feeding once weekly saved labour, but the animals could not eat sufficient grain (almost 10 kg) in one meal, the residues of which attracted vermin. The results for the twice weekly fed group are presented here. The experiment continued for 38 weeks (Table 3.9) of which the first 3 were spent introducing the cattle to the grain diet, the next 26 on the main experiment, and then in the recovery phase, 1 week on lucerne hay and the following 8 on pasture. In the first 3 weeks of the experiment with the change from roughage to an all-grain diet, the weight of cattle fell by 38 kg, presumably largely reflecting reduced weight of rumen contents. During the experimental period, cattle lost weight at approximately 380 g/day, to weigh 177 kg during the first 84 days, but during the next 98 days, weight loss declined by 133 g/day so that final weight was 165 kg. Weight lost during the whole experiment was thus 83 kg, or about 36 % of original weight, whereas during grain feeding the loss of 45 kg was equivalent to approximately 21 %.

The average maintenance ME required by these cattle between days 0 and 84, and 84 and 182 of the grain feeding regime was calculated from Morris (1968) at 388 kJ ME per unit metabolic body size. The ME provided by the diet was, when averaged over the first 84 days, 3.9 MJ/day below maintenance, but with calculated ME required for maintenance falling with declining body weight, the estimated mean deficit in the next 98 days was only 2.2 MJ/day. The accuracy of these calculated energy deficits and the interpretation of weight loss data depend on a number of assumptions, including the amount of digesta in the tract. However, they indicate that the ME System can be used to indicate the quantities of feed adequate to sustain a controlled weight loss in animals (Table 3.9). In the recovery period, the cattle immediately resumed normal feed intake. When offered lucerne hay for a week, the cattle consumed about 3 kg/day, and when turned out to pasture regained their pre-experimental weights within 8 weeks. During this phase,

Table 3.9 Live weight and metabolizable energy (ME) requirements and deficit in Hereford heifers fed 16.1 MJ ME/day as 1.35 kg cracked sorghum grain offered twice weekly (Ryley et al. 1960)

Time (days)	Live weight (kg)	ME requirement (MJ/day)	ME deficit (MJ/day)
0	210	21.3	5.1
84	178	18.9	2.7
182	165	17.9	1.7

although the amount of digesta included in weight gain was not known, the animals would have benefited, in terms of both level of feed intake and efficiency of use of ME, from the period of compensatory weight gain. During such gain, the cattle would have deposited a higher proportion of protein and a lower proportion of fat in gain.

3.11 Managing Starvation

Ruminant animals are kept as production units of commercial enterprises and are subject to commercial decisions. Some decisions are based on the assessment of the macro-environment and of the factors limiting animal production described in Table 3.1. These will probably determine the nature of the enterprise and which animal species and breeds are suitable. Essential to decisions of a biological nature is risk assessment, and in particular the reliability of rain. Many aspects of risk assessment can be made on the basis of large regions, but rainfall affects individual properties. The prediction of rainfall has advanced greatly in recent years but can provide only a broad guide for management decisions on individual properties (Vizard et al. 2005). The assessment of the probability that the dry season will extend into drought in pan-pacific countries of the Southern Hemisphere has been helped in recent years by the development of the Southern Oscillation Index to predict length of the wet season. Drought management plans implemented by livestock producers should take this into effect.

Essential to any strategy is the retention of young breeding females to rebuild stock numbers after the drought (Whitmore 2000). This implies that older and potentially less productive animals will be sold, providing that a market is available, because other producers may be pursuing the same strategy. Of the remaining stock, some may be sent to other properties (agisted) or to feedlots. With a wool-producing venture, castrate males and non-pregnant females are often allowed access to the entire property with occasional inspections to ensure their welfare, both for attack by ectoparasites like blowflies and to ensure adequate water. If other sources of feed exist on the property, such as edible trees or shrubs, a programme of tree pruning may be undertaken to give stock access to this source of feed. For those animals given highest priority, it is recommended that the drought feeding techniques to conserve the animals' energy and to ensure a share of the feed for every animal be followed.

3.12 Animal Welfare Regulations

An important development in welfare legislation governing starvation of animals has been introduced in several developed countries in the last decade. Livestock producers are now subject to a duty of care for their animals, so that they can be prosecuted, for example, for not providing sufficient care for animals in their charge. This includes not having sufficient feed reserves to provide for their stock during an anticipated feed shortage, such as over a dry season or winter. Under the *Duty of Care Act* in Queensland, for example, producers must provide adequate food and water and also facilities whereby animals can display normal behaviour. Formerly livestock producers could only be prosecuted if there was evidence of starvation and suffering by the animals. In general, legislation has required livestock producers to avoid 'unnecessary suffering'. Extreme weather events, including drought, are usually regarded as outside the producer's control and it is rare for them to be prosecuted for the deaths of their stock. The actions of a reasonably competent and compassionate person are the benchmark, against which producers actions must be judged. An animal can be deemed to be suffering if people omit to do something, such as providing feed, or by actually doing something, such as offering an inappropriate feed, although there is greater recognition of the latter. Potential penalties for infringement are usually monetary, but can be prevention of care or ownership of animals in future. Animals kept for other purposes usually are given better protection legally, for example, in Australia the *Code of Practice for the Care and Use of Animals for Scientific Purposes* (NHMRC 2013) determines that animals used for scientific purposes should not be subjected to withholding or severe restriction of food or water if it produces a continuing detrimental effect. If any studies could potentially induce such an effect, live weight or fluid balance must be monitored and maintained within predetermined limits. Similarly, allowing companion animals to starve to death would bring rapid prosecution in most jurisdictions.

Laws may conflict in controlling farm livestock feeding, for example, whereas the duty of care acts usually require that stock are not allowed to suffer excessive hunger, environmental protection legislation may require that feed is withheld before and during transport to reduce distribution of excreta.

3.13 Recovering from Starvation

At the end of a long period of weight loss, animals 'are of dejected appearance, reluctant to undertake unnecessary movements, and with an observed reduction in rectal temperature' (Morris 1968). The ending of a drought by rain may find such animals in a state too weak to withstand the cold stress of rain accompanied by wind. Further, there may be enough rain to cause flooding, which brings further problems for weak animals. There is a need to provide shelter for affected animals. Pasture plants begin to grow again after rain, but the early green shoots are sparse and not very nutritious, and unrestrained animals must spend a great deal of energy seeking food. If such feed is eaten in large quantities it can cause serious digestive upsets. Hence drought-affected stock should be kept in their enclosures for several weeks and given access to immature pasture plants only gradually.

Health issues also arise as infective helminth larvae, excreted by grazing animals, begin to inhabit pasture plants and infect other grazing animals. Some toxic pasture plants such as the South African Gifblaar (*Dichapetalum apetalum*) send out sprouts quickly and their content of fluoroacetate causes casualties.

3.14 Short-Term Feed Restrictions

Temporary feed restrictions regularly occur before and during transport of cattle and sheep, resulting in the negative affective state of hunger. The physiology of hunger is well understood, in particular involving ghrelin peptide hormone production in the abomasum and the associated reduced immunocompetence (Hogan et al. 2007). However, the *extent* of hunger in cattle and sheep experiencing short-term feed restrictions, and its welfare implications, are not well understood. These are likely to differ from monogastric species because the rumen acts as a feed reservoir, but will also take longer than the gastrointestinal tract of monogastric species to recover following disturbance to the microbial digestion process. Where regulations exist for cattle and sheep, the maximum permissible time off water (and hence feed) is particularly long in Australia (48 h), as a result of the significant distances that animals have to be transported from remote breeding properties. During this period, cattle and sheep will typically experience weight loss of 10–15 %; hunger; stress-related increased nitrogen output by 25–30 %, some of which derives from increased tissue protein catabolism; a major reduction in rumen microbial activity; a transition of energy sourcing from volatile fatty acids (VFA) produced in the rumen to non-esterified fatty acids mobilized from adipose depots; and a potential increase in enteropathogenic bacteria (Hogan et al. 2007). Repletion of glycogen reserves that have been seriously depleted by feed restriction or transport will take many days or even weeks.

Another serious case of short-term feed restriction occurs when unwanted male 'bobby' calves are transported for long periods before slaughter. These calves, some as young as 5 days old, would naturally be provided with a milk feed approximately five times a day if they were being suckled by their mother. During the transport process, changes in blood glucose concentration suggest that they start to feel hungry approximately 9 h after their last feed and this increases exponentially thereafter (Phillips and Hogan 2011). Maximum journey times for calves, which are sometimes presented as times off feed, are highly variable between jurisdictions and range from a 6 h transportation limit in Switzerland to the currently proposed 30-h time-off feed limit in Australia.

3.15 Toxic Compounds and Pollutants

Extensively grazed cattle and sheep are more likely to encounter toxic plants than those fed a more controlled diet, resulting in effective starvation if toxins interfere with normal digestive processes. However, careful management and regular exposure can help them to learn to avoid the toxic plants, though naive and often hungry livestock may browse these toxic plants. Fatalities may occur, for example, following relocation of animals. Furthermore, the increased spread of weed species around the world is enabling them to colonize novel regions and pose new threats to grazing livestock (Hogan and Phillips 2011). Frequent overgrazing and climatic extremes are supporting increased weed colonization of existing grazing lands.

Industrial pollution sometimes contaminates grazing lands for livestock and especially poses problems in drought periods, when the animals may consume significant quantities of soil with their herbage. Planned or accidental release of heavy metals, radioactive chemicals, and carcinogens may accumulate in animals, mainly in the liver, kidney, and muscle tissues. This can pose a risk to humans consuming them, as the impact of chronic accumulation of toxic agents on livestock is usually less than it is for humans because of the relatively short lives of the former. The toxic elements cadmium and lead may infect agricultural land and interfere with the normal processes of digestion (Phillips et al. 2005, 2011).

3.16 Future Management of Starvation of Livestock

Problems of undernutrition of livestock are likely to increase, as a result of increasing climate variability in marginal land, human, and animal population pressure eroding their grazing lands and farmers overstocking their land to make a profit. The increasing prosperity in many developing countries will maintain a strong demand for livestock products for those who can afford them, despite the associated problems of some inefficient use of good quality land and increased pollution potential from livestock emissions. Those in developing countries whose livelihood remains poor will be faced with low food availability and, in the case of graziers, diminishing land resources for their herds and flocks. In a scenario of increasing food demand worldwide and competing interests for energy crops, there will be pressure on marginal land to be more productive. It is expected, therefore, that some livestock production will be further intensified in feedlots, with the attendant inefficient use of land for production of high-quality animal feed, as well as pollution and waste disposal problems. At the same time, although livestock undernutrition in marginal areas often currently goes unnoticed because of the remote nature of the farms, future development of these regions and increased public awareness will bring increased scrutiny. Although the public tend to consider drought and associated cattle feeding problems as ‘natural’ and therefore of less concern than husbandry procedures (Phillips et al. 2009), evidence of the manmade nature of climate change could rapidly change that perception. This could lead to increased public concern about livestock feeding practices and their relation to animal welfare.

3.17 Knowledge Gaps and Future Research Needs

Our ability to manage starvation in ruminants will depend on the willingness and ability of livestock producers to introduce measures to feed their cattle appropriately and the prevalence of risk factors, which are anticipated to increase, as outlined above. A better understanding of the impact of starvation on the welfare of livestock would assist in knowing how to allocate resources to control the various welfare issues challenging ruminant livestock. Obviously, if an animal dies, its welfare was probably severely compromised in the last stages of its life, but when did it become affected and how can we recognize it? Better forecasting of weather patterns that will affect pasture production will help producers prepare for drought, as would an improved understanding of how global climate change is likely to have an impact on rangeland systems. In response to climate change and reduced tolerance of the impact of drought on livestock welfare, more drought-tolerant crops are likely to be required. These may not be as productive in periods of high rainfall, but will offer greater economic and ethical sustainability for the production system, as well as maintaining greater consistency for the producer. Indeed, greater certainty in the entire system will facilitate management of feed resources; recently there have been major perturbations in the price of cereals, for example. Understanding, modelling, and ultimately controlling long-term future trends will enable better planning in livestock enterprises.

There will always be producers who do not manage their properties well, operating at too high stocking densities, being unprepared for drought, keeping the wrong type of animals for the land type, for example. The best method to control these practices is little understood. Legislation, advisory services, and codes of practice are all possible, but the provision of assistance from fellow farmers may ultimately be the most respected option. Knowing how to foster and encourage such mutual support should be a priority in times when farms are getting larger and contact between farmers is potentially infrequent and competitive, rather than supportive. The advantages of internet communication potentially offer opportunities to improve communications with farmers, but little is done to organize or understand the benefits of this.

3.18 Conclusions

Starvation in domestic ruminants is of two types: either sudden or prolonged onset. Sudden starvation, which occurs when normally fed animals are abruptly deprived of feed, generally affects animals with adequate body reserves of glycogen and fat. The animal is able to withdraw nutrients from these reserves by mechanisms that are well understood, and unless the animal has specific nutrient demands, such as in late pregnancy or lactation, with accompanying physiological or psychological stressors, it survives well and responds rapidly to re-alimentation. Prolonged-onset starvation occurs following a period of undernutrition in which the body reserves have been largely exhausted. In this scenario, body weight typically declines by about one-third of the standard reference weight and reaches a critical live weight where the animal is close to death. Such an animal can recover if re-alimented, but if no additional feed is available, the owner has to consider humane disposal.

Between these two extremes is undernutrition, a consequence of depletion of feed quality and quantity, often exacerbated by the annual cyclical change of pasture nutritional value from germination through growth to maturity. In consequence, the animal experiences weight changes ranging from slightly positive through zero to negative. The ME and protein rationing systems available provide a useful framework with which to calculate nutrient requirements of animals and to ration available feed to control weight loss. Husbandry systems have been developed to keep the maximum number of animals alive at minimum cost in times of feed scarcity. Undernutrition appears to be largely reversible, but recent reports of harmful effects of undernutrition on the embryo early in development suggest that permanent reduction in growth potential of offspring is possible.

The welfare impact of starvation is profound, as this poses a specific threat to the survival of the animal. Yet, its signs may be difficult to detect, with the animal entering a quiescent phase and changes in behaviour not being obvious. Changes in body condition are slow to eventuate and for all but the most careful of observers they go undetected. Correct feeding of livestock is not an easy task, requiring skill, resources, knowledge, and empathy. Whenever possible, a choice of feedstuffs should be provided, allowing the animal to make the decision as to what is best for it. However, when feed is limited, early action to move stock, purchase supplementary feed, or even sell some animals, will be rewarded by the preservation of as many animals in good condition as possible and the maintenance of their welfare.

References

Agenäs S, Heath MF, Nixon RM, Wilkinson JM, Phillips CJC (2006) Indicators of undernutrition in cattle. Anim Welf 15(2):149–160

Allden WG (1968) Undernutrition of Merino sheep and its sequelae. I. The growth and development of lambs following prolonged periods of nutritional stress. J Agric Res 19(4):621–638

Barker DJP, Clark PM (1997) Fetal undernutrition and disease in later life. Rev Reprod 2 (2):105–112

Blackstone C, Chang C-R (2011) Mitochondria unite to survive. Nat Cell Biol 13(5):521–522

Blair HT, Jenkinson CMC, Peterson SW, Kenyon PR, van der Linden DS, Davenport LC, Mackenzie DDS, Morris ST, Firth EC (2010) Dam and granddam feeding during pregnancy in sheep affects milk supply in offspring and reproductive performance in grand-offspring. J Anim Sci 88(Suppl 13):E40–E50

Briggs PK (1968) Drought feeding of sheep. Proc Aust Soc Anim Prod 7:13–19

Broom DM, Fraser AF (2007) Domestic animal behaviour and welfare, 4th edn. CABI, Wallingford

Brownlee H (1973) Wool production, ground cover and botanical composition of pastures grazed by Merino wethers in central western New South Wales. 1. Natural pasture. Barrel medic pasture. Aust J Exp Agric Anim Hus 13:502–515
Buckley F, O'Sullivan K, Mee JF, Evans RD, Dillon P (2003) Relationships among milk yield, body condition, cow weight and reproduction in spring-calved Holstein-Friesians. J Dairy Sci 86(7):2308–2319
Butler-Hogg BW (1984) Growth-patterns in sheep: changes in the chemical composition of the empty body and its constituent parts during weight-loss and compensatory growth. J Agric Sci (Camb) 103(1):17–24
Cheema UB, Sultan JI, Javaid A, Akhtar P, Shahid M (2011) Chemical composition, mineral profile and in situ digestion kinetics of fodder leaves of four native trees. Pak J Bot 43 (1):397–404
Chiy PC, Phillips CJC, Omed HM (1993) Sodium fertilizer application to pasture. 3. Rumen dynamics. Grass Forage Sci 48(3):249–259
Cooper MA, Arney DR, Phillips CJC (2010) The motivation of high and low yielding dairy cows to obtain supplementary concentrate feed. J Vet Behav Clin Appl Res 5(2):75–81
CSIRO [Commonwealth Scientific and Industrial Research Organisation] (2007) In: Freer M, Dove H, Nolan JV (eds) Nutrient requirements of Australia's ruminants. CSIRO, Melbourne
Denney GD, Hogan JP, Lindsay JR (1979) The digestion of barrel medic *Medicago truncatula* hay and seed pods by sheep. Aust J Agric Res 30(6):1177–1184
Doyle PT, Egan PK (1987) The palatability of mature subterranean clover. In: Wheeler JL, Pearson CJ, Robards JL (eds) Temperate pastures. CSIRO, Melbourne, pp 328–330
Ernst AJ, Wythes JR, Gulbransen B, Nicol DC, Venamore PC, Beasley RC, Knight JL, Dodt RM, Smith PC, Powell EE, Daly JJ, Schmidt PJ (1984) Molasses as a drought feed. Proc Aust Soc Anim Prod 15:224–226
Franklin MC (1950) The influence of diet on dental development in sheep. Bulletin 252. CSIRO, Melbourne
Gardener CJ, Freire LCL, Murray RM (1988) Effect of superphosphate application on the nutritive value of *Stylosanthes* spp.: native grass pasture for cattle. 1. Composition of the diet selected. Proc Aust Soc Anim Prod 17:190–193
Gearhart MA, Curtis CR, Erb HN, Smith RD, Sniffen CJ, Chase LE, Cooper MD (1990) Relationship of changes in condition score to cow health in Holsteins. J Dairy Sci 73 (11):3132–3140
Goldstein JL, Zhao T-J, Li RL, Sherbet DP, Liang G, Brown MS (2011) Surviving starvation: essential role of the ghrelin-growth hormone axis. In: Proceedings of the Cold Spring Harbour Symposium on Quantitative Biology 2011.http://symposium.cshlp.org/content/early/2011/07/22/sqb. 2011 Abstract Accessed 28 Oct 2013
Gomes LC, Di Benedetto G, Scorrano L (2011) During autophagy, mitochondria elongate, are spared from autophagy and maintain cell viability. Nat Cell Biol 13(5):589–598
Gutteridge RC, Shelton HM (eds) (1998) Forage tree legumes in tropical agriculture. Tropical Grassland Society of Australia, St Lucia
Hailey DW, Rambold AS, Satpute-Krishnan P, Mitra K, Souqrat R, Kim PK, Lippincott J (2010) Mitochondria supply membranes for autophagosome biogenesis during starvation. Cell 141 (4):656–667
Harding JE, Jaquiery JL, Hernandez CE, Oliver MH, Derriak JGB, Bloomfield FH (2011) Animal studies on the effects of early nutrition on long-term health. In: van Goudoever H, Guandalini S, Kleinman R (eds) Early nutrition impact on short and long-term health. The 68th Nestlé Nutrition Institute Workshop, pp 1–16
Hayder AM, Bakhiet AO, Mohammed AA (2006) Retrospective study on the prevalence of foreign body in goats' rumen: Omdurman Province, Khartoum State, Sudan (1998–2002). J Anim Vet Adv 5(6):449–451
Healey WB, Ludwig TB (1965) Wear of sheep's teeth. I. The role of ingested soil. N Z J Agric Res 8(4):73–752

Higgs ARB, Norris RT, Richards RB (1991) Season, age and adiposity influence death rates in sheep exported by sea. Aust J Agric Res 42(2):205–214
Hogan JP (1996) Options for manipulating nutrition if feed supply is immutable. Aust J Agric Res 47(2):289–305
Hogan JP, Phillips CJC (2008) Nutrition and the welfare of ruminants. In special issue, theme topic "unraveling animal welfare". Ann Rev Biomed Sci 10:T33–T50
Hogan JP, Phillips CJC (2011) Transmission of weed seed by livestock: a review. Anim Prod Sci 51(5):391–398
Hogan JP, Kennedy PM, McSweeney CS, Schlink AC (1989) Quantitative studies of the digestion of tropical and temperate forages by sheep. Aust J Exp Agric 29(3):333–337
Hogan JP, Petherick JC, Phillips CJC (2007) The physiological and metabolic impacts on sheep and cattle of feed and water deprivation before and during transport. Nutr Res Rev 20(1):17–28
Hunter RA, Siebert BD, Breen MJ (1976) The botanical and chemical composition of the diet selected by steers grazing Townsville stylo-grass pastures during a period of liveweight gain. Proc Aust Soc Anim Prod 11:457–460
Jaquiery AL, Oliver MH, Bloomfield MH, Harding JE (2011) Periconceptional events perturb postnatal growth regulation in sheep. Pediatr Res 70(3):261–266
Kamphues J (1998) Problems in feeding food producing animals to meet the energy and nutrient requirements as well as to enable animals' well-being. Dtsch Tierarztl Wochenschr 3:105
Kojima M, Kangawa K (2005) Ghrelin: structure and function. Physiol Rev 85(2):495–522
Leng RA (1970) Glucose synthesis in ruminants. Adv Vet Sci 14:209–260
Leng RA (1984) The potential of solidified molasses-based blocks for the correction of multinutritional deficiencies in buffaloes and other ruminants fed low quality agro-industrial by-products. In: The use of nuclear techniques to improve buffalo production in Asia. International Atomic Energy Agency, Vienna, pp 135–150
Lindsay DP, Setchell BP (1976) The oxidation of glucose, ketone bodies and acetate in the brain of normal and ketonaemic sheep. J Physiol 254:801–823
Lindsay JA, Mason GWJ, Toleman MA (1982) Supplementation of pregnant cows with protected proteins when fed tropical forage diets. Proc Aust Soc Anim Prod 14:67–68
Lindström T, Redbo I (2000) Effect of feeding duration and rumen fill on behaviour in dairy cows. Appl Anim Behav Sci 70(2):83–97
Lowry JB, Prinsen JH, Burrows DM (1994) *Albizia lebbeck*. A promising forage tree for semi-arid regions. In: Gutteridge RC, Shelton HM (eds) Forage tree legumes in tropical agriculture. CABI, Wallingford, pp 75–83
McGregor BA (2005) Nutrition and management of goats in drought. A report for the Rural Industries Research and Development Corporation, Australia. RIRDC Publ. No. 05/188. RIRDC Project No. DAV 217A. http://www.acga.org.au/goatnotes/drought_full_report.pdf. Accessed 28 Oct 2013
Minson DJ (1987) Plant factors affecting intake. In: Snaydon RW (ed) Managed grasslands. B. Analytical studies. Elsevier, Amsterdam, pp 37–144
Morris JG (1968) The survival feeding of beef cattle during drought. Proc Aust Soc Anim Prod 7:20–39
NHMRC [National Health and Medical Research Council] (2013) Australian code for the care and use of animals for scientific purposes, 8th edn. http://www.nhmrc.gov.au/guidelines/publications/ea28. Accessed 28 Oct 2013
Pell JM, Bergman EN (1983) Cerebral metabolism of amino acids and glucose in fed and fasted sheep. Am J Physiol 244:E282–E288
Phillips CJC, Hogan J (2011) Independent assessment of Dairy Australia Project No. TIG 124. Determining a suitable time off feed for bobby calf transport under Australian conditions, by Andrew Fisher, Peter Mansell, Bronwyn Stevens, Melanie Conley, Ellen Jongman, Mariko Lauber & Sue Hides. Report to Animals Australia, unpublished

Phillips CJC, Kitwood SE (2003) The effects of feed supplements on the production, metabolism and behaviour of dairy cows fed low levels of concentrates during lactation. J Appl Res Vet Med 1(3):187–195
Phillips CJC, Phillips AP (2010) Attitudes of Australian sheep farmers to animal welfare. J Int Farm Manag 5(2):17–43
Phillips CJC, Chiy PC, Zachou E (2005) The effects of cadmium in herbage on the apparent absorption of elements by sheep, in comparison with inorganic cadmium added to their diet. Environ Res 99(2):224–234
Phillips CJC, Wojciechowska J, Meng J, Cross N (2009) Perceptions of the importance of different welfare issues in livestock production in Australia. Animal 3(8):1152–1166
Phillips CJC, Mohamed MO, Chiy PC (2011) Effects of duration of exposure to lead in the diet on rumen metabolism and the accumulation of heavy metals in sheep. Small Rumin Res 100 (2–3):113–121
Purser DB, Hogan JP (1993) The sheep's environment. Effects upon wool. In: Doyle PT, Fortune JA, Adams NR (eds) Management for wool quality in Mediterranean environments. Western Australia Department of Agriculture, Perth
Rambold A, Lippincott-Schwartz J (2010) Starved cells use mitochondria for autophagosome biogenesis. Cell Cycle 9(18):3633–3634
Reid RL (1960) Studies on the carbohydrate metabolism of sheep. XI. The role of the adrenals in ovine pregnancy toxaemia. Aust J Agric Res 11(3):364–382
Reid RL, Hinks NT (1962) Studies on the carbohydrate metabolism of sheep. XVII. Feed requirements and voluntary intake in late pregnancy, with particular reference to prevention of hypoglycaemia and hyperketonaemia. Aust J Agric Res 13(6):1092–1111
Richards RB, Norris RT, Dunlop RH, Mcquade NC (1989) Causes of death in sheep exported live by sea. Aust Vet J 66(2):33–38
Ryley JW, Gartner RJW, Morris JG (1960) Drought feeding studies with cattle. 5. The use of sorghum grain as a drought fodder for non-pregnant heifers. Qld J Agric Sci 17:339–359
Schlink AC, Nguyen M-L, Viljoen GJ (2010) Water requirements for livestock production: a global perspective. Rev Sci Tech 29(3):603–619
Strydom S, Agenäs S, Heath MF, Phillips CJC, Rautenbach GH, Thompson PN (2008) Evaluation of biochemical and ultrasonographic measurements as indicators of undernutrition in cattle. Onderstepoort J Vet Res 75(3):207–213
Thornton RF, Hood RL, Jones PN, Re VM (1979) Compensatory growth in sheep. Aust J Agric Res 30(1):525–531
Underwood EJ, Suttle NF (1999) The mineral nutrition of livestock. CABI, Wallingford
Vasseur S, Paull DR, Atkinson SJ, Colditz IG, Fisher AD (2006) Effects of dietary fibre and feeding frequency on wool biting and aggressive behaviours in housed Merino sheep. Aust J Exp Agric 46(6–7):777–782
Vizard AL, Anderson GA, Buckley DJ (2005) Verification and value of the Australian Bureau of Meteorology township seasonal rainfall forecasts in Australia, 1997–2005. Meteorol Appl 12 (4):343–355
Warriss PD, Brown SN, Bevis EA, Kestin SE, Young S (1987) Influence of feed withdrawal at various times pre-slaughter on carcass yields and meat quality in sheep. J Sci Food Agric 39 (4):325–344
Weston RH (1996) Some aspects of constraint to forage consumption by ruminants. Aust J Agric Res 47(2):175–197
Weston RH, Hogan JP (1968) Rumen ammonia in relation to characteristics of the diet and parameters of nitrogen metabolism. Proc Aust Soc Anim Prod 7:359–363
Weston RH, Hogan JP (1986) Nutrition of herbage-fed ruminants. In: Alexander G, Williams OB (eds) The pastoral industries of Australia. Sydney University Press, Sydney, pp 183–210
Whitmore JS (2000) Drought management on farmlands. Kluwer Academic, Dordrecht
Whyte RO (1958) Plant exploration, collection and introduction. FAO Agricultural Studies, No. 41. FAO, Rome, Italy, p 117

Wilson AD (1977) The digestibility and voluntary intake of the leaves of trees and shrubs by sheep and goats. Aust J Agric Res 28(3):501–508

Wilson JR, Kennedy PM (1996) Plant and animal constraints to voluntary feed intake associated with fibre characteristics and particle breakdown and passage in ruminants. Aust J Agric Res 47 (2):199–225

Winter WH, Coates DB, Henricksen RE, Kerridge PC, McLean RW, Miller CP (1990) Phosphorus and beef production in northern Australia. 1. The response of cattle to fertiliser and supplementary P. Trop Grassl 24(3):170–184

Wu G, Bazer FW, Wallace M, Spencer TE (2006) Intrauterine growth retardation: implications for the animal sciences. Board-invited review. J Anim Sci 84(9):2316–2337

Yeates NTM, Schmidt PJ (1974) Beef cattle production. Butterworths, Sydney

Chapter 4
The Use of Feeding Behaviour in the Assessment of Animal Welfare

Birte L. Nielsen, Ingrid C. de Jong, and Trevor J. DeVries

Abstract Feeding behaviour is an important aspect of animal production, as it constitutes the link between the feed provided and that which is consumed. Measures of feeding behaviour can be used as a tool with which to gauge how an animal perceives the diet offered, as well as its motivation to feed, i.e. its level of hunger. The feed intake of an animal may also depend on the presentation of the food, the previous experience of the animal with a given food, and to what extent other competing motivations affect the behaviour of the animal. In this chapter, examples from ruminants, pigs, and poultry are provided on how measures of feeding behaviour can be used to elucidate the constraints on feeding imposed by the nutritional, physical, and social environment of our production animals. For ruminants, the relationship between meal patterns and rumen fermentation is described, together with the effects of feed delivery and feeding frequency on the feeding behaviour, in particular the sorting of feed components. Further examples are provided on how competition at the feed trough results in higher feeding rates and can lead to uneven distribution of the feed between animals in a group, dependent on space availability, trough design, and partitioning. We describe how the provision of fibre may alleviate hunger and aggression in parent stock, such as gestating sows and broiler breeders that are restrictively fed. Examples are provided of how automatic feeders can be used to detect changes in the feeding patterns of individual animals, which in turn can be used in the detection of health problems. It is described how grazing ruminants show behavioural adaptability not only to changes in sward composition and height but also to hot and cold environments. Other examples include studies of feeding behaviour that have explored

B.L. Nielsen (✉)
INRA, UR1197 Neurobiologie de l'Olfaction, Université Paris-Saclay, Jouy-en-Josas, France

Faculty of Science and Technology, Department of Animal Science, Aarhus University, Tjele, Denmark
e-mail: birte.nielsen@jouy.inra.fr

I.C. de Jong
Livestock Research, Wageningen University and Research Centre, Wageningen, The Netherlands

T.J. DeVries
Department of Animal Biosciences, University of Guelph, Guelph, ON, Canada

C.J.C. Phillips (ed.), *Nutrition and the Welfare of Farm Animals*, Animal Welfare 16, DOI 10.1007/978-3-319-27356-3_4

ways to reduce the stress experienced by piglets in connection with weaning, and how feather pecking in laying hens may be prevented by provision of roughage. Overall, the understanding of species-specific feeding behaviour and nutritional requirements is important for the assessment of animal welfare. It may help solve many feeding problems and allow animals to maximize the potential of the nutritional value of the feed provided to them.

4.1 Introduction

Feeding behaviour is an important aspect of animal production, as it constitutes the link between the feed and the intake. It consists of a number of different aspects, including finding the food supply, choosing the food, gaining and maintaining access to the food, as well as the amount of food eaten at any one time (meal size) and the speed of ingestion (feeding rate).

Measures of feeding behaviour can be used as a tool with which to gauge how an animal perceives the diet offered, as well as its motivation to feed, i.e. level of hunger. We expect a highly palatable feed to be eaten faster and in larger amounts than a feed with less nutritional value or with a suboptimal composition. In addition, animals that are hungry will eat faster than more satiated conspecifics (Nielsen et al. 2011). The feed intake of an animal may also depend on the presentation of the food, the previous experience of the animal with the particular food, and to what extent other competing motivations affect the behaviour of the animal. Research in rats has shown that if a given feed is first encountered when the animal is hungry, this feed is perceived as being more desirable than if the animal was satiated the first time it tasted the feed, a phenomenon known as state-dependent learning (Balleine 1992; Balleine and Dickinson 1998).

Animals on the farm usually feed in groups, involving competition between group members for access to the feed as well as social facilitation, where an animal begins to feed because it sees another animal feeding. The latter does not necessarily affect intake in the long term, but may add to the level of competition already present, if several group members attempt to eat simultaneously from limited trough space. Differences between animals in their feeding behaviour may reflect different levels of stress when feeding (Nielsen 1999a, b, 2004), and changes within an animal over time may also relate to changes in its health (Bach et al. 2007; Goldhawk et al. 2009; Desnoyers et al. 2011).

It is often difficult to identify feeding problems amongst group-housed animals. Individual feeding behaviour may be overlooked in farming environments where feed intake is measured at group level. However, the increased accessibility of technologies such as transponders, which automatically register individual visits to the food trough, will enable automatic, real-time, within- and between-animal comparisons of feeding behaviour, thus allowing producers to prioritize inspection of those animals, which show signs of aberrant feeding within a group as flagged by such systems (Shabi et al. 2005; Borderas et al. 2009).

In this chapter, we demonstrate the importance of feeding behaviour when considering the nutritional welfare of various livestock species. We focus on ruminants (in particular dairy breeds), poultry and pigs, and each species is dealt with in turn. This is not a comprehensive literature review, but a collection of examples of how we can use measures of feeding behaviour to elucidate the constraints on feeding imposed by the nutritional, physical, and social environment of the animal. In addition, we will show how measurements of temporal changes in feeding behaviour add valuable information relating to the health and welfare of the animal. The solution to many feeding problems can be found not only in the understanding of species-specific feeding behaviour and requirements but also by taking into account other behaviours, unrelated to feeding, that the animal may be more motivated to perform.

The notion that one can use feeding behaviour to gain information on the welfare of the animal is therefore the essence of this chapter, and this is independent of the economics of the feeding methods employed. Although economically viable solutions are important, a thorough analysis of the costs and benefits of different behavioural monitoring methods is beyond the scope of this chapter. It is, however, worth keeping in mind that what may appear technically impossible or too expensive to implement today often will become common practice a decade later. Apps on modern smart phones are examples of this type of technology.

4.2 Ruminants in Intensive and Extensive Production Systems

4.2.1 Feeding Behaviour Patterns and Management

Under natural grazing conditions, cattle will engage in foraging behaviour from 4 to 9 h/day (Hafez and Bouissou 1975). This feeding time is split into a number of smaller meals occurring throughout the day, with the largest meals occurring in the early morning and late afternoon. Modern, intensively housed ruminants have modified feeding patterns compared with these extensive situations. For example, dairy cattle fed a conserved ration typically consume their daily dry matter (DM) intake in up to 6 h/day, divided into more than six meals per day (DeVries et al. 2003). Management practices that cause ruminants to eat fewer and larger meals more quickly have been associated with large postprandial depressions in rumen pH and an associated increased incidence of subacute ruminal acidosis (SARA) (Krause and Oetzel 2006; Desnoyers et al. 2011). The reason for this increased risk is that ruminal pH declines following the ingestion of a meal; the rate of this pH decline increases with the size of the meal and as dietary effective fibre concentration decreases (Allen 1997). It has been shown in dairy cattle that as they spend less overall time feeding and increase their rate of feed consumption, daily salivary secretion is reduced (Beauchemin et al. 2008), decreasing the buffering capacity of the rumen and reducing rumen pH. This type of feeding pattern is related to greater depressions in rumen pH in goats (Desnoyers et al. 2011). Alternatively, when ruminants slow down their rate of DM consumption and have more frequent, smaller meals throughout the day, rumen buffering is maximised, large within-day depressions in pH are avoided, and the risk of SARA is decreased. Similar problems occur in some milk-fed calves, where consumption of milk infrequently and in large meals increases the risk of abomasal ulceration (Ahmed et al. 2002). Related to these changes in rumen conditions, greater feed digestibility has been found in dairy cows which feed more evenly across the 24-h period and spend more time eating and ruminating per unit of ingested feed (Aikman et al. 2008). Thus, to maximize health and welfare of ruminants, it is important to utilize feeding management strategies that promote the frequent consumption of feed in small meals throughout the day.

Intensively raised ruminants are commonly fed a total mixed ration (TMR), which is a homogenous mixture designed to minimise the selective consumption of individual feed components. Unfortunately, even when providing feed as a TMR, ruminants have been shown to preferentially select the grain component and discriminate against the longer forage components (Leonardi and Armentano 2003). Sorting of the diet has been shown to result in dairy cows consuming an inconsistent ration, whereby the ration actually consumed has more fermentable carbohydrates than intended and less effective fibre, thus increasing the risk of SARA (DeVries et al. 2008). Related to this, in two recent studies, it has been observed that such sorting of a TMR is associated with producing milk with lower fat percentage (DeVries et al. 2011; Fish and DeVries 2012). Sorting of the TMR can reduce the nutritive value of that remaining in the feed bunk in the last hours before feed delivery (DeVries et al. 2005; Hosseinkhani et al. 2008). For group-fed ruminants, this may be detrimental for those animals that do not have access to feed at the time when it is delivered, such as when there is intense competition at the feed bunk. In such cases, their welfare may be compromised due to inadequate nutrient intake to meet requirements for milk production (Krause and Oetzel 2006). Thus, at a herd level, imbalanced nutrient intake and altered rumen fermentation, as a result of sorting, have been shown to impact negatively the efficiency of feed conversion and production levels (Sova et al. 2013). The likelihood of sorting is largely affected by the composition of the ration, especially the proportion, type, and length of forage (Leonardi and Armentano 2003). A greater proportion of forage in the ration reduces feed sorting (DeVries et al. 2008) but also has other behavioural consequences: Nielsen et al. (2000) found that a high level of silage in the TMR (75 % silage DM) resulted in cows eating for longer, ruminating whilst standing more often, and spending less time lying inactive, compared with cows fed the low silage diet (50 % silage DM).

When grazing, cattle often synchronize their behaviour so that many animals in the group feed, ruminate, and rest at the same time (Miller and Wood-Gush 1991). Cattle in groups indoors also synchronize their behaviour, particularly at feeding (Curtis and Houpt 1983). Further, it has been reported that the circadian feeding patterns of lactating dairy cattle are mostly influenced by the time of feed delivery and feed push-up between feedings and time of milking (DeVries et al. 2003). Feed delivery is the primary stimulus to affect the daily feeding pattern of dairy cattle (DeVries and von Keyserlingk 2005). Feeding patterns are not influenced to the same degree by feed push-up, milking activity, or, as seen in extensive grazing cattle, the time of day. As a result, even though dairy cattle may still spread their meals throughout the day, the largest ones will occur right after the delivery of fresh feed. Since the delivery of fresh feed is such a strong stimulus to initiate feeding behaviour, increased frequency of feed delivery can influence short-term feeding behaviour and also affect cattle health, productivity, and welfare. When cattle are offered feed only once daily, there is a significant peak in feeding activity in the immediate time period following feed delivery compared with two-times feeding (DeVries et al. 2005). The behavioural response elicited by the delivery of fresh feed provided once a day could result in slug feeding (i.e. eating a large meal quickly) and predispose cows to SARA due to large circadian fluctuations in ruminal pH (Shabi et al. 1999). Conversely, cows fed more frequently (twice or more daily) tend to consume feed more evenly after each feed delivery, increasing, and distributing, their feeding time more consistently throughout the day (DeVries et al. 2005; Mantysaari et al. 2006; Hart et al. 2014). Further, providing feed twice a day or more often has also been demonstrated to reduce the amount of feed sorting as compared with feeding once daily (DeVries et al. 2005; Sova et al. 2013), which would further contribute to more consistent nutrient intakes over the course of the day. Several researchers have similarly reported that more frequent feeding of beef cattle also leads to more stable ruminal pH, with fewer daily fluctuations (Soto-Navarro et al. 2000; Robles et al. 2007). Similarly, increasing the frequency of milk replacer provision reduces the risk of abomasal ulceration in calves (Ahmed et al. 2002). Thus, frequent provision of feed to ruminant species, both young and mature, appears to be beneficial for their health and welfare.

Another common feeding management practice believed to stimulate feeding activity in floor-fed ruminants is feed push-up. When fed a TMR, cattle have a natural tendency to continually sort through the feed and toss it forward where it may no longer be within reach. DeVries et al. (2003) demonstrated that feed push-ups may not have the same stimulatory effect on feeding activity as does the delivery of fresh feed. However, it must be stressed that pushing up feed does play a vital role in ensuring that feed is accessible when cattle want to eat. Related to this is the need for dairy cattle to have feed available to them upon the return from milking. The presence of feed, particularly fresh feed, will promote feeding activity after milking, which helps to prevent dairy cows from lying down immediately after milking and as a consequence reduces the risk of subclinical mastitis (DeVries et al. 2010; Watters et al. 2014).

4.2.2 *Feeding Competition*

As social animals, ruminants tend to synchronize their feeding behaviour, including a strong desire to access the feed bunk as a group. Reduced space availability increases aggressive behaviour in cattle (Kondo et al. 1989). When feed bunk space is limited, increases in aggressive behaviour limit the ability of some cattle to access feed at times when feeding motivation is high, particularly after the delivery of fresh feed (DeVries et al. 2004; Huzzey et al. 2006; González et al. 2008). Competition at the feed bunk increases the rate at which dairy cows feed (Hosseinkhani et al. 2008), with fewer but larger and longer meals. Competition changes the distribution of dry matter intake over the course of the day, resulting in higher intakes during the later hours after feed delivery when much of the feed sorting has already occurred. Thus, increased competition promotes feeding behaviour that forces subordinate cows to consume more of their feed after the dominant cows have sorted (i.e. selected and eaten the most preferred items from) the TMR. Subordinate animals within groups of young and adult cattle therefore usually experience the greatest alteration of their feeding behaviour as a result of competition, such as greater eating rate (Harb et al. 1985; Olofsson 1999; DeVries and von Keyserlingk 2009). As a result, such feeding patterns that occur as a result of competition may increase ruminal acidosis and as a consequence increase the prevalence of lameness and liver abscesses (Nagaraja and Chengappa 1998; Cook et al. 2004).

Milk-fed calves show similar responses to increases in feeding competition. When milk-fed calves are unable to feed simultaneously, they are likely to have more frequent competitive interactions, especially around time of fresh feed delivery (when ad libitum-fed calves typically consume their largest meal; Appleby et al. 2001). Restricting teats available for group-housed calves increases frequency of competitive interactions and rate of intake, with reduced feeding time and milk intake (Jensen 2004; von Keyserlingk et al. 2004).

Similar results have also been observed for intensively fed feedlot cattle. González et al. (2008) found that when competition for access to concentrate feeders was increased, a higher number of aggressive displacements from these feeders were seen, together with large modifications in feeding behaviour. In that study, competition had a detrimental impact on the digestive function of heifers and subordinate animals, who exhibited the greatest increases in SARA and blood haptoglobin concentration. Interestingly, it was suggested that dominant animals may also be at greater risk of SARA because their priority of access to feeders allows them to eat the most desirable ingredients first, and eat uninterrupted, which results in larger meals with high proportions of highly fermentable carbohydrates (González et al. 2008).

Providing greater amounts of feed bunk space or feeding places will improve feed access; this increases feeding times and decreases competition, with subordinate animals showing the greatest responses (DeVries et al. 2004; von Keyserlingk et al. 2004; Huzzey et al. 2006; DeVries and von Keyserlingk 2009; Proudfoot et al. 2009). Competition for feed can also be reduced through improved feeding trough design and area, in particular the access barrier separating animals from their feed. Feed barrier types which allow animals to have a good backward view, to leave easily, and which provide physical separation between single feeding places confer the least stress on a feeding ruminant. A metal palisade fulfils all these characteristics in goats, whereas other feed barrier types (neck bar, wooden

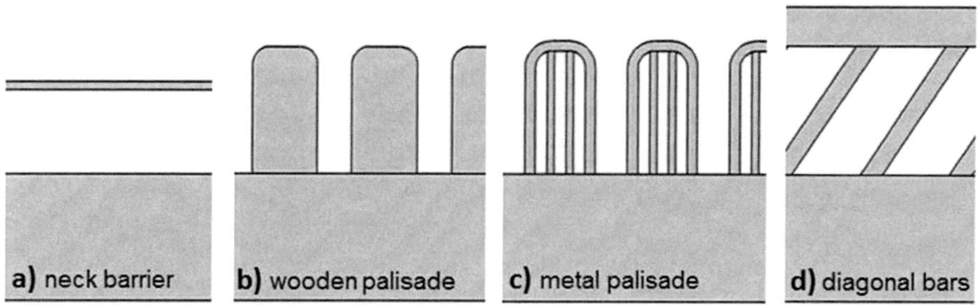

Fig. 4.1 Different types of feed barriers used for ruminants: (**a**) neck barrier, (**b**) wooden palisade, (**c**) metal palisade, and (**d**) diagonal bars. Only barrier (**c**) provides both a good backward view and easy withdrawal of the head as well as physical separation between feeding animals (adapted from Nordmann et al. 2011)

palisade, diagonal bars; see Fig. 4.1) were unfavourable in one or more characteristics (Nordmann et al. 2011). A headlock system for lactating dairy cows reduces competition at the feed bunk compared with a post-and-rail system (Endres et al. 2005; Huzzey et al. 2006). Another option to reduce competition is the use of partitions (feed stalls) between adjacent dairy cows at the feed bunk. DeVries and von Keyserlingk (2006) demonstrated that feed stalls resulted in increased feeding time and decreased competition, particularly for subordinate cows. Milk-fed calves also benefit (Jensen et al. 2008), with competition reduced at increased barrier sizes. Thus, providing increased feed access through more feeding space and feeding places, particularly when combined with a physical partition between adjacent feeding ruminants, will increase their welfare, particularly those that are socially subordinate.

4.2.3 Feeding Behaviour on Pasture

Pasturing of ruminants, which often exposes them to a variety of plant species, may be beneficial to both their health and welfare. Lambs exposed to diverse diets early in life have increased initial acceptance of new flavours in novel environments compared with lambs exposed only to monotonous diets early in life (Villalba et al. 2012). In a review by Rutter (2010), grazing sheep and cattle are described as having 70 % preference for clover when offered as a monoculture sward alongside grass and that dairy cattle and sheep when given the ability to select their own diet show higher levels of production than animals grazing mixed swards. There is a need for more research to investigate whether ruminants prefer grazing or eating prepared rations indoors, as results are equivocal (Charlton et al. 2011a, b). In one study, cows given a choice spent more than 90 % of their time indoors with access to a TMR (Charlton et al. 2011a). However, they had no previous experience of pasture and the distance to the grazing area from the indoor housing was approximately 100 m. When cows with grazing experience were used and the distance reduced to 40 m, cows spent more than 70 % of their time on pasture, which was not influenced by provision of a TMR on the pasture (Charlton et al. 2011b). Motivation to go outside to forage may depend on the relative quality of the TMR and the grass, the physiological state of the animal, and hence its needs for nutrients, but also on non-nutritional characteristics such as indoor floor quality, outdoor temperature, and level of precipitation. It is very difficult to disentangle these various aspects, and a more relevant question would be whether conditions exist under which provision of a TMR indoors without access to pasture is better for the welfare of the animal?

There can be considerable welfare issues associated with keeping sheep in hill flocks, and Morgan-Davies et al. (2008) demonstrated that good nutritional management during winter is a key component to good welfare and productivity. They also confirmed that body condition score (BCS) can be used as an easily applied, quantitative predictor of animal welfare and that poor mid-winter scores indicate a high risk of mortality, both at the flock and individual ewe level. In order to prevent mortality during cold spells for extensively kept ruminants, it is pertinent to ensure grazing on pasture that provides sufficient feed to allow a certain BCS to develop. For example, McGregor and Butler (2008) found no mortality due to hypothermia amongst Angora goats in good condition (BCS $\geq$ 2.5), but did find that mortality increased sharply when mean BCS was below 2. However, these researchers also found that mortality of animals with BCS $\leq$ 2.0 could be prevented, provided they grazed at the lowest stocking rate (7.5 animals/ha). In a 5-year study (Hejcmanova et al. 2009), heifers balanced their intake by increasing grazing time to counteract the decrease in grazing rate as the season progressed under both high and low grazing intensities. The grazing rate increased by 2 bites/min for each 1 cm decrease in sward height.

When assessing the welfare challenges to out-wintered pregnant suckler cows, Morgan et al. (2009) found that restricting the ME intake by 17 % had no effect on the behaviour of the cows, calving ease, or calf birth weight. Winter weather may be energetically demanding; however, the cows sought shelter from the wind to counteract the potentially adverse climatic conditions. There are indications that young cattle with no experience of winter range are less efficient in their use of the forage on the pasture and consequently lose more body fat and live weight than older, more experienced conspecifics (Beaver and Olson 1997).

In hot environments, the ability to feed in the shade—such as provided by trees—increases grazing time in sheep and leads to a less variable behaviour pattern when compared with sheep without access to shade (Ferreira et al. 2011). In many African countries, limited access to grazing land close to the village forces the pastoralists to walk with the livestock to distant pastures, where the cattle are often allowed to graze only during the daytime and where the daily walk to pasture limits time for foraging. Jung et al. (2002) recorded the behaviour and productivity of Zebu cattle when having 0.5 h daily walk (15 min each way; or 0.8 km twice) and 9 h on pasture compared with 4.5 h daily walk (2.25 h each way; or 10.8 km twice) and 5 h on pasture. The pastures were of low nutritional value and had a low density of grazed and browsed plant species. During the first 3 h of grazing, both groups spent a large proportion of the time grazing, but subsequently, the cows that had walked for more than 2 h to the pasture decreased their grazing activity, and spent more time standing, often in the shade of bushes or trees. This was probably due to this coinciding with the hottest time of the day. As a consequence, the animals grazed less overall compared with the first 5 h of the 9-h treatment, and the daily live weight gain was higher when 9 h were allowed on pasture. These results illustrate that animal welfare and performance are likely to be compromised when walking distance is long and time is reduced on pastures of poor quality. Bayer (1990) found that cattle with limited access to grazing during the day are able to postpone ruminating until night, but this was not confirmed by Jung et al. (2002).

4.3 Pigs in Intensive and Extensive Production Systems

4.3.1 Gestating Sows

In modern husbandry systems, gestating (pregnant) sows are typically fed a restricted food allowance just above their maintenance requirements allocated in one or two daily concentrate meals. The level of feed provided is well below the voluntary intake of the animal (Martin and Edwards 1994) and is associated with negative animal welfare aspects in the form of prolonged hunger (D'Eath et al. 2009). The restricted feeding of parent stock is described in more detail in Chap. 2 of this book by Tolkamp and D'Eath.

Allowing these animals to meet their energy requirements, by feeding more energy-rich diets, may simply change the nature of the problem, as the associated increase in live weight can have severe negative effects on both health and production. Another option is to make the diet more bulky to increase gut-fill and prolong (or induce) the feeling of satiety. A number of investigations of the effects of increased dietary fibre on animal welfare have been carried out in gestating sows. The general consensus is that although inclusion of high levels of dietary fibre to restrictedly fed sows may have satiating properties, it provides only transient reductions in feeding motivation, i.e. the animals are still feeling hungry. However, depending on the types of fibre used, bulky feed can reduce aggression and the incidence of stereotypies (Robert et al. 1993; Stewart et al. 2010).

The welfare of restrictively fed animals may be improved, at least in theory, if they spend more time obtaining their food. It may be possible to exploit the natural motivation of pigs to explore and engage in appetitive foraging, especially if the behaviour is occasionally rewarded by the discovery of edible morsels. One way of obtaining this is to provide the daily amount of feed in smaller, more frequent meals. Feeding the ration in two daily portions may increase night time satiety in gestating sows (Jensen et al. 2012). Another way of supplementing a quantitatively unsatisfying diet is to give these non-ruminant animals access to roughage. This can be in the form of litter or supplied from feeders or racks. Access to straw, although of limited value for satiety (Pedersen and Jensen 2009), allows the animals to engage in species-specific foraging.

The feeding of gestating sows is a good example to illustrate how the necessity for restricted feeding impinges on other aspects of the sow environment, which in turn affect their welfare (Meunier-Salaün et al. 2001). Until recently, many sows in modern production systems were kept in individual stalls with little possibility for movement in order to ensure that each sow got its allocated ration. The development of automatic feeding systems, such as the electronic sow feeder, allows controlled feeding of gestating sows kept in groups, as only one sow at a time can enter the feeder. However, because the sows are hungry, their attention is often centred on the entrance to the feeder, which can result in aggression and vulva-biting of the sow occupying the feeder (Olsson et al. 2011). On the positive side, data from electronic sow feeders can be used to detect health problems, such as lameness, when the feeding pattern of individual sows changes from that usually observed (Cornou et al. 2008; Hinrichs and Hoy 2010).

One attempt to reduce sow competition for access to an electronic feeder has been to equip it with a loudspeaker programmed to play an individual sound immediately before each sow receives a portion of feed (Manteuffel et al. 2010). In this study, animals were called six times daily to be fed the respective fraction of the daily feed allowance. On average, the animals entered the feeder following 80 % of their calls after 8 days of conditioning. At the start of the training, dominant sows blocked the entrance of the feeder, but this behaviour decreased significantly as these sows learned to recognize their own signal for entering the feeder. Subsequent work has shown that sows are able to remember their individual sounds in the following gestation period (Kirchner et al. 2014).

4.3.2 Lactating Sows and Their Piglets

In feral pigs, the weaning process is gradual, with suckling ceasing when the piglets are around 17 weeks of age (Petersen 1994). Piglets in most commercial production systems are weaned abruptly and at an early age (3–4 weeks). The young piglet is separated from the sow, and the regular intake of warm milk from a teat is suddenly replaced by a trough with solid food and cold water available from a nipple- or cup-drinker. This leads to a delay in feeding post-weaning and a high incidence of diarrhoea (Dybkjær et al. 2006). One way of ameliorating the transition between feeds at weaning is to allow piglets access to post-weaning feed whilst still with the sow, so-called creep feeding. Creep feeding initiated 5–10 days from birth has been found to reduce significantly the post-weaning diarrhoea scores in piglets (Yan et al. 2011).

In a series of experiments, a Dutch research group investigated whether piglets were able to learn aspects of eating solid feed from their mother and whether this would benefit the welfare of the piglets after weaning. They found that the presence of the mother during encounters with novel food decreased food neophobia in piglets, as they approached the novel food sooner, more piglets sampled it, and more of the novel food was eaten (Oostindjer et al. 2011a). Piglets were also found to eat more of a solid food before weaning if they could observe or participate when their mother was eating the same food, and they would eat more from a trough position previously used by the dam (Oostindjer et al. 2011b). Finally, when sows were fed a flavoured food during late gestation, piglets were found to eat more, weigh more, and have a lower occurrence of diarrhoea post-weaning when given access to similarly flavoured food that they had been exposed to prenatally (Oostindjer et al. 2010). These experiments demonstrate the importance of the sow in the vertical transfer of information about feed to her offspring and that this may be exploited in commercial settings to reduce the post-weaning stress experienced by piglets (Oostindjer et al. 2014).

4.3.3 Growing and Fattening Pigs

Growing and fattening pigs are most often fed ad libitum, and the concentrated form of the feed means that ingestion of the daily amount is very rapid. In nature, pigs will spend a considerable amount of time searching for food, and when this is no longer required, pigs may redirect their appetitive foraging towards their pen-mates in the form of rooting, chewing, or biting the ears, tails, and bellies of conspecifics. This may be at least partially prevented by supplying small amounts of food, which are not readily accessible, but which can be acquired through rooting, biting, or other manipulation. Provision of such rooting materials to growing pigs has been obligatory in the European Union (EU) since 2003 through legislation:

> "... pigs must have permanent access to a sufficient quantity of material to enable proper investigation and manipulation activities, such as straw, hay, wood, sawdust, mushroom compost, peat or a mixture of such, which does not compromise the health of the animals." (EC 2001)

Access to straw has a number of beneficial effects for the general welfare of pigs, as it allows natural rooting behaviour to occur (Fig. 4.2), reduces chewing of pen-mates, and the incidence of tail biting (Fraser et al. 1991; Moinard et al. 2003). Morgan et al. (1998) found more activity when straw was provided and, as a consequence, a higher frequency of aggressive encounters, but also a higher growth rate.

Cost of purchasing, manure handling constraints, and concern for hygiene and biosecurity are the main reasons that otherwise suitable roughage materials are not used more widely in production systems for growing pigs and gestating sows. Therefore, other measures, such as objects or devices, are often employed, although many of them, such as chains and rope, do not resemble the description found in the EU legislation above. For any enrichment measure, there is a risk of habituation when the animal becomes accustomed to the device resulting in declined usage over

Fig. 4.2 Provision of straw to growing pigs increases rooting and explorative activity, reduces the incidence of tail biting, and may improve growth (Photos: Liat Romme Thomsen)

time. In a study of environmental enrichment, van de Weerd et al. (2003) examined no less than 74 different objects to assess their potential enrichment value for pigs of different ages. They found that objects with high initial attractiveness were characterized by being odorous, deformable, and chewable, whereas sustained interest was maintained for objects, which were ingestible and destructible. From a feeding point of view, it is interesting that two of the most intensively and persistently used enrichment objects were lavender straw with whole peanuts and coconut halves suspended on a string.

The accessibility of the feed will affect the feeding behaviour of an animal, and this is important to consider for growing pigs kept in groups, as this impinges on aspects such as the maximum animal to trough ratio. If it is difficult for growing pigs to gain access to the feed (e.g. when a door is placed in front of the trough to protect the feeding animal) this will result in fewer, but longer meals, most often without affecting the level of daily feed intake (Morrow and Walker 1994; Nielsen et al. 1995a, b, 1996). If, in contrast, it is difficult to retain access to the feed (if competition is fierce), this may result in shorter and more frequent visits to the feed trough. Although enclosing the feeder makes access more difficult, it also protects the feeding animal, which is of importance when animal-to-trough ratio is high. It is interesting that feeding behaviour variables such as feeding rate and daily number of visits are now being considered in genome studies (Guo et al. 2015).

One aspect of feeding behaviour that changes in a social environment is the speed with which the food is ingested. Animals fed in a group-housed environment will increase their rate of eating even when this is not necessary to sustain a given level of intake. Pigs kept in groups of 20 have been found to eat faster than pigs in smaller groups (Nielsen et al. 1995b). This was not due to an overall time constraint, as there were times when the feeder was not occupied, even during daylight hours. Feeding rate may therefore be used as a tool to quantify welfare aspects of the social constraint affecting an individual pig within a group (Nielsen 1999a). It also highlights the motivation of social animals to synchronize their behaviour, including feeding, which is not possible if trough spaces are limited. Behavioural synchrony may lead to a pig resting with its pen-mates instead of feeding from an otherwise available trough. This may contribute to the reduced feed intake found in group housed compared with individually housed pigs (De Haer and Merks 1992; Morgan et al. 1999).

4.4 Poultry in Intensive and Extensive Production Systems

4.4.1 Laying Hens

The behaviour of modern laying hens is still very similar to that of their ancestors (Dawkins 1989); however, feeding-related behaviour is strongly affected by current housing and management conditions. Jungle fowl housed in a semi-natural environment spend about 60 % of their active time on feeding-related behaviours (Dawkins 1989), whereas for laying hens in cages and aviaries this is around 40 % (Hansen 1994). It has been shown that there is a relationship between the time spent feeding and foraging and feather pecking behaviour in laying hens (Blokhuis 1986; Nørgaard-Nielsen et al. 1993; Aerni et al. 2000; Rodenburg et al. 2013). Feather pecking behaviour (i.e. pecking at and pulling out the feathers of conspecifics) is one of the major welfare problems in laying hens. It not only causes damaged feathers but may also lead to injuries, increase the risk of cannibalism, and cause serious economic losses for the farmer (Blokhuis and Wiepkema 1989).

The time budget for feeding has been shown to be dependent on the interaction between feed form and litter substrate. Laying hens fed pellets and housed in floor pens without access to straw spent 29 % of the observed time on feeding-related behaviour, whereas laying hens fed mash feed and housed in floor pens with access to long-cut straw spent 52 % of the observed time on feeding behaviour. Time spent on feather pecking was very low in the latter group, but very high in the group fed pellets without access to straw (Aerni et al. 2000; El-Lethey et al. 2000).

Van Krimpen et al. (2008, 2009, 2011) added sand to the diet of laying hens and studied the effect on eating time and eating rate. Adding up to 20 % sand resulted in a proportional increase in feed intake during rearing as well as laying (van Krimpen et al. 2008, 2009, 2011). Surprisingly, with respect to eating behaviour, laying hens responded differently to the diet dilution in rearing and laying. During rearing, hens reduced the time spent eating when fed sand-diluted diets, thus increasing their eating rate (van Krimpen et al. 2009). In the laying phase, eating time and eating rate increased with increasing sand level in the diet (van Krimpen et al. 2008, 2011).

Fig. 4.3 Providing maize silage to prevent feather pecking in laying hens (Photo: Thea van Niekerk)

Interestingly, the hens that were fed the rearing diets diluted with 15 % sand had a significantly improved feather condition at 49 weeks of age. The authors suggest that the pullets were more focused on their feed during rearing and therefore less oriented towards the feathers of conspecifics, reducing the risk for feather pecking at a later age (van Krimpen et al. 2009).

Supplementing the diet of laying hens with roughage such as carrots, maize silage (Fig. 4.3), and barley–pea silage in the laying period reduced feather pecking behaviour compared with the control groups that received no additional roughage (Steenfeldt et al. 2007). It can be expected that hens fed roughage increase their time spent on feeding behaviour, but the study did not provide data on behavioural time budgets. Laying hens do not show a preference for a diet rich in antioxidants over a non-supplemented diet (Loetscher et al. 2014). However, chicks have been found to accept unfamiliar food more readily if it has the same flavour as that fed to the mother hen before the eggs were laid (Aigueperse et al. 2013). Chicks also adjust their pecking rate towards mobile food, such as meal worms, as a function of catch efficiency and protein requirements (Murphy et al. 2014).

Van Krimpen et al. (2009) investigated whether adding non-starch polysaccharides (NSP) to the diet of laying hens in the rearing and laying periods affected the behaviour of the hens. NSP content of the diet was increased by adding 10 % oat hulls. Adding NSP to rearing and laying diets did not increase eating time and feeding-related behaviour, in contrast to diet dilution with sand. However, feather condition of the hens was significantly improved when fed NSP-diluted diets in the laying period. Interestingly, the hens fed NSP-diluted diets during laying spent more time walking and ground searching (van Krimpen et al. 2009). Although the effect of NSP-diluted diets on laying hen behaviour merits further study, the examples here show that the nutrient content of the diet affects not only feeding behaviour but may also have effects on other behaviours and even reduce the risk of feather pecking.

4.4.2 Broiler Breeders

To prevent health and reproduction problems due to fast growth, broiler breeders of fast-growing breeds are fed restricted quantities of feed, especially during the rearing period (De Jong and Guemene 2011; van Krimpen and de Jong 2014). The restriction level is most severe between 8 and 16 weeks of age, and feed intake at these ages is between 25 and 33 % of the intake of ad libitum fed birds (De Jong and Jones 2006). There is substantial evidence that the feed restriction during rearing has negative effects on broiler welfare. As a consequence of the feed restriction, broiler breeders show abnormal behaviours, such as stereotypic pecking at non-feed objects, over-drinking, and pacing (De Jong et al. 2002; Mench 2002). It has been shown that the proportion of time spent in stereotypic spot-pecking increases with increasing level of feed restriction (Hocking et al. 1996). Also, feed-restricted broiler breeders will work for access to a foraging substrate, even when no food is provided in the substrate, and this high motivation to perform the appetitive component of feeding behaviour is indicative of hunger (Dixon et al. 2014).

In the period of most severe restriction during rearing, broiler breeders may consume their daily ration in as little as 15 min. The restricted quantities of feed provided lead to severe feed competition between the birds (Fig. 4.4). As a consequence, birds show aggressive pecking behaviour around the feeder (Jones et al. 2004; Hocking and Jones 2006), and they may damage each other when trying to reach the feeder when feed is provided (De Jong, pers. obs.). In this respect, it is very important to provide sufficient feeder space to enable each bird to reach the feeder and consume its daily portion.

Although the level of restriction in the laying period is much lower as compared with the rearing period, the motivation to feed is still very high in the laying period. Thus, female broiler breeders abandon the sitting phase of pre-laying behaviour (an important aspect of nesting behaviour) when the feed is provided during nesting (Sheppard and Duncan 2011).

Scattering the feed in the litter by using a so-called spin feeder instead of providing the feed in pans or chains (feed distributed over a conveyor belt) is a common practice during rearing in several countries. This practice has several advantages over providing feed in pans or chains. There is less competition around the feeders and foraging behaviour of the birds is stimulated. According to farmers, the increased foraging leads to less leg problems in the birds. Time to consume the daily ration is prolonged compared with feeding with pan feeders or feeding chains

Fig. 4.4 Sixteen-week-old broiler breeders at the moment their daily portion of feed is provided (Photo: Ingrid de Jong)

(De Jong et al. 2005a). It has even been suggested that by promoting a more natural feeding behaviour, the development of stereotypic behaviour due to feed restriction might be prevented, as found in sows (Lawrence and Terlouw 1993; Spoolder et al. 1995). However, although the pecking behaviour of broiler breeders when fed with the spin feeder is directed at the litter, it has been shown that there are no indications that this feeding method is less stressful in terms of reducing frustration or hunger due to feed restriction (De Jong et al. 2005a).

Skip-a-day feeding programmes are widely used in broiler breeding rearing in some parts of the world. There are various skip-a-day feeding programmes, such as 6 days of feed and 1 day without, or 5 days of feed and 2 days without. The reason for using these feeding programmes is that the amount of feed provided on a daily basis is very small, and by dividing the weekly amount over fewer days the feed is more likely to reach all birds and provide an adequate ration. It is not clear yet from the literature if these feeding programmes are better, worse, or equal as compared with daily feeding in terms of bird welfare. In some EU countries, such as the United Kingdom and Sweden, a daily feed allocation is required by law (EFSA 2010).

As for restricted-fed sows (see previous section), a number of investigations have studied the effect of increasing the fibre content of broiler breeder feed to promote gut-fill and satiety, thereby reducing the hunger and frustration from restricted feeding. Results of these studies differed with respect to the effects of diet dilution on indicators of hunger and frustration. Diet dilution reduced stereotypic object pecking in some studies (Hocking et al. 2004; De Jong et al. 2005b), but not in others (Jones et al. 2004; Hocking 2006). More recently, it has been shown that feed diluted with insoluble fibres was successful in reducing stereotypic pecking behaviour and promoting foraging, dust bathing, and comfort behaviours, whereas birds fed a diet with a high proportion of soluble fibre showed signs of discomfort, such as a hunched posture indicative of intestinal malaise (Nielsen et al. 2011). This indicates that the type of fibre used is important in terms of improving bird welfare. In a study of De Jong et al. (2005b), diluted feeds were also provided during the laying period. Surprisingly, the feed with the lowest energy density and the most positive effects on broiler breeder welfare during the rearing period seemed to be perceived as stressful by the birds in the laying period. This was probably due to birds fed this diet in laying spending more time feeding, thereby reducing the time available for other (essential) behaviours (De Jong et al. 2005b). Diets have been developed that allow for ad libitum feeding of broiler breeders without excessive live weight gain (Sandilands et al. 2006); however, it is necessary to include an appetite suppressant (calcium propionate) in these diets, which may create new and different welfare concerns.

4.4.3 Broilers

Broilers of fast-growing strains spend between 60 and 90 % of their time being inactive from an age of 3 weeks and older (Arnould and Faure 2003; Bokkers and Koene 2003). Promoting activity of broilers has been suggested to be effective in reducing leg disorders, one of the major welfare problems in fast-growing broilers. A possible method to stimulate activity of broilers and thereby reducing leg problems is applying feed restriction to stimulate foraging behaviour, either by quantitative or qualitative restriction (Nielsen et al. 2003). Qualitative restriction by diluting the feed with 15 % oat hulls was not effective in stimulating broiler activity. Quantitative feed restriction indeed stimulated bird activity, but it also induced hunger which had negative effects on bird welfare.

A recent study determined whether providing either whole grains or pelleted feed in the litter was effective in stimulating locomotor activity in broilers of a fast-growing strain. It was shown that providing whole grains in the litter did not stimulate foraging activity of the broilers at all. Broilers seemed not to be motivated to perform any foraging behaviour as long as they could meet their nutrient requirements through easily available feed in the feeders (Jordan et al. 2011), which is in line with an earlier study (Bizeray et al. 2002). Only when the entire diet was scattered in the litter did foraging activity increase significantly. A disadvantage of scattering the whole diet in the litter was a reduction in the growth rate of the birds (Jordan et al. 2011), and it has to be ensured that the birds are able to find sufficient feed to compensate for the increased activity.

4.5 Knowledge Gaps and Future Research Needs

4.5.1 Ruminants

- Little is known of the behavioural need of ruminants for grazing activity. Further research is needed to identify whether conditions exist under which provision of a TMR indoors without access to pasture is better for the welfare of the animal, taking into account also non-nutritional aspects.
- Ruminants are naturally selective in their feeding habits and have been shown under certain circumstances to select feed components to maximize health and productivity. Selective consumption (sorting) of prepared mixed rations has also been shown to be detrimental to dairy cattle, and thus many nutritional and management practices are employed to reduce this behaviour. Further research is needed to determine if ruminants are frustrated when fed mixed rations that are difficult to sort (i.e. selectively consume).
- Previous research has demonstrated that the timing and frequency of ration provision can affect ruminants' feeding behaviour patterns. Despite this knowledge, further research is needed on the optimal daily timing and frequency of ration provision to promote feeding behaviour patterns that ensure maximal rumen health, efficiency, and productivity.
- Reducing feeding competition is important for group-fed ruminants to ensure that they can consume their rations with little social pressure. Despite much research in this area, the optimal feeding space requirements for ruminants to maximise health and productivity and to minimise social pressure are not known; thus further research in this area is needed.

4.5.2 Pigs

- Although progress has been made, we still do not have methods to feed gestating sows without prolonged periods of hunger. Re-calculations of the optimum feeding level in terms of balancing productivity and welfare may be needed to ensure appropriate nutrition, combined with novel ways to ensure sufficient foraging.
- The weaning of piglets is one of the most abrupt and radical food changes imposed on a production animal. Changes to management that makes this transition more gradual will improve both the welfare and the performance of newly weaned pigs.
- Increased opportunities for fattening pigs to perform foraging behaviour will improve their welfare, reduce aggression, and may support faster growth. Better ways to supply roughage or similar foraging materials are needed, which also take into account the biosecurity of the production as well as concurrent adjustment of the manure handling.

4.5.3 Poultry

- Although diet dilution in laying hens seems promising as a means to reduce feather pecking behaviour, more research is needed into the composition of such diets and the effect on the (damaging) behaviour during rearing and laying.
- The effects of different feeding programmes (skip-a-day feeding) in broiler breeder rearing on bird welfare are largely unknown and require further research.
- Although some studies indicate that diet dilution may alleviate hunger in broiler breeders during rearing, the effect of such diets during the laying period needs further study. In addition, results of the different studies are not consistent, and therefore more research is needed before diet dilution can be implemented in practice.
- Different methods have been used to stimulate broiler activity to reduce leg problems, such as diet dilution, feed restriction, or scattered feeding. However, these methods have not yet been successful, and further research is needed into feeding methods and feed composition and their influence on broiler welfare.

4.6 Conclusions

Feeding behaviour can be used as a tool to detect or prevent health problems. Specifically, the relationships between meal patterns and rumen fermentation and health are critical. Automatic feeders can be used to detect changes in the feeding patterns of individual animals in a group, and these changes can be used to identify health problems, often before clinical symptoms are present.

Appropriate feeding management and feeder design can improve animal welfare. In relation to this, feeding method (delivery time, frequency) affects the time course and composition of feed consumed, which in turn can have impacts on animal health and productivity. Competition at the feed trough leads to higher feeding rates, an indicator of stress, and can lead to uneven distribution of the feed between animals in a group, where particularly the lower ranking individuals lose out. Space availability, scatter feeding, trough design, and partitioning between animals whilst feeding can reduce the negative effects of competition.

The provision of roughage is beneficial for both ruminants and monogastric animals. In the case of the latter, the provision of specific fibre types may in part alleviate hunger and reduce aggression in restrictively fed parent stock, such as gestating sows and broiler breeders. Optimal feeding methods, including provision of roughage, may prevent the development or outbreaks of feather pecking in laying hens. Environmental stimulation of growing pigs requires odorous, deformable, or chewable materials, which are ingestible or destructible over time.

The natural behavioural pattern of farm animals may be better exploited to improve welfare as well as productivity. Behavioural synchrony in group-housed animals affects their feeding behaviour and may affect feed intake. Grazing animals show behavioural adaptability not only to sward composition and height but also to hot and cold environments, as well as distance to pasture. Post-weaning stress in piglets may be lessened by utilising the innate curiosity and the sensory modalities of this species to the animals' advantage.

Acknowledgements The authors thank Liat Romme Thomsen and Thea van Niekerk for their photographs.

References

Aerni V, El-Lethey H, Wechsler B (2000) Effect of foraging material and food form on feather pecking in laying hens. Br Poult Sci 41(1):16–21

Ahmed AF, Constable PD, Misk NA (2002) Effect of feeding frequency and route of administration on abomasal luminal pH in dairy calves fed milk replacer. J Dairy Sci 85(6):1502–1508

Aigueperse N, Calandreau L, Bertin A (2013) Maternal diet influences offspring feeding behavior and fearfulness in the precocial chicken. PLoS One 8(10):e77583

Aikman PC, Reynolds CK, Beever DE (2008) Diet digestibility, rate of passage, and eating and rumination behavior of Jersey and Holstein cows. J Dairy Sci 91(3):1103–1114

Allen MS (1997) Relationship between fermentation acid production in the rumen and the requirement for physically effective fiber. J Dairy Sci 80(7):1447–1462

Appleby MC, Weary DM, Chua B (2001) Performance and feeding behaviour of calves on ad libitum milk from artificial teats. Appl Anim Behav Sci 74(3):191–201

Arnould U, Faure JM (2003) Use of pen space and activity of broiler chickens reared at two different densities. Appl Anim Behav Sci 84(4):281–296

Bach A, Dinares M, Devant M, Carre X (2007) Associations between lameness and production, feeding and milking attendance of Holstein cows milked with an automatic milking system. J Dairy Res 74(1):40–46

Balleine B (1992) Instrumental performance following a shift in primary motivation depends on incentive learning. J Exp Psychol Anim Behav Process 18(3):236–250

Balleine WB, Dickinson A (1998) The role of incentive learning in instrumental outcome revaluation by sensory-specific satiety. Anim Learn Behav 26(1):46–59

Bayer W (1990) Behavioural compensation for limiting grazing time by herded cattle in central Nigeria. Appl Anim Behav Sci 27(1–2):9–19

Beauchemin KA, Eriksen L, Nørgaard P, Rode LM (2008) Salivary secretion during meals in lactating dairy cattle. J Dairy Sci 91(5):2077–2081

Beaver JM, Olson BE (1997) Winter range use by cattle of different ages in south-western Montana. Appl Anim Behav Sci 51(1–2):1–13

Bizeray D, Estevez I, Leterrier C, Faure JM (2002) Influence of increased environmental complexity on leg condition, performance, and level of fearfulness in broilers. Poult Sci 81 (6):767–773

Blokhuis HJ (1986) Feather-pecking in poultry—its relation with ground-pecking. Appl Anim Behav Sci 16(1):63–67

Blokhuis HJ, Wiepkema PR (1989) Studies of feather pecking in poultry. Vet Q 20(1):6–9

Bokkers EAM, Koene P (2003) Behaviour of fast- and slow growing broilers to 12 weeks of age and the physical consequences. Appl Anim Behav Sci 81(1):59–72

Borderas TF, Rushen J, von Keyserlingk MAG, de Passillé AMB (2009) Automated measurement of changes in feeding behavior of milk-fed calves associated with illness. J Dairy Sci 92 (9):4549–4554

Charlton GL, Rutter SM, East M, Sinclair LA (2011a) Preference of dairy cows: indoor cubicle housing with access to a total mixed ration vs. access to pasture. Appl Anim Behav Sci 130 (1–2):1–9

Charlton GL, Rutter SM, East M, Sinclair LA (2011b) Effects of providing total mixed rations indoors and on pasture on the behavior of lactating dairy cattle and their preference to be indoors or on pasture. J Dairy Sci 94(8):3875–3884

Cook NB, Nordlund KV, Oetzel GR (2004) Environmental influences on claw horn lesions associated with laminitis and subacute ruminal acidosis in dairy cows. J Dairy Sci 87 (Suppl):E36–E46

Cornou C, Vinther J, Kristensen AR (2008) Automatic detection of oestrus and health disorders using data from electronic sow feeders. Livest Sci 118(3):262–271

Curtis SE, Houpt KA (1983) Animal ethology: its emergence in animal science. J Anim Sci 57 (2):234–247

D'Eath RB, Tolkamp BJ, Kyriazakis I, Lawrence AB (2009) 'Freedom from hunger' and preventing obesity: the animal welfare implications of reducing food quantity and quality. Anim Behav 77(2):275–288

Dawkins MS (1989) Time budgets in Red Junglefowl as a baseline for the assessment of welfare in domestic fowl. Appl Anim Behav Sci 24(1):77–80

De Haer LCM, Merks JWM (1992) Patterns of daily food intake in growing pigs. Anim Prod 54 (1):95–104

De Jong IC, Guemene D (2011) Major welfare issues in broiler breeders. Worlds Poult Sci J 67 (1):73–81

De Jong IC, Jones B (2006) Feed restriction and welfare in domestic birds. In: Bels V (ed) Feeding in domestic vertebrates. CABI, Wallingford, pp 120–135

De Jong IC, van Voorst S, Ehlhardt DA, Blokhuis HJ (2002) Effects of restricted feeding on physiological stress parameters in growing broiler breeders. Br Poult Sci 43(2):157–168

De Jong IC, Fillerup M, Blokhuis HJ (2005a) Effect of scattered feeding and feeding twice a day during rearing on parameters of hunger and frustration in broiler breeders. Appl Anim Behav Sci 92(1–2):61–76

De Jong IC, Enting H, Van Voorst S, Ruesink EW, Blokhuis HJ (2005b) Do low density diets improve broiler breeder welfare during rearing and laying? Poult Sci 84(2):194–203

Desnoyers M, Giger-Reverdin S, Sauvant D, Duvaux-Ponter C (2011) The use of a multivariate analysis to study between-goat variability in feeding behavior and associated rumen pH patterns. J Dairy Sci 94(2):842–852

DeVries TJ, von Keyserlingk MAG (2005) Time of fresh feed delivery affects the feeding and lying patterns of dairy cows. J Dairy Sci 88(2):625–631

DeVries TJ, von Keyserlingk MAG (2006) Feed stalls affect the social and feeding behavior of lactating dairy cows. J Dairy Sci 89(9):3522–3531

DeVries TJ, von Keyserlingk MAG (2009) Competition for feed affects the feeding behavior of growing dairy heifers. J Dairy Sci 92(8):3922–3929

DeVries TJ, von Keyserlingk MAG, Beauchemin KA (2003) Diurnal feeding pattern of lactating dairy cows. J Dairy Sci 86(12):4079–4082

DeVries TJ, von Keyserlingk MAG, Weary DM (2004) Effect of feeding space on the inter-cow distance, aggression, and feeding behavior of free-stall housed lactating dairy cows. J Dairy Sci 87(5):1432–1438

DeVries TJ, von Keyserlingk MAG, Beauchemin KA (2005) Frequency of feed delivery affects the behavior of lactating dairy cows. J Dairy Sci 88(10):3553–3562

DeVries TJ, Dohme F, Beauchemin KA (2008) Repeated ruminal acidosis challenges in lactating dairy cows at high and low risk for developing acidosis: feed sorting. J Dairy Sci 91 (10):3958–3967

DeVries TJ, Dufour S, Scholl DT (2010) Relationship between feeding strategy, lying behavior patterns, and incidence of intramammary infection in dairy cows. J Dairy Sci 93(5):1987–1997

DeVries TJ, Holsthausen L, Oba M, Beauchemin KA (2011) Effect of parity and stage of lactation on feed sorting behavior of lactating dairy cows. J Dairy Sci 94(8):4039–4045
Dixon LM, Brocklehurst S, Sandilands V, Bateson M, Tolkamp BJ, D'Eath RB (2014) Measuring motivation for appetitive behaviour: food-restricted broiler breeder chickens cross a water barrier to forage in an area of wood shavings without food. PLoS One 9(7):e102322
Dybkjær L, Jacobsen AP, Tøgersen FA, Poulsen HD (2006) Eating and drinking activity of newly weaned piglets: effects of individual characteristics, social mixing, and addition of extra zinc to the feed. J Anim Sci 84(3):702–711
EC [European Commission] (2001) Commission Directive 2001/93/EC of 9 November 2001 amending Directive 91/630/ EEC laying down minimum standards for the protection of pigs. Off J Eur Communities L 316:36–38, 1.12.2001
EFSA [European Food Safety Authority] (2010) Scientific opinion on welfare aspects of the management and housing of the grand-parent and parent stocks raised and kept for breeding purposes. EFSA J 8(7):1667 [81 p]
El-Lethey H, Aerni V, Jungi TW, Wechsler B (2000) Stress and feather pecking in laying hens in relation to housing conditions. Br Poult Sci 41(1):22–28
Endres MI, DeVries TJ, von Keyserlingk MAG, Weary DM (2005) Effect of feed barrier design on the behavior of loose-housed lactating dairy cows. J Dairy Sci 88(7):2377–2380
Ferreira RA, Estrada LHC, Thiebaut JTL, Granados LBC, de Souza VR (2011) Evaluation of the behaviour of Santa Ines sheep in a silvopastoral system in the north of Rio de Janeiro state. Cienc Agrotecnol 35(2):399–403
Fish JA, DeVries TJ (2012) Varying dietary dry matter concentration through water addition: effect on nutrient intake and sorting of dairy cows in late lactation. J Dairy Sci 95(2):850–855
Fraser D, Phillips PA, Thompson BK, Tennessen T (1991) Effect of straw on the behaviour of growing pigs. Appl Anim Behav Sci 30(3–4):307–318
Goldhawk C, Chapinal N, Veira DM, Weary DM, von Keyserlingk MAG (2009) Pre-partum feeding behavior is an early indicator of subclinical ketosis. J Dairy Sci 92(10):4971–4977
González LA, Ferret A, Manteca X, Ruíz-de-la-Torre JL, Calsamiglia S, Devant M, Bach A (2008) Performance, behavior, and welfare of Friesian heifers housed in pens with two, four and eight individuals per concentrate feeding place. J Anim Sci 86(6):1446–1458
Guo YM, Zhang ZY, Ma JW, Ai HS, Ren J, Huang LS (2015) A genomewide association study of feed efficiency and feeding behaviors at two fattening stages in a White Duroc * Erhualian F population. J Anim Sci 93(4):1481–1489
Hafez ESE, Bouissou MF (1975) The behaviour of cattle. In: Hafez ESE (ed) The behaviour of domestic animals, 3rd edn. Bailliere Tindall, London, pp 203–245
Hansen I (1994) Behavioral expression of laying hens in aviaries and cages—frequencies, time budgets and facility utilization. Br Poult Sci 35(4):491–508
Harb M, Reynolds VS, Campling RC (1985) Eating behavior, social dominance and voluntary intake of silage in group-fed milking cattle. Grass Forage Sci 40(1):113–118
Hart KD, McBride BW, Duffield TF, DeVries TJ (2014) Effect of frequency of feed delivery on the behavior and productivity of lactating dairy cows. J Dairy Sci 97(3):1713–1724
Hejcmanova P, Stejskalova M, Pavlu V, Hejcman M (2009) Behavioural patterns of heifers under intensive and extensive continuous grazing on species-rich pasture in the Czech Republic. Appl Anim Behav Sci 117(3–4):137–143
Hinrichs B, Hoy S (2010) Sick sows come later to the electronic feeding station—results of a first statistical analysis. Landtechnik 65(4):272–275
Hocking PM (2006) High-fibre pelleted rations decrease water intake but do not improve physiological indexes of welfare in food-restricted female broiler breeders. Br Poult Sci 47(1):19–23
Hocking PM, Jones EKM (2006) On-farm assessment of environmental enrichment for broiler breeders. Br Poult Sci 47(4):418–425
Hocking PM, Maxwell MH, Mitchell MA (1996) Relationships between the degree of food restriction and welfare indices in broiler breeder females. Br Poult Sci 37(2):263–278

Hocking PM, Zaczek V, Jones EKM, McLeod MG (2004) Different concentrations and sources of dietary fibre may improve the welfare of female broiler breeders. Br Poult Sci 45(1):9–19

Hosseinkhani A, DeVries TJ, Proudfoot KL, Valizadeh R, Veira DM, von Keyserlingk MAG (2008) The effects of feed bunk competition on the feed sorting behavior of close-up dry cows. J Dairy Sci 91(3):1115–1121

Huzzey JM, DeVries TJ, Valois P, von Keyserlingk MAG (2006) Stocking density and feed barrier design affect the feeding and social behavior of dairy cattle. J Dairy Sci 89(1):126–133

Jensen MB (2004) Computer-controlled milk feeding of dairy calves: the effects of number of calves per feeder and number of milk portions on use of feeder and social behavior. J Dairy Sci 87(10):3428–3438

Jensen MB, de Passillé AMB, von Keyserlingk MAG, Rushen JP (2008) A barrier can reduce competition over teats in pair-housed milk-fed calves. J Dairy Sci 91(4):1607–1613

Jensen MB, Pedersen LJ, Theil PK, Yde CC, Knudsen KEB (2012) Feeding motivation and plasma metabolites in pregnant sows fed diets rich in dietary fiber either once or twice daily. J Anim Sci 90(6):1910–1919

Jones EKM, Zaczek V, McLeod M, Hocking PM (2004) Genotype, dietary manipulation and food allocation affect indices of welfare in broiler breeders. Br Poult Sci 45(6):725–737

Jordan D, Stuhec I, Bessei W (2011) Effect of whole wheat and feed pellets distribution in the litter on broilers' activity and performance. Arch Geflugelkunde 75(2):98–103

Jung J, Yngvesson J, Jensen P (2002) Effects of reduced time on pasture caused by prolonged walking on behaviour and production of Mpwapwa Zebu cattle. Grass Forage Sci 57 (2):105–112

Kirchner J, Manteuffel C, Manteuffel G, Schrader L (2014) Learning performance of gestating sows called to the feeder. Appl Anim Behav Sci 153:18–25

Kondo S, Sekine J, Okubo M, Asahida Y (1989) The effect of group size and space allowance on the agonistic spacing behaviour of cattle. Appl Anim Behav Sci 24(2):127–135

Krause KM, Oetzel G (2006) Understanding and preventing subacute ruminal acidosis in dairy herds: a review. Anim Feed Sci Technol 126(3–4):215–236

Lawrence AB, Terlouw EMC (1993) A review of behavioral factors involved in the development and continued performance of stereotypic behaviors in pigs. J Anim Sci 71(10):2815–2825

Leonardi C, Armentano LE (2003) Effect of quantity, quality, and length of alfalfa hay on selective consumption by dairy cows. J Dairy Sci 86(2):557–564

Loetscher Y, Hospenthal AR, Kreuzer M, Messikommer RE (2014) Do antioxidants induce feed selection behavior in laying hens? Preference for, and palatability of, different antioxidant sources. Eur Poult Sci 78:1–11. doi:10.1399/eps.2014.65

Manteuffel G, Mannewitz A, Manteuffel C, Tuchscherer A, Schrader L (2010) Social hierarchy affects the adaption of pregnant sows to a call feeding learning paradigm. Appl Anim Behav Sci 128(1–4):30–36

Mantysaari P, Khalili H, Sariola J (2006) Effect of feeding frequency of a total mixed ration on the performance of high-yielding dairy cows. J Dairy Sci 89(11):4312–4320

Martin JE, Edwards SA (1994) Feeding behaviour of outdoor sows: the effects of diet quantity and type. Appl Anim Behav Sci 41(1–2):63–74

McGregor BA, Butler KL (2008) Relationship of body condition score, live weight, stocking rate and grazing system to the mortality of Angora goats from hypothermia and their use in the assessment of welfare risks. Aust Vet J 86(1–2):12–17

Mench JA (2002) Broiler breeders: feed restriction and welfare. Worlds Poult Sci J 58(1):23–29

Meunier-Salaün MC, Edwards SA, Robert S (2001) Effect of dietary fibre on the behaviour and health of the restricted-fed sow. Anim Feed Sci Technol 90(1–2):53–69, Special Issue

Miller K, Wood-Gush DGM (1991) Some effects of housing on the social behaviour of dairy cows. Anim Prod 53(3):271–278

Moinard C, Mendl M, Nicol CJ, Green LE (2003) A case control study of on-farm risk factors for tail biting in pigs. Appl Anim Behav Sci 81(4):333–355

Morgan CA, Deans LA, Lawrence AB, Nielsen BL (1998) The effects of straw bedding on the feeding and social behaviour of growing pigs fed by means of single-space feeders. Appl Anim Behav Sci 58(1–2):23–33

Morgan CA, Nielsen BL, Lawrence AB, Mendl MT (1999) Describing the social environment and its effects on food intake and growth. In: Kyriazakis I, Forbes M (eds) Quantitative biology of the pig. CABI, Wallingford, pp 99–125

Morgan CA, McIlvaney K, Dwyer CM, Lawrence AB (2009) Assessing the welfare challenges to out-wintered pregnant suckler cows. Animal 3(8):1167–1174

Morgan-Davies C, Waterhouse A, Pollock ML, Milner JM (2008) Body condition score as an indicator of ewe survival under extensive conditions. Anim Welf 17(1):71–77

Morrow ATS, Walker N (1994) A note on changes to feeding behaviour of growing pigs by fitting stalls to single-space feeders. Anim Prod 59(1):151–153

Murphy KJ, Hayden TJ, Kent JP (2014) Chicks change their pecking behaviour towards stationary and mobile food sources over the first 12 weeks of life: improvement and discontinuities. PeerJ 2:e626

Nagaraja TG, Chengappa MM (1998) Liver abscesses in feedlot cattle: a review. J Anim Sci 76 (1):287–298

Nielsen BL (1999a) On the interpretation of feeding behaviour measures and the use of feeding rate as an indicator of social constraint. Appl Anim Behav Sci 63(1):79–91

Nielsen BL (1999b) Perceived welfare issues in dairy cattle, with special emphasis on metabolic stress. In: Oldham JD, Simm G, Groen AF, Nielsen BL, Pryce JE, Lawrence TLJ (eds) Metabolic stress in dairy cows. British Society of Animal Science Occasional Publication, No. 24, Edinburgh, pp 1–7

Nielsen BL (2004) Behavioural aspects of feeding constraints: do broilers follow their gut feelings? Appl Anim Behav Sci 86(3–4):251–260

Nielsen BL, Lawrence AB, Whittemore CT (1995a) Effects of single-space feeder design on feeding behaviour and performance of growing pigs. Anim Sci 61(3):575–579

Nielsen BL, Lawrence AB, Whittemore CT (1995b) Effect of group size on feeding behaviour, social behaviour, and performance of growing pigs using single-space feeders. Livest Prod Sci 44(1):73–85

Nielsen BL, Lawrence AB, Whittemore CT (1996) Feeding behaviour of growing pigs using single or multi-space feeders. Appl Anim Behav Sci 47(3–4):235–246

Nielsen BL, Veerkamp RF, Lawrence AB (2000) Effects of genotype, feed type and lactational stage on the time budget of dairy cows. Acta Agric Scand Sect A Anim Sci 50(4):272–278

Nielsen BL, Litherland M, Nøddegaard F (2003) Effects of qualitative and quantitative feed restriction on the activity of broiler chickens. Appl Anim Behav Sci 83(4):309–323

Nielsen BL, Thodberg K, Malmkvist J, Steenfeldt S (2011) Proportion of insoluble fibre in the diet affects behaviour and hunger in broiler breeders growing at similar rates. Animal 5 (8):1247–1258

Nordmann E, Keil NM, Schmied-Wagner C, Graml C, Langbein J, Aschwanden J, von Hof J, Maschat K, Palme R, Waiblinger S (2011) Feed barrier design affects behaviour and physiology in goats. Appl Anim Behav Sci 133(1–2):40–53

Nørgaard-Nielsen G, Vestergaard K, Simonsen HB (1993) Effects of rearing experience and stimulus enrichment on feather damage in laying hens. Appl Anim Behav Sci 38(3–4):345–352

Olofsson J (1999) Competition for total mixed diets fed for ad libitum intake using one or four cows per feeding station. J Dairy Sci 82(1):69–79

Olsson A-C, Andersson M, Botermans J, Rantzer D, Svendsen J (2011) Animal interaction and response to electronic sow feeding (ESF) in 3 different herds and effects of function settings to increase capacity. Livest Sci 137(1–3):268–272

Oostindjer M, Bolhuis JE, van den Brand H, Roura E, Kemp B (2010) Prenatal flavor exposure affects growth, health and behavior of newly weaned piglets. Physiol Behav 99(5):579–586

Oostindjer M, Munoz JM, van den Brand H, Kemp B, Bolhuis JE (2011a) Maternal presence and environmental enrichment affect food neophobia of piglets. Biol Lett 7(1):19–22

Oostindjer M, Bolhuis JE, Mendl M, Held S, van den Brand H, Kemp B (2011b) Learning how to eat like a pig: effectiveness of mechanisms for vertical social learning in piglets. Anim Behav 82(3):503–511

Oostindjer M, Kemp B, van den Brand H, Bolhuis JE (2014) Facilitating 'learning from mom how to eat like a pig' to improve welfare of piglets around weaning. Appl Anim Behav Sci 160:19–30

Pedersen LJ, Jensen MB (2009) Hunger in pregnant sows: the effect of fibre diet and free access to straw. In: Proceedings of the 43rd International Congress of the International Society for Applied Ethology, Cairns, Australia, p 122

Petersen V (1994) The development of feeding and investigatory behavior in free-ranging domestic pigs during their first 18 weeks of life. Appl Anim Behav Sci 42(2):87–98

Proudfoot KL, Veira DM, Weary DM, von Keyserlingk MAG (2009) Competition at the feed bunk changes the feeding, standing, and social behavior of transition dairy cows. J Dairy Sci 92 (7):3116–3123

Robert S, Matte JJ, Farmer C, Girard CL, Martineau GP (1993) High-fiber diets for sows—effects on stereotypies and adjunctive drinking. Appl Anim Behav Sci 37(4):297–309

Robles V, González LA, Ferret A, Mantexa X, Calsamiglia S (2007) Effects of feeding frequency on intake, ruminal fermentation, and feeding behavior in heifers fed high-concentrate diets. J Anim Sci 85(10):2538–2547

Rodenburg TB, van Krimpen MM, de Jong IC, de Haas EN, Kops MS, Riedstra BJ, Nordquist RE, Wagenaar JP, Bestman M, Nicol CJ (2013) The prevention and control of feather pecking in laying hens: identifying the underlying principles. Worlds Poult Sci J 69(2):361–373

Rutter SM (2010) Grazing preferences in sheep and cattle: Implications for production, the environment and animal welfare. Can J Anim Sci 90(3):285–293

Sandilands V, Tolkamp BJ, Savory CJ, Kyriazakis I (2006) Behaviour and welfare of broiler breeders fed qualitatively restricted diets during rearing: are there viable alternatives to quantitative restriction? Appl Anim Behav Sci 96(1–2):53–67

Shabi Z, Bruckental I, Zamwell S, Tagari H, Arieli A (1999) Effects of extrusion of grain and feeding frequency on rumen fermentation, nutrient digestibility, and milk yield and composition in dairy cows. J Dairy Sci 82(6):1252–1260

Shabi Z, Murphy MR, Moallem U (2005) Within-day feeding behavior of lactating dairy cows measured using a real-time control system. J Dairy Sci 88(5):1848–1854

Sheppard KC, Duncan IJH (2011) Feeding motivation on the incidence of floor eggs and extraneously calcified eggs laid by broiler breeder hens. Br Poult Sci 52(1):20–29

Soto-Navarro SA, Krehbiel CR, Duff GC, Galyean ML, Brown MS, Steiner RL (2000) Influence of feed intake fluctuation and frequency of feeding on nutrient digestion, digesta kinetics, and ruminal fermentation profiles in limit-fed steers. J Anim Sci 78(8):2215–2222

Sova AD, LeBlanc SJ, McBride BW, DeVries TJ (2013) Associations between herd-level feeding management practices, feed sorting, and milk production in freestall dairy farms. J Dairy Sci 96 (7):4759–4770

Spoolder HAM, Burbidge JA, Edwards SA, Simmins PH, Lawrence AB (1995) Provision of straw as a foraging substrate reduces the development of excessive chain and bar manipulation in food restricted sows. Appl Anim Behav Sci 43(4):249–262

Steenfeldt S, Kjaer JB, Engberg RM (2007) Effect of feeding silages or carrots as supplements to laying hens on production performance, nutrient digestibility, gut structure, gut microflora and feather pecking behaviour. Br Poult Sci 48(4):454–468

Stewart CL, Boyle LA, McCann MEE, O'Connell NE (2010) The effect of feeding a high fibre diet on the welfare of sows housed in large dynamic groups. Anim Welf 19(3):349–357

van de Weerd HA, Docking CM, Day JEL, Avery PJ, Edwards SA (2003) A systematic approach towards developing environmental enrichment for pigs. Appl Anim Behav Sci 84(2):101–118

van Krimpen MM, de Jong IC (2014) Impact of nutrition on welfare aspects of broiler breeder flocks. Worlds Poult Sci J 70(1):139–150 van Krimpen MM, Kwakkel RP, van der Peet-Schwering CMC, den Hartog LA, Verstegen MWA (2008) Low dietary energy concentration, high non-starch polysaccharide concentration, and coarse particle sizes of non-starch polysaccharides affect the behavior of feather-pecking-prone laying hens. Poult Sci 87(3):485–496

van Krimpen MM, Kwakkel RP, van der Peet-Schwering CMC, den Hartog LA, Verstegen MWA (2009) Effects of nutrient dilution and non-starch polysaccharide concentration in rearing and laying diets on eating behavior and feather damage of rearing and laying hens. Poult Sci 88 (4):759–773

van Krimpen MM, Kwakkel RP, van der Peet-Schwering CMC, den Hartog LA, Verstegen MWA (2011) Effects of dietary energy concentration, nonstarch polysaccharide concentration, and particle sizes of nonstarch polysaccharides on digesta mean retention time and gut development in laying hens. Br Poult Sci 52(6):730–741

Villalba JJ, Catanese F, Provenza FD, Distel RA (2012) Relationships between early experience to dietary diversity, acceptance of novel flavors, and open field behavior in sheep. Physiol Behav 105(2):181–187

von Keyserlingk MAG, Brusius L, Weary DM (2004) Competition for teats and feeding behavior by group-housed dairy calves. J Dairy Sci 87(12):4190–4194

Watters MEA, Barkema HW, Leslie KE, von Keyserlingk MAG, DeVries TJ (2014) Relationship between post-milking standing duration and risk of intramammary infection in free-stall housed dairy cows milked three times per day. J Dairy Sci 97(6):3456–3471

Yan L, Jang HD, Kim IH (2011) Effects of varying creep feed duration on pre-weaning and post-weaning performance and behavior of piglet and sow. Asian Aust J Anim Sci 24 (11):1601–1606

Chapter 5
Welfare Is Affected by Nutrition Through Health, Especially Immune Function and Inflammation

Giuseppe Bertoni, Erminio Trevisi, Jos Houdijk, Luigi Calamari, and Spiridoula Athanasiadou

Abstract Welfare, for an animal, means the possibility to cope with the environment of life and to avoid, as much as possible, pain and suffering. It is obvious that good health is of paramount importance for this objective; nevertheless, not only clinical conditions but also any mild malaise status should be avoided for better results. The relationship between nutrition and health is well known; in fact, as well as offering nutrients for maintenance, growth, and reproduction, diet can influence health in a variety of ways: it can directly affect the fitness of infectious pathogens, detrimentally alter the environment in which they reside, and improve host resistance to the pathogens. Since host resistance to infection is mediated to a large extent through involvement of its immune system, the focus here is also on the effects of nutritional environment on immune responses, particularly to gastrointestinal parasites. We consider a variety of nutritional compounds, such as macro- and micronutrients and also secondary plant metabolites with evidence available of the effects on immunoregulation.

Good nutrition can affect animal health, avoiding tissue damage or metabolic disorders associated with deficiencies and excesses of nutrients. It also provides resilience against poisonous compounds found in toxic plants or produced by fungi growing on feed. Good nutrition works against bacteria growing in the digestive tract in the case of diets that are too highly fermentable. Nevertheless, nutrition is also important for inflammatory response, which is a major cause of reduced welfare; in fact, good nutrition not only reduces the frequency of such events but also minimizes their intensity and duration. In particular, we consider the possible modes of action of nutrients in immune function and propose directions on how we can further advance our knowledge on interactions between nutrition, immunity

G. Bertoni (✉) • E. Trevisi • L. Calamari
Istituto di Zootecnica, Facoltà di Agraria, Università Cattolica del Sacro Cuore, Via Emilia Parmense 84, 29122 Piacenza, Italy
e-mail: giuseppe.bertoni@unicatt.it

J. Houdijk • S. Athanasiadou
Disease Systems, Animal and Veterinary Sciences, SAC, Roslin Institute Building, Easter Bush, Midlothian EH25 9RG, UK

C.J.C. Phillips (ed.), *Nutrition and the Welfare of Farm Animals*, Animal Welfare 16, DOI 10.1007/978-3-319-27356-3_5

and inflammatory response, and thus welfare (as well an important factor of better performance and efficiency).

5.1 Introduction

In this chapter, underlining the relationship between welfare and health, we briefly remind the reader of a few general concepts. From the interpretation of them can follow different definitions of welfare and therefore different modalities to obtain it, as well as different models for its assessment.

According to Johnson (2009), but also Von Keyserlingk et al. (2009), there are three ways of thinking about welfare judgement:

- feeling based, which refers to the importance of ascertaining what an animal feels in terms of pleasure, suffering, distress, and pain;
- functioning based, which focuses on the fitness, performance, and health of animals;
- nature based, which considers the current state of breeding compared with natural behaviour of animals under natural conditions.

Often people, particularly scientists, tend to consider these different approaches as facets of the same issue (Sejian et al. 2011); Johnson (2009) suggests that welfare includes several disciplines: performance, physiology, anatomy, health, and behaviour. In the same direction is the attempt to consider—although in captivity—positive emotions of animals, as suggested by Broom and Fraser (2007, p. 14) '... it is also important to recognize and assess the degree of good welfare, i.e. happiness, contentment, control of interactions with the environment and possibilities of exploiting abilities'. Data are now available, as showed by Rousing and Wemelsfelder (2006), Boissy et al. (2007), and Napolitano et al. (2009), but a further issue becomes critical: to what extent each multiple aspect of welfare must be contemporary fulfilled? The question is not new because Webster (1994) has written, arguing about the 'Five freedoms', that 'absolute attainment of all five freedoms is unrealistic, indeed they are to some extent incompatible. Complete behavioural freedom, for example, is unhygienic for all us animals! In fact, all commercial husbandry systems have their strengths and weaknesses; therefore the five freedoms make it more difficult to sustain a sense of absolute outrage against any particular breeding system'.

Obviously, this does not mean that some animal needs can be fully neglected and compensated by others; on the contrary, according to Von Keyserlingk et al. (2009): 'Animal welfare science attempts to address all three types of concerns—feeling, functioning, nature—by identifying problems in production systems and developing solutions to these problems. The best solutions are win-win, improving the lives of cattle and the people who work with them'. Some difficulties can, however, arise from some recent suggestions for welfare assessment (Thompson 2010) that perhaps do not have much following among scientists, but are often supported by philosophers and citizens '... more consensus on the ultimate limits to constraining natural behaviours, even when no elements of animal health or affective states seem to be adversely affected' (Thompson 2010). Thompson suggests that the possibility to express natural behaviours covers any other negative effect on welfare (for instance, the hygienic ones suggested by Webster 1994). Thompson (2010) admits that '... it can lead one to take a more favorable view of the welfare of animal in traditional extensive setting such as pastures and barnyards, even though animals kept in such environments are vulnerable to predators and to infectious disease. Mortality and morbidity may then be viewed as "natural" in the sense of being normal for animals of the species living under conditions more closely resembling those in which behavioural drives and capabilities evolved. Given this way of interpreting the role of natural behaviours in animal welfare, extensive systems will be viewed as ethically better and more consistent with an animal's welfare, irrespective of objective measures in the animal bodies and animal minds categories'. For Thompson, the natural living criterion is clear: simply allowing animals to live as naturally as possible, but we see this approach as naïve; in fact, some natural conditions such as exposure to climatic extremes, diseases, parasite infections, and predator attacks cannot be good for the animals, as suggested by Von Keyserlingk et al. (2009). Furthermore, given the genetic changes that have occurred among cattle (and other animals), because of selection, there is great difficulty in identifying what their natural life is (Rushen et al. 2008). Thus, we need to know which aspects of the natural life are really important to animals, how this might be evaluated, and whether any benefits can be achieved in a more practical way. Allowing cattle to live a reasonably natural life could seem to require practices perceived as impractical or uneconomical in farm husbandry, such as leaving calves with their cows so that they can suckle. Thus, a key role for research in this area is also to discover ways of providing key features to the animals that are practical for producers (von Keyserlingk et al. 2009).

Furthermore, we do believe that the welfare of animals is most objectively measured by animal-based indicators (Calamari and Bertoni 2009; de Vries et al. 2011), which shows that a good level can be obtained in well-managed intensive farms. The models to evaluate welfare in farm animals can be considered to be objective because we and others carried out validation (Calamari et al. 2004); nevertheless, the current animal-based parameters mainly evaluate the animal response to negative 'stressors'. For the future, it is hoped that welfare evaluation could be possible by 'examining states of pleasure with a view to promoting these states in farm animals' (Sejian et al. 2011), as previously suggested by Boissy et al. (2007); this could in fact give greater assurances about the real welfare status of animals, bred in either intensive or extensive systems.

5.2 Welfare and Health

Health is a major component of welfare, though not the only one, as suggested by the OIE (2015), which defines animal welfare as (1) how well an animal is coping with the conditions in which it lives, (2) an animal having good welfare if, as indicated by scientific evidence, it is healthy, comfortable, well nourished, safe, able to express key aspects of behaviour, and if it is not suffering from unpleasant states, such as pain, fear, and distress, and (3) good animal welfare requires disease prevention and veterinary treatment for illness and injuries, appropriate shelters, management, nutrition, humane handling, and humane slaughter or killing.

Among the factors causing negative experiences to the animals (pain, discomfort, or distress and therefore low welfare), 'disease' is of major importance (Fregonesi and Leaver 2001). Even better, the third freedom '... from pain, injury and disease', suggests that health impairment is caused by physical pain as well as by psychological depression due to pro-inflammatory cytokines (Johnson and Finck 2001), as recently confirmed in humans by Dantzer et al. (2008). Therefore, a healthy animal cannot be merely considered as free from diseases, but a body in optimal homeostatic conditions.

Also, Broom (2006) showed that good health is central to good welfare, because any kind of pathology involves some degree of poor welfare. Nevertheless, pathological conditions can be caused by genetics; physical, thermal, and chemical injuries; infections and infestations; metabolic abnormalities; and nutritional disorders. All these causes are in fact associated with a status of irritation, injury, or infection that result in the induction of macrophages, and other cells, to release inflammatory mediators and eventually in a tissue inflammatory response.

Therefore, if inflammation is the common event of any health disorder, the relatively recent finding of a strong relationship between sickness behaviour and pro-inflammatory cytokines (IL-1 and TNFα) should be noted. Sickness behaviour comprises a set of nonspecific symptoms occurring in the case of infection and inflammation, including fever, weakness, malaise, listlessness, and inability to concentrate. This is of importance for two reasons. First it is one of the mechanisms that connect health disorders and low welfare (physical and mind effects), as the pro-inflammatory cytokines cause a negative impact on the brain (depression, irritability, and cognitive disorders), as suggested by Dantzer et al. (2008). Second, there is an exaggerated pain response, through eicosanoids release, which is a more physical effect (Watkins and Maier 2000). At the same time, the likely effects of cytokines at a subclinical level can be assessed objectively through the blood changes of acute phase proteins (positive and negative) (Trevisi et al. 2011a, 2015; Bertoni and Trevisi 2013); these proteins can be therefore a further tool to objectively evaluate animal welfare.

5.2.1 Nutrition, Health, and Welfare: Immune Function

Within the well-known effects of nutrition on health status (thus welfare) of farm animals, besides the obvious metabolic or infectious diseases, subclinical conditions are of major interest because they are relatively frequent in intensive farms and likely to reduce welfare, as shown in pigs (Fossum et al. 1998). The latter are also supposed to be the cause of malaise, observed, for example, in the postpartum period of dairy cows, which do not show clear symptoms, but are detectable in the blood through reduced negative acute phase proteins, indices of inflammation (Bertoni et al. 2008). This could explain (Bertoni et al. 2009) the frequent occurrence of cows without obvious problems, but listless, apathetic, depressed, and with little appetite. In the end, inflammation could therefore be one of the possible linkages between health and welfare, also in connection with nutrition. In fact, this relationship is not limited to infections, parasitism, metabolic diseases, digestive upsets, toxicosis, etc., but it should be extended to any situation characterized by a significant release of pro-inflammatory cytokines (with or without clinical symptoms) (Bertoni et al. 2015).

However, a large literature is also available on the effects of diet on host resistance, i.e. the ability of the host to regulate establishment, development, reproduction, and survival of the pathogen. Host resistance is in turn regulated and mediated through immunity, which is affected by nutrition. The immunomodulatory effects of dietary components on the immune responses to pathogens are not yet fully understood. A variety of approaches have been used to investigate them. In medical research, for example, most of the evidence derives from studies conducted in the absence of a specific disease or from in vitro studies (Kroner et al. 2015; Spinas et al. 2015; Hale et al. 2015). In animal science, many examples arise from studies where hosts have been infected with specific pathogens, and the relevant responses investigated (Sakkas et al. 2012; Tan et al. 2014). The dietary components may affect immunity in a variety of ways. Recent studies have suggested a potential role for dietary amino acids by regulating lymphocyte and macrophage activation, lymphocyte proliferation, antibody production, and also hormone excretion (Stephen and Avenell 2006; Osorio et al. 2014). Other studies have monitored the consequences of nutrient supplementation on antibody titres and cytokine profiles. For example, dietary supplementation with branched chain amino acids, such as leucine and valine, has resulted in changes in the blood cytokine profile in healthy subjects, such as an increase in interferon-γ production and a decrease in interleukin-4 release (Bassit et al. 2002), leading towards an improved Th1 immune response.

In this chapter, we draw most of our examples from the immune-modulatory effects of nutrition on gastrointestinal parasitism, though some evidence may indicate that the same principles are relevant for immunity towards bacterial and viral pathogens (Bikle 2008; Watzl et al. 2005; Spears and Weiss 2008; Volman et al. 2008). Here, we present examples of studies where direct measurements on immune responses (i.e. effector molecules, involving cells, or immunoregulating compounds) have been performed. It is noted that these are, almost by definition, correlated responses and thus they may not necessarily be protective (Maizels et al. 2004). We first focus attention on the effects of macro- and micronutrients on immune responses to gastrointestinal nematodes. We then examine effects of Plant Secondary Metabolites (PSM) on immune regulation and discuss the implications of these interactions. We finally look into possible modes of action of nutrients in immune functions and propose directions on how to advance our knowledge of interactions between nutrition and immunity.

5.2.1.1 Effects of Macro- and Micronutrients on Immunity

A considerable body of research has focused on the relationship between protein nutrition and gastrointestinal nematode parasitism. This seems sensible, as the immune system can be expected to draw on protein resources due to the highly proteinaceous nature of its effector molecules (Lewis and Austen 1981; MacRae 1993; Coop and Holmes 1996), which would support the view that protein supply can be limiting for immune functions. However, like any function, the immune system would also have requirements for energy and micronutrients. It has been shown that an increased availability of each of these resources can be expected to increase immune responses provided that they were initially scarce (Steinbrenner et al. 2015). For example, improved circulating and intestinal antibodies, circulating eosinophils, intestinal granulocyte numbers, in vitro proliferation of lymphocytes, increased cytokine levels, and neutrophil and lymphocyte functions have been observed following zinc, iron, selenium, vitamin E, vitamin A, and molybdenum supplementations (e.g. McClure et al. 1999; Finch and Turner 1996; Boulay et al. 1998; Koski and Scott 2001; Calder et al. 2002). On the contrary opposite, dietary energy supplementation does not appear to affect immunity to parasites. This inconsistency is not further discussed here, but has been the subject of a recent review (Houdijk 2012).

A large body of evidence shows that increased dietary protein supply improves expression of immunity to nematodes in growing animals (Table 5.1). In general, protein supplementation to growing hosts results in an increased concentration of circulating and local inflammatory cells, mast cell proteases, and circulating antibodies. Apparent inconsistencies may be attributed to the timing of sampling in relation to the development of the infection, the duration of nutrient supplementation, and the type of immune response assessed. For example, effects of protein nutrition on immune responses are often more pronounced during the expression phase of immunity (van Houtert et al. 1995). This would be related to the fact that immune responses will differ in relation to the development of the infection, with eosinophilia occurring at later stages in the infection. Protein supplementation also affects the various types of effector arms differently. Detailed studies in mice infected with intestinal nematodes show that protein supplementation increases worm expulsion by promoting gut-associated Th2 responses, whilst reducing pro-inflammatory Th1 responses (Ing et al. 2000).

Protein supplementation also affects effector responses in periparturient hosts (Table 5.2). Although such responses have been rather variable for some effector responses, e.g. circulating antibodies, effects on globule leukocytes appear consistent. The latter may reflect the long known importance of this type of inflammatory cell in expression of immunity to gastrointestinal nematodes in ruminants (O'Sullivan and Donald 1973; Huntley et al. 2004). Effects of protein nutrition on plasma antibodies may be more variable, since plasma antibody concentration depends on the balance between antibody production, utilization, and excretion in the milk. As protein supply increases milk production, effects of protein nutrition

Table 5.1 Effects of protein supplementation on immune responses in parasitized growing hosts

	Response	Effect[a]	Reference
Sheep			
Oesophagostomum columbianum	Goblet cells	↑	Dobson and Bawden (1974)
	Mast cells	↑	
	Eosinophils	↑	
	Globule leukocytes	↑	
Trichostrongylus colubriformis	Circulating eosinophils	↑, =	van Houtert et al. (1995), Kambara and McFarlane (1996)
	Plasma antibodies	=	Kahn et al. (2000)
	Lymphocyte stimulation	↑	
	Circulating T-cells	↑	
	Sheep mast cell proteases	↑	
Nematodirus battus	Mucosal mast cells	=	Israf et al. (1996)
	Sheep mast cell proteases	=	
	Globule leukocytes	↑	
	Mucosal eosinophils	↑	
	Circulating eosinophils	=	
	Plasma antibodies	↑	
Teladorsagia circumcincta	Mucosal mast cells	↑	Coop and Holmes (1996), Coop et al. (1995)
	Globule leukocytes	↑	
	Sheep mast cell proteases	↑	
Haemonchus contortus	Plasma antibodies	↑	Strain and Stear (2001)
Mice			
Heligmosomoides bakeri	Cytokines	↑, =, ↓	Coltherd et al. (2011), Tu et al. (2007a), Boulay et al. (1998)
	Circulating antibodies	↑, =, ↓	Ing et al. (2000), Tu et al. (2007b)
	Serum mast cell proteases	↑	
	Pro-inflammatory cytokines	↓	
	Intestinal mast cells	↑	
	Circulating eosinophils	↑	
	Intestinal eosinophils	↑	
Trichuris muris	Circulating antibodies	↓	Michael and Bundy (1992a, b)
Pigs			
Trichuris suis + *Ascaris suum*	Circulating eosinophils	↑	Pedersen et al. (2002)

[a]Relative to the unsupplemented group, ↓: decrease, ↑: increase, =: no change

Table 5.2 Effects of protein supplementation on immune responses in parasitized periparturient hosts

	Response	Effect[a]	Reference
Ewes			
Teladorsagia circumcincta	Mucosal mast cells	=	Houdijk et al. (2000, 2001, 2003, 2005, 2006)
	Globule leukocytes	↑	Huntley et al. (2004)
	Circulating eosinophils	=	
	Mucosal eosinophils	=	
	Plasma antibodies	↑, =, ↓	
Trichostrongylus colubriformis	Mucosal mast cells	=	Houdijk et al. (2009)
	Globule leukocytes	↑	
T. circumcincta + *T. colubriformis*	Mucosal mast cells	↑	Sykes et al. (2007)
	Mucosal eosinophils	=	
	Globule leukocytes	↑, =	
	Serum interleukin-5	↓	
	Plasma antibodies	=, ↓	
Rats			
Nippostrongylus brasiliensis	Mucosal mast cells	↑	Jones et al. (2009, 2011)
	Goblet cells	↑, =	
	Eosinophils	↑, =	
	Mucosal antibodies	↑	
	Plasma antibodies	=	

[a]Relative to the unsupplemented group, ↓: decrease,↑: increase, =: no change

on circulating antibodies may not always follow expected dose–response relationships (Houdijk et al. 2003).

A wide range of protein sources have been used to improve immunity to parasites, including soybean meal (e.g. Houdijk et al. 2003), fish meal (e.g. Donaldson et al. 1998), cottonseed meal (e.g. Shaw et al. 1995), casein (e.g. Jones et al. 2009), ovalbumin (e.g. Boulay et al. 1998), sunflower meal (e.g. van Houtert et al. 1995), and urea, increasing microbial protein (Strain and Stear 2001). Relatively high levels of protein supplementation were used in most of these studies. We suggest that if levels of protein supplementation are closer to host protein requirements, then protein quality (i.e. amino acid composition) becomes important to consider. This hypothesis was recently tested (Sakkas et al. 2012); the results demonstrated that Metabolizable Protein (MP) supplementation from xylose-treated soybean—higher rumen undegradable protein-based rations—is more effective at reducing parasitism than the same amount of MP from faba bean-based rations. The effect observed was probably attributable to the superiority of soya protein in terms of amino acid composition, as both diets were formulated to provide the same amount of MP. Thus, protein source is an important factor to consider when nutritional strategies are designed to improve the control of disease.

5.2.1.2 Consequences of Plant Secondary Metabolites on Immune Regulation

PSM are organic compounds that do not have an obvious function in plant growth and reproduction. More than 80,000 different PSM compounds have been described (Acamovic and Brooker 2005) and although they are present in all plants in small amounts, only a small proportion of plant species contain PSM in high concentration (Gershenzon 2002). Saponins, alkaloids, lactones, glycosides, and polyphenols are all PSMs. For some, immunomodulatory effects have been reported, for example, caffeine is a well-known alkaloid that may reduce neutrophil and monocyte chemotaxis and suppress pro-inflammatory cytokine Tumour Necrosis Factor alpha (TNF-α) production (Horrigan et al. 2006). Similarly digoxin, which is a cardiac glycoside, also suppresses production of TNF-α and plays a role in regulating the nuclear factor NF-κB pathway, which is a central regulator of inflammation (Yang et al. 2005). Saponins, in contrast, are known for their immunogenic abilities; an example is Quil A, which has been used as adjuvant for a variety of veterinary vaccines for a number of years (Potter et al. 2008).

A particular class of PSM are the polyphenols, which are the most abundant antioxidants in human diets and have long been known for their immunomodulatory activity (Spears and Weiss 2008). This is of relevance to immune cells as they are particularly susceptible to oxidative stress (Neyestani 2008). A specific class of polyphenols are the condensed tannins (CT), which are polymers of flavonoid units. The consumption of CT has resulted in a reduction of the level of gastrointestinal nematode parasitism in sheep and goats (Athanasiadou et al. 2000, 2001; Paolini et al. 2003a, 2005; Marley et al. 2006; Niezen et al. 2002). The administration of a CT extract to infected goats resulted in an increase in small intestinal mucosal mast cell numbers (Paolini et al. 2003a), cells associated with expulsion of nematodes from the gastrointestinal tract (Huntley et al. 1992; Harrison et al. 1999). Such effects may not be notable in other gut compartments; the same CT extract did not affect either the number of mast cells in the abomasum (true stomach) of infected sheep (Paolini et al. 2003b) or the parasite burden (Athanasiadou et al. 2001). This may be related to reduced bioavailability of CT, as CT are in complexes with dietary protein or other macromolecules in the abomasum (Mueller-Harvey 2006). Appropriate pH and other conditions in the small intestine may enable the disassociation of CT from such complexes and probably allow them to exert their antiparasitic activity.

The effects of CT have also been studied under grazing conditions. Infected sheep grazing on the CT-rich legume sulla (*Hedysarum coronarium*) showed increased numbers of mucosal mast cells and globule leukocytes in their abomasum compared to sheep grazing on conventional pastures (Tzamaloukas et al. 2006). Antibody titres against specific parasite antigens were higher in sheep grazing on sulla compared to conventional legumes (Niezen et al. 2002). Furthermore, the consumption of CT-rich sainfoin (*Onobrychis viciifolia*) increased ovine small intestinal concentrations of mucosal eosinophils, mast cells, and Paneth cells, when compared to those grazing on conventional pastures (Ríos-de Alvarez et al. 2008). However, such immunomodulatory effects may not necessarily be attributed to direct activity of CT. As mentioned before, CT form stable complexes with dietary protein and protect it from rumen degradation, which results in higher availability of protein for absorption from the small intestine (Min and Hart 2003). Thus, CT consumption has the ability to increase host protein availability and immunogenic effects observed may at least in part arise from improved protein nutrition, as discussed above, rather than the CT per se.

5.2.1.3 Implications and Future Research Directions

Although it is well recognized that host nutritional environment affects different manifestations of immunity to pathogens, in many cases it is not clear how this is modulated. Most of the evidence available to date is supportive of a quantitative relationship between nutrition and immunity. Hosts aim to satisfy their nutritional requirements for maintenance, growth and reproduction, and effective immune response through nutrition. As a consequence, under nutrient scarcity, immunity may be penalized, and nutritional supplementation may improve immunity (Coop and Kyriazakis 1999). However, a more qualitative relationship between certain nutrients and effector arms of immunity may also be implicated, i.e. specific nutrients or their deficiency may affect the expression of genes that are responsible for specific immune traits. For example, branched chained amino acids may act as regulators of gene expression via modulation of the initiation phase of mRNA translation (Kimball and Jefferson 2006). Also, protein supplementation resulted in the downregulation of pro-inflammatory pathways in parasitized hosts, which aids parasite expulsion (Athanasiadou et al. 2011). Although it is unclear how this is mediated, it appears that there is a relationship between nutrient availability and the synthesis of certain components of the immune response which may consequently result in an enhanced immune response to parasites. The evidence presented above describes the relationship between nutrient supply and immune response during the chronic parasitic infections; it is unclear whether this is the same during acute infections. There is already abundant evidence on the effects of nutrition on gene expression and such investigations have been greatly advanced by the exploitation of high-throughput technologies (Athanasiadou and Huntley 2008). Gene expression profiling has revealed canonical pathways related to ageing, metabolic disease, and diabetes that can be affected by host nutrition. Similar advances will enable the investigation of nutritional regulation of expression of genes responsible for the immune responses controlling diseases in mammals, which could ultimately lead to effective dietary interventions to prevent disease. By bringing together basic immunology and nutrition (in terms of both nutrient supplementation and deficiency) and by exploring the novel technologies that can assure an unbiased approach, we can advance our understanding of the immunomodulatory effects of nutrition (Athanasiadou and Huntley 2008). Such advances promise to make significant contributions to nutritional recommendations alone, or when nutrition is used in combination with other treatments, to improve protection from infectious disease, at prevention or treatment level.

5.2.2 *Nutrition and Health (Welfare): Metabolic Diseases and Gastrointestinal Disorders*

As shown in the previous section, undernutrition and nutrient deficiency affect the innate, adaptive, and cellular immune responses and suppress immune functions, leading to deregulation of a coordinated host response to infection, thereby enhancing the virulence of pathogens (Wintergerst et al. 2007) and thus compromising health (and welfare). However, according to Calder and Jackson (2000), inadequate dietary intake and disease per se are both immediate causes of malnutrition and they reinforce one another synergistically (Scrimshaw et al. 1968). In fact, malnutrition makes the individual more susceptible to infection and decreases immune defences against invading pathogens. In turn, pathogens influence nutritional status, mediated by changes in dietary intake, absorption, and nutrient requirements and losses of endogenous nutrients. Malnutrition takes several forms that often appear in combination, such as protein-energy malnutrition and deficiencies in micronutrients such as vitamin A, Fe, Zn, and I (Calder and Jackson 2000). Nevertheless, our major interest is on the effects of malnutrition on tissue damage, metabolic diseases, and digestive disorders: all causes of cytokine release and, consequently, inflammation and welfare impairment (Fig. 5.1). Only to give an idea of them, we suggest some examples selected within the two major ruminant systems: extensive and intensive. In theory, both systems could guarantee satisfactory nutrition conditions; nevertheless, it appears reasonable to suggest that:

- **extensive systems** are typically affected by irregular deficiency problems: primarily energy and proteins. These problems occur in hot-dry or in cold seasons because of low feed intake, as a consequence of the unavailability of feed or of the availability only of feed of very low quality. In extensive systems, vitamin deficiencies could be a minor problem; otherwise mineral deficiencies (both macro and trace) can be important if there is a low soil availability, or the spoilage of mature grass and difficulty in providing mineral supplements. Moreover, extensive systems do not exclude excess problems (i.e. minerals in some areas) or the effects of toxicants (i.e. phytoestrogens with subterraneous clover; mycotoxins from Claviceps spp. in fungi-infected grasses, especially fescues; or toxic weed intake in some cases of pasture shortage). A well-known case is 'fescue toxicosis' because fescue can be infected by an endophyte producing ergotamine. This infected forage not only depresses milk yield and dry matter intake (DMI) but also causes an increase in respiration rate and suppresses the immune system. Tissue damage, metabolic or infectious diseases, and the consequences of any kind of malnutrition are undoubtedly responsible for pain

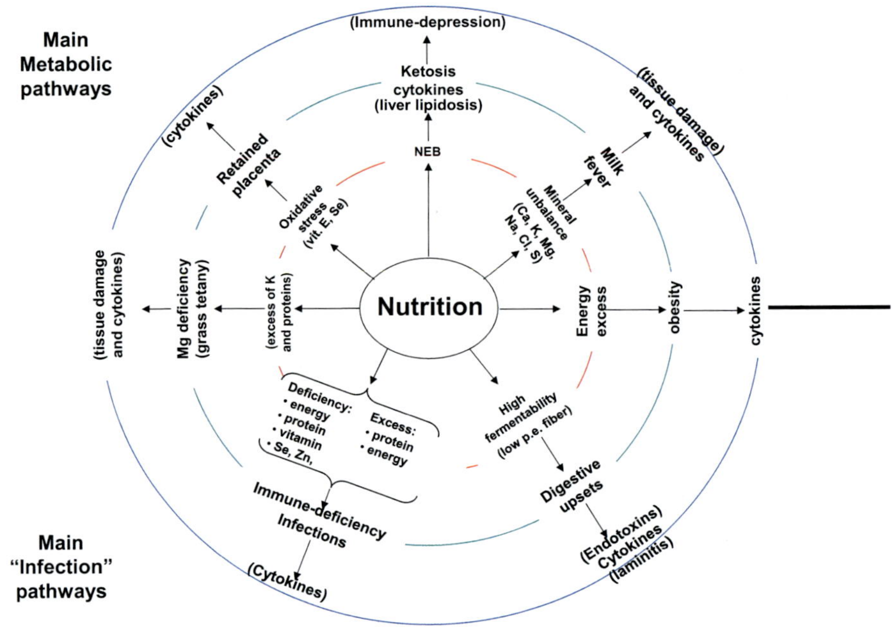

Fig. 5.1 Different pathways from nutrition to health disorders, pro-inflammatory cytokines, and inflammation

and suffering (Yates 1962; Thompson and Stuedemann 1993). In extensive conditions, some different metabolic or infectious diseases can also occur depending on species (sheep or cattle), production type (meat or milk), and yield levels. A specific case of metabolic disease occurring as a consequence of energy deficiency is the pregnancy toxaemia of ewes and goats at the end of pregnancy (particularly when carrying twins and triplets), as a result of lack of concentrates or good forages (LeValley 2010). In our experience, at least in late pregnant goats, hyperketonemia is preceded by changes of blood indices of inflammation (Trevisi et al. 2005), suggesting that a reduced voluntary intake, induced by cytokines also without clinical symptoms, could contribute to the energy shortage and thus to the metabolic disease. In cattle, in particular in early lactation of high-yielding dairy cows on pasture, low availability of nutrients from forages causes high BCS losses (Roche et al. 2006) and low fertility (Horan et al. 2006).

In extensive systems, mineral disorders are more frequent because feeding is mainly based on local forage production without proper supplements. We distinguish:

- primary deficiency: i.e. areas with low cobalt (vitamin B_{12} deficiency with anaemia and low propionic acid utilization), low selenium (white muscle), or low phosphorus (pica and bone disorders); and

– excesses that in a few cases can cause toxicological phenomena (e.g. molybdenosis), but more often can interfere with the utilization of other minerals causing a secondary deficiency. This is the case, for example, with potassium (and NH_3) excess which can depress the Mg absorption, leading to a metabolic disease (grass tetany) in ruminants. For copper as well, a secondary deficiency can be caused by an excess of sulphur and molybdenum; the consequences are not only anaemia, bone disorders, neonatal ataxia, cardiovascular disorders, and diarrhoea but also higher susceptibility to infections (Underwood and Suttle 1999).

Therefore, any effort to avoid—especially in extensive systems—the negative consequences of malnutrition on health is useful to improve animal welfare.

- **Intensive systems** are usually characterized by high feed intake of well-balanced diets. Deficiencies are rare, while the excesses are infrequent if the diets are calculated to cover the corresponding requirements of high-yielding animals. This, however, does not exclude any unsuitable feeding: occasionally of micronutrients but more often pertinent to energy and proteins. We give a few examples regarding dairy cows. Of major interest is the case of high-yielding dairy cows that, despite DMI almost doubling in the first 2–3 months of lactation compared with the dry period, still remain for several weeks in Negative Energy Balance (NEB), mainly because of physical limit of the gut size (we previously showed that also inflammation can reduce appetite, Bertoni et al. 2009). For this reason, it is difficult to properly judge the welfare situation of these animals: they may have fulfilled the intake capacity (the satiety centre is therefore satisfied), so it appears reasonable to suppose that they do not suffer hunger. However, nutrients may be in deficit and therefore very early lactation is characterized by increased risk of metabolic (ketosis) or infectious (e.g. mastitis) diseases, also due to the nutrient deficiency or unbalance (Drackley 1999).

  Nevertheless, with this kind of production system, two further aspects of nutrition–health–welfare relationship are noteworthy:

  – defining the best energy (and nutrient) level for the dry period to avoid metabolic diseases and susceptibility to infectious diseases in the transition period; and
  – managing the diets at the end of pregnancy and at the beginning of lactation to reduce the risks of highly fermentable diets, which are useful to avoid the NEB, but are a possible cause of rumen–intestine fermentation disorders.

The dry period has an opposite situation with respect to early lactation: low requirements and relatively high DMI because the gastrointestinal tract remains over-developed. The hypothesis to 'fill' the reserves to be utilized after calving has been shown to be wrong at least for calcium and phosphorus (excesses increase the risk of milk fever, as suggested by Martin-Tereso and Vestegen 2011), for protein (excess of NH_3), and particularly for energy, which causes overweight cows. This last aspect is an old issue since the discovery of the 'fat cow syndrome' (Morrow 1976): obese cows at calving time show a very high frequency of retained placenta, milk fever, ketosis, and steatosis, as well as an increased mortality. More recent data showed that cows fed a high-energy diet in the dry period, no matter if they are obese or not, appear more susceptible to puerperal metabolic diseases and high liver triglycerides levels (Van den Topp et al. 1995). Moreover, Janovick Guretzky et al. (2007) also observed that these cows show the highest frequency of inflammatory conditions (clinical or subclinical). This last observation tends to support the idea that, in ruminants, a relatively prolonged energy excess could trigger a metabolic syndrome like condition (similar to the overweight or obese human situation, characterized by a 'low level' of inflammation with insulin resistance, type 2 diabetes, high blood pressure, fatty liver, etc.). Therefore, cows fed a high-energy diet in the last part of pregnancy, or which are too fat at calving time, will be at risk of dystocia, retained placenta, ketosis, etc., as well as more susceptible to infectious diseases (Janovick and Drackley 2010).

The risk of high fermentability of the diet—that here will be considered for early lactating cows—is, however, typical of any kind of intensive animal husbandry, no matter whether ruminant or monogastric. In both cases, the excesses of starch (and sugars) fermented in the rumen or escaping into the large intestine are responsible for substantial pH reduction (besides an increased and modified microbial population) that tends to increase the endotoxins (lipopolysaccharides, LPS) or microbe translocation (Khafipour et al. 2009a, b). These phenomena are responsible inter alia for liver abscesses (Nagaraja and Chengappa 1998), laminitis and sole ulcers (Nocek 1997). Particularly in horses, LPS can cause large intestine colics as well as laminitis (Frape 1986). Nevertheless, the presence of LPS in the rumen and intestine is not always the cause of the appearance of the above health problems, *provided* that the mucosa is wholesome. An increase of mucosa permeability, due to acidosis itself (Wannemuehler 1995; Minuti et al. 2014) or to 'external' causes (e.g. hypoxia, heat stress, heavy physical work, etc.; Alexander 1998), or to a combination of these factors, can contribute to the above translocation and consequent inflammation with several negative effects (Fig. 5.2).

It is obvious that pain and suffering are strictly linked to all the above health disorders associated with energy overnutrition and consequent clinical or subclinical inflammations, as well as digestive disorders, because LPS and other antigens are responsible of pro-inflammatory cytokine release (Klasing 1988).

5.2.3 Nutrition and Health (Welfare): Inflammatory Response

According to the previous discussion on the relationships between malnutrition and health-welfare impairment, the need for a special care to obtain a proper nutrition, either in extensive or intensive systems, is obvious:

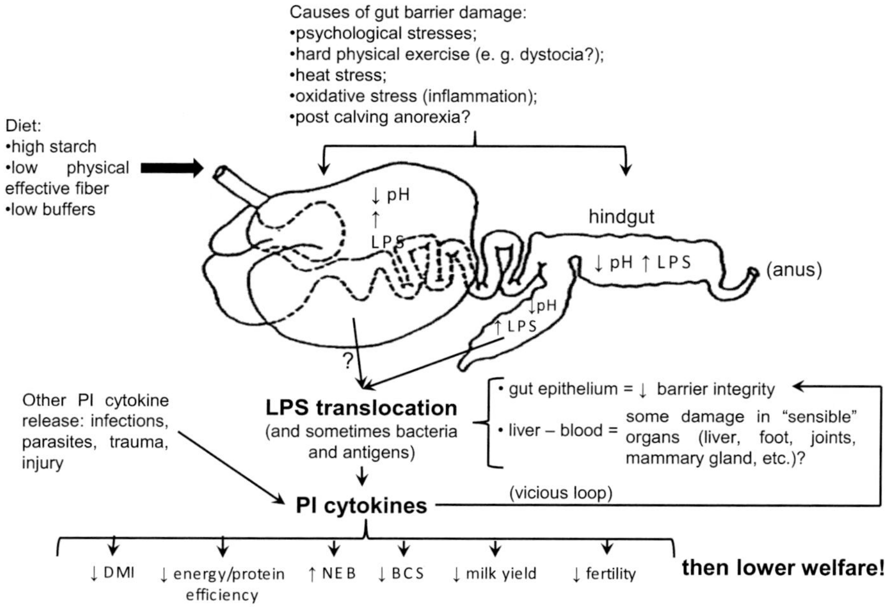

Fig. 5.2 Factors relating to digestive disorders and gut permeability increase in periparturient dairy cows and their possible consequences. *LPS* lipopolysaccharides, *DMI* dry matter intake, *NEB* negative energy balance, *BCS* body condition score, *PI* pro-inflammatory

- to avoid disorders and/or diseases (hence inflammation), and
- to reduce the inflammatory response when the phenomenon occurs.

The first aspect needs only few practical references; otherwise, the inflammatory response and its relationship with nutrition deserve further considerations. For a proper nutrition in extensive systems, besides the adoption of genetically suitable breeds and strains and the proper management of land and stocking rate, the main achievable tool is the administration of supplements to prevent serious deficiencies of proteins (i.e. with urea, such as by giving molasses-urea blocks) and of minerals, namely, macro (i.e. magnesium and phosphorus) and trace elements (cobalt, copper, iodine, selenium, and zinc). A further feasible tool is the reduction or avoidance of toxins: especially those in toxic weeds and mycotoxin-producing moulds.

In intensive systems, and especially for high-yielding dairy cows, two major conditions—both around calving—have to be properly managed:

- before calving to ensure constant (relatively high) DMI until the delivery and to minimize the risk of metabolic and infectious diseases around calving (e.g. retained placenta, milk fever, displaced abomasum, mastitis, lameness). Although it needs further study, prevention of overweight conditions appears important. According to recent suggestions of Drackley (personal communication) and corresponding to the feeding rules developed in our Institute (Bertoni and Trevisi 1997), the proper diet for the last 40–50 days of pregnancy in dairy cows must be at 'low' concentrations of energy, Ca and P, but with high fibre content. Only in the last few days of pregnancy can more fermentable carbohydrates be added;

- after calving to ensure a quick rise in DMI (and energy-protein intake too), but remembering that some body condition loss is unavoidable. Therefore, the opposite risks of ketosis (shortage of energy intake) and of acidosis (excess of fermentable energy intake) must be avoided, while special care has to be taken to prevent diseases and to optimize gut–liver function as well as immune system activity.

  Particularly in case of Total Mixed Rations (TMR) for high-yielding cows, a diet with high energy and protein contents is required, although safe for rumen and intestinal bacteria. Furthermore, a proper amount and type of rumen (and intestine) buffers is needed to avoid the acidosis, and therefore the risk of digestive diseases, with their potentially dangerous consequences.

Before and after calving it is noteworthy that not only feeding but other aspects of animal management (termed 'subclinical stressors' by Drackley et al. 2005) are also important to consider with the aim of achieving the above-mentioned objectives, namely good health and welfare. Thus, the non-feeding tools are good hygiene, but also prophylaxis (vaccines, antiparasitic treatments, good mastitis treatments at drying off, feet care, etc.), as well as the reduction of dystocia (e.g. by using sires that generate small calves) and the avoidance of stress conditions at calving time. From a nutritional point of view, it not only means maximizing the capacity of the immune system (through appropriate nutrient supply) but also preventing metabolic and digestive diseases which occur more frequently around calving (milk fever, retained placenta, rumen acidosis, etc.) (Trevisi et al. 2011a).

5.2.3.1 Inflammatory Response and Its Consequences

Inflammation is an adaptive response to that triggered by noxious stimuli and conditions, such as infections and tissue injury (Medzhitov 2008). It is started by immune system cells (mainly macrophages) which produce pro-inflammatory cytokines, which in turn activate the release of eicosanoids. The inflammatory event is often needed for the induction of a strong immune response: recruitment of more phagocytic cells and specific immune response against the invader if present (Goddeeris 2010). Because inflammatory response is per se responsible for health problems and thus for suffering (low welfare), it merits a special attention also with reference to nutrition. In fact, the inflammatory cascade is not always triggered by a 'pathogen', but it can start from any cause of tissue damage, and that can be also of nutritional origin: vitamin and mineral deficiencies with their pathological consequences; pro-oxidative conditions allowing the rise of Reactive Oxygen Metabolites (ROM); digestive disorders associated with

LPS or bacteria translocation; or toxins contained in feeds, such mycotoxins. More recently it has been observed that the energy excess, which is per se the cause of obesity, can be also included in the risky nutritional factors; indeed it is responsible for a 'low-grade systemic inflammation, called metaflammation (referring to metabolically triggered inflammation) associated with obesity and chronic disease' (Egger and Dixon 2009). This risk factor has been clearly demonstrated for humans and experimental animals and recently has been suggested to create a similar condition for dairy cows fed high-energy diets before calving (Janovick Guretzky et al. 2007; Janovick and Drackley 2010).

Besides the above inflammatory conditions that are not justified by usual diseases: infectious or metabolic (Grimble 1998), the inflammation mechanism is per se of great interest because it is self-triggering (Bertoni et al. 2015) and has the potential to damage the host. For example, the oxidant molecules produced as part of the inflammatory response may damage the still healthy tissues. Morbidity and mortality in a number of infectious diseases and conditions are related to excessive production of pro-inflammatory cytokines. This can occur because the acute phase of the inflammatory process normally leads to the recovery and to the reinstatement of the usual functions within a few days. However, if the response is not properly functioning, the process can develop into a chronic low-grade inflammatory state, which may trigger different diseases (Medzhitov 2008; Wärnberg et al. 2010).

All of the above-described negative consequences of uncontrolled inflammation suggest the need for a mechanism to reduce its side effects. In fact, a number of protective systems exist naturally to limit the inflammatory activities of the immune system (Grimble 1998). These activities include an increase of glucocorticoids, which are known to have an immunosuppressive effect, but they also show an anti-activation effect on the Nuclear Factor-kappa B (NF-kB), a nuclear factor responsible for expression of pro-inflammatory genes (Barnes 1997). Another protective system is the production of natural antagonists (like the anti-inflammatory cytokines, e.g. IL-10 and TGFbeta) towards the action of Interleukin-1 (IL-1) and Tumour Necrosis Factor alpha (TNFα), thus reducing the acute phase response. The rise of antioxidant defence components, which reduce ROM concentration, is also against the NF-kB release and therefore suppresses the pro-inflammatory cytokine production (Rimbach et al. 2002). Moreover, other important intrinsic anti-inflammatory systems could be adrenomedullin, as showed by Elsasser et al. (2006), and adiponectin as described by Ouchi and Walsh (2007).

5.2.3.2 Nutrition and Inflammation

If the inflammatory phenomena can be partly independent from usual diseases and, much more important, can have different pathways within the same 'disease', it means that immune functions, mainly aimed to fight 'infections', can be partly separated from inflammatory responses, as suggested by Uauy (2007). Moreover, it could be a reasonable specific approach to nutritional aspects:

- to avoid inflammation events, because a proper nutrition guarantees tissue integrity and reduces metabolic diseases, independent from infections; and
- to reduce the level (and length) of the inflammation process, minimizing its negative effects.

The first aspect is per se obvious and we merely mention that several nutrients exert an immune-modulating function (Suchner et al. 2000), and this would contribute to the reduction of infections and parasitizations. Some of them are macronutrients such as amino acids (glutamine, arginine, cysteine, branched amino acids, and taurine), nucleotides, and lipids (monounsaturated and polyunsaturated fatty acids), but energy and protein malnutrition must be also included (Uauy 2007). Others are micronutrients such as vitamins (A, D, E, and C) and trace elements (Fe, Zn, and Se) (Maggini et al. 2007), but their respective carrier proteins in plasma Retinol Binding Protein (RBP), transferrin, ceruloplasmin, albumin, etc., can be also important (Uauy 2007). Moreover, Moore et al. (2006) showed in humans that nutritional deficiencies during prenatal or early postnatal life may disrupt the development of immune system. This confirms the suggestion of Koutsos and Klasing (2001) for farm animals: a nutrient deficiency or excess during embryonic development may affect immunocompetence and may result in increased susceptibility to disease.

Nevertheless, our major interest is not limited to the improvement of the immune cell function—which is in practice the objective of good nutrition as demonstrated—but involves the reduction of inflammatory responses, when they occur. This latter possibility is in fact much less known and scarcely utilized in animal husbandry. The attempt to control inflammation by nutritional tools is not new (Grimble 1992); nevertheless, the tremendous increase in knowledge of useful nutrients and of their mechanisms in the last few years justifies a more decisive use of this strategy (Fig. 5.3).

The interaction between genes and the environment (nutrition) can also be important, as emerging from the new science of epigenetics around early stages of life. This interaction suggests the possibility that some genes (and in some specific cells) could be temporarily or for a long time depressed by methylation of DNA or post-translational changes in histones. Of special interest is, for instance, the case of reduced Peroxisome Proliferator-Activated Receptor alfa (PPARα)

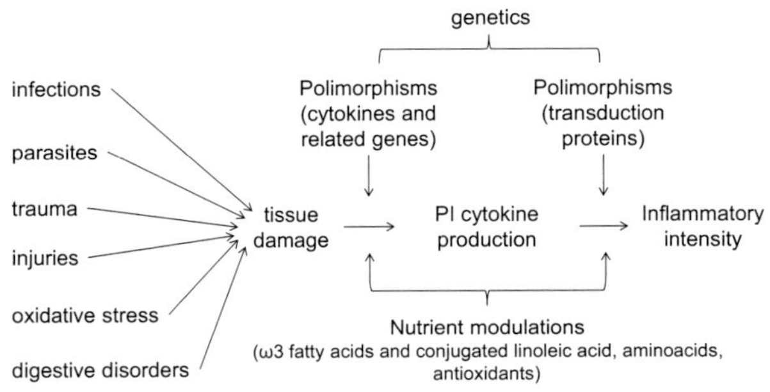

Fig. 5.3 Genetic and nutritional influence on pro-inflammatory (PI) cytokine production and inflammatory response. Adapted from Grimble (2001)

expression occurring in liver of rats whose mothers were fed a protein-restricted diet during pregnancy (Burdge et al. 2007). Nevertheless, PPARα is known to be involved in lipid metabolism and ketosis risk in dairy cows (Drackley 1999) but also in anti-inflammatory phenomena (Moraes et al. 2006). PPARα, when associated with its ligands Polyunsaturated Fatty Acids (PUFA), can inhibit the release of NF-kB (Rimbach et al. 2002), and thus the expression of pro-inflammatory genes.

The first nutrients suggested to reduce the inflammatory response in humans have been ω3 PUFA and antioxidants (vitamin C, vitamin E, β-caroten, polyphenols, Se, etc.). Besides the reduction of the release of pro-inflammatory eicosanoids by the competition of Eicosapentaenoic Acid (EPA) versus the arachidonic acid (Simopoulos 2002), ω3 PUFA appears to modulate several nuclear transcription factors (the PPARα, the NF-kB, the sterol regulatory element 1, and the PUFA regulatory element) which influence the expression of some genes affecting inflammation (Carpentier 2001). Other PUFA and conjugated linoleic acid (CLA) have also been included within the anti-inflammatory nutrients (Williams 2000); they are in fact ligands of PPARα and therefore act by inhibiting NF-kB, as demonstrated by Ringseis and Eder (2009). Some preliminary results of our Institute regarding transition dairy cows seem promising in animal husbandry as well (Trevisi et al. 2008, 2011b; Bertoni et al. 2012).

The importance of antioxidants as meaningful nutrients for inflammation depression is self-evident because ROMs are known to release NF-kB—a nuclear activator of pro-inflammatory gene expression—from its inhibitory unit (Inhibitor-kB) (Rimbach et al. 2002). Well known are the effects of vitamin E, vitamin C, and carotene, but selenium has also been included in this mechanism (McKenzie et al. 2002). The importance of ROMs and their relationship with nutrition have been emphasized by Bernabucci et al. (2005): over-conditioned cows tend to have more oxidative stress at calving and consequently (see above) increased TNFα release, which means more inflammation, as confirmed by O'Boyle et al. (2006), that in turn causes a new release of ROMs. The high levels of ROMs are therefore a cause of inflammation (but they can also be the consequence of inflammation; Bertoni and Trevisi 2013), and this might suggest an attempt to manipulate antioxidant enzymes to reduce the inflammatory status as well as the incidence of some diseases in periparturient cows (Aitken et al. 2009), which are followed by inflammation. Nevertheless, this could determine some misunderstanding, because oxidative stress would be originally the consequence of active metabolic processes in early lactation, while the inflammation would follow later. The results of Wullepit et al. (2009) in early lactating heifers are therefore of great interest: '... there were no clear indications that this increased milk yield causes an alteration of plasma oxidative status and makes heifers sensitive to oxidative stress'. In fact, this does not totally exclude the possibility that oxidative stress could occur and therefore cause some inflammation damage (Conner and Grisham 1996); nevertheless, the frequent inflammations observed at calving time (Bertoni et al. 2008) suggest that these occur first and that anti-inflammatory efforts are of major importance in also preventing some oxidative stress.

Other anti-inflammatory nutrients include some amino acids, namely glutamine, which improves the intestinal barrier function (Calder and Newsholme 2002). Sulphur-containing amino acids (glutathione source) are involved in the antioxidant systems; thus, they are also suggested to have an anti-inflammatory effect (Grimble 2002). Quite recently, the active form of vitamin D3 has been recognized as a downregulator of pro-inflammatory cytokine production by Th_1 cells in favour of Th_2 which produce anti-inflammatory cytokines (Flores 2005; Hewison 2012).

However, a new concept of nutritional involvement—beyond the anti-inflammatory effects—is the possibility to improve the inflammation resolution (Serhan et al. 2008). The molecules involved provide potent signals that selectively stop neutrophil and eosinophil infiltrations (with anti-inflammatory effects) and stimulate non-phlogistic recruitment of monocytes (which occurs without pro-inflammatory mediator release). Simultaneously, these molecules activate the macrophage phagocytosis of microorganisms and of apoptotic neutrophils, as well as the increased exit of phagocytes from the inflamed site through the lymphatic vessel system; furthermore, they stimulate the expression of molecules involved in antimicrobial defence. This is what can be considered the pro-resolution phenomenon, being prodromic to the return to homeostasis and nutrition is involved for the precursors of pro-resolution mediators: arachidonic acid for lipoxin and EPA/DHA for protectins and resolvins. Furthermore, their synthesis can be potentiated by salicylic acid (Calder 2008; Serhan et al. 2008); unfortunately, some preliminary studies of our Institute aimed to stimulate the synthesis of the pro-resolution mediators were unsuccessful in dairy cows.

To conclude, inflammation is a costly physiological process (cause of health and welfare impairment, besides nutrient losses) that implies natural counter-regulation mechanisms to modulate it (glucocorticoids, some cytokines, antioxidant systems, etc.). Nevertheless, there is a strong individual variability that suggests a genetic background and the possibility of improvement; moreover, there are important nutritional factors (ω3 and CLA PUFA, as well as antioxidants, are of major importance, but not alone) which can epigenetically modify pro-inflammatory gene expression or can slow down the pro-inflammatory cytokines release, or facilitate the resolution of the inflammation. Thus, the possibility to have animals with a different inflammatory phenotype, such as with different response to inflammatory stimuli, as suggested by Calder (2008), could become a reality (Fig. 5.3), even though further studies are needed. This could possibly reduce the malaise risks for the animals and therefore positively affect their welfare and efficiency.

5.3 Conclusions

Welfare is without doubt closely related to the status of health because such problems cause pain and suffering, besides feeling bad: all conditions characterizing a low welfare. Simultaneously, nutrition is a very important contributing factor for good health through several different mechanisms:

(1) proper nutrition is essential to ensure a high level of body defences against bacteria, virus, fungi, and parasites (tissue integrity and immune cell activity);
(2) avoidance of nutrient deficiencies, but also excesses, which can be responsible for tissue damage or metabolic disorders, which cause health troubles; and
(3) care to utilize safe feeds and feeding techniques minimizing the intake of toxic compounds (or their production by commensal microbes along the gastrointestinal tract): all causing several types of health problems.

Nevertheless, there is a growing awareness that an important part of better health conditions is played by lower inflammatory conditions. They are in fact essential within the immune response, but can also be dangerous (tissue damage, energy and nutrient waste, cause of suffering and pain, etc.), thus more negative than useful (metaflammation, chronic inflammation, autoimmune diseases, etc.).

Therefore, it could be useful not only to avoid inflammatory conditions (i.e. prevention of any kind of diseases, injuries, tissue damage, oxidative stress etc.) but also to diminish and to mitigate as much as possible the inflammatory phenomena occurring. Also, in this respect nutrition has an important role, because several nutrients (fatty acids, amino acids, micronutrients) have a direct or indirect (antioxidant) effect on inflammation. This will be greatly facilitated by investing in the exploration of the underlying mechanisms of nutritional immunomodulation, to achieve the optimum use of the nutritional resources for increased disease resistance and resilience (which include 'reasonable' inflammatory response). Possible applications of the advantages of understanding the immunomodulatory effects of nutrients and PSM include the development of personalized nutrition in humans and the matching of breed/genotypes with appropriate nutritional environments in livestock for health and immunity.

It can be therefore concluded that better knowledge in the field of relationships between nutrition and different facets of health can be fundamental for better animal welfare, as well as better performance and efficiency, in any animal production system.

References

Acamovic T, Brooker JD (2005) Biochemistry of plant secondary metabolites and their effects in animals. Proc Nutr Soc 64:403–412

Aitken SL, Karcher EL, Rezamand P, Gandy JC, VandeHaar MJ, Capuco AV, Sordillo LM (2009) Evaluation of antioxidant and proinflammatory gene expression in bovine mammary tissue during the periparturient period. J Dairy Sci 92:589–598

Alexander JW (1998) Bacterial translocation during enteral and parenteral nutrition. Proc Nutr Soc 57:389–393

Athanasiadou S, Huntley JF (2008) Emerging technologies and their applications in interactions between nutrition and immunity to gastrointestinal parasites in sheep. Parasite Immunol 30:101–111

Athanasiadou S, Kyriazakis I, Jackson F, Coop RL (2000) Effects of short-term exposure to condensed tannins on adults *Trichostrongylus colubriformis*. Vet Rec 146:728–732

Athanasiadou S, Kyriazakis I, Jackson F, Coop RL (2001) Direct anthelmintic effects of condensed tannins towards different gastrointestinal nematodes of sheep: *in vitro* and *in vivo* studies. Vet Parasitol 99:205–219

Athanasiadou S, Jones LA, Burgess STG, Kyriazakis I, Pemberton AD, Houdijk JGM, Huntley JF (2011) Genome-wide transcriptomic analysis of intestinal tissue to assess the impact of nutrition and a secondary nematode challenge in lactating rats. PLoS One 6:e20771

Barnes PJ (1997) Nuclear factor-kB. Int J Biochem Cell Biol 29:867–870

Bassit RA, Sawada LA, Bacurau RF, Navarro F, Martins E Jr, Santos RV, Caperuto EC, Rogeri P, Costa Rosa LF (2002) Branched-chain amino acid supplementation and the immune response of long-distance athletes. Nutrition 18:376–379

Bernabucci U, Ronchi B, Lacetera N, Nardone A (2005) Influence of body condition score on relationships between metabolic status and oxidative stress in periparturient dairy cows. J Dairy Sci 88:2017–2026

Bertoni G, Trevisi E (1997) Le principali malattie della produzione delle lattifere. L'Informatore Agrario 53(Suppl 47):1–34

Bertoni G, Trevisi E (2013) Use of the liver activity index and other metabolic variables in the assessment of metabolic health in dairy herds. Vet Clin N Am Food Anim Pract 29(2):413–431

Bertoni G, Trevisi E, Han X, Bionaz M (2008) Effects of inflammatory conditions on liver activity in the puerperium and consequences for performance in dairy cows. J Dairy Sci 91:3300–3310

Bertoni G, Lombardelli R, Trevisi E (2009) Some new aspects of nutrition, health conditions and fertility of intensively reared dairy cows. Ital J Anim Sci 8:491–518

Bertoni G, Grossi P, Trevisi E (2012) A possible connection among CLA, inflammation and fertility in dairy cow. Proceedings of the 244th American Chemical Society National Meeting (Spencer Award Symposium), pp 50

Bertoni G, Minuti A, Trevisi E (2015) Immune system, inflammation and nutrition. Anim Prod Sci 55(6):354–360

Bikle DD (2008) Vitamin D and the immune system: role in protection against bacterial infection. Curr Opin Nephrol Hypertens 17:348–352

Boissy A, Manteuffel G, Jensen MB, Oppermann Moe R, Spruijt B, Keeling LJ, Winckler C, Forkman B, Dimitrov I, Langbein J, Bakken M, Veissier I, Aubert A (2007) Assessment of positive emotions in animals to improve their welfare. Physiol Behav 92:375–397

Boulay M, Scott ME, Conly SL, Stevenson MM, Koski KG (1998) Dietary protein and zinc restrictions independently modify a *Heligmosomoides polygyrus* (Nematoda) infection in mice. Parasitology 116:449–462

Broom DM (2006) Behaviour and welfare in relation to pathology. Appl Anim Behav Sci 97:73–83

Broom DM, Fraser AF (2007) Introduction and concepts. In: Domestic animal behaviour and welfare, 4th edn. CABI, Wallingford, pp 1–16

Burdge GC, Hanson MA, Slater-Jefferies JL, Lillycrop KA (2007) Epigenetic regulation of transcription: a mechanism for inducing variations in phenotype (fetal programming) by differences in nutrition during early life? Br J Nutr 97:1036–1046

Calamari L, Bertoni G (2009) Model to evaluate welfare in dairy cow farms. Ital J Anim Sci 8 (Suppl 1):301–323

Calamari L, Bionaz M, Trevisi E, Bertoni G (2004) Preliminary study to validate a model of animal welfare assessment in dairy farms. In: 5th Congress of the European Society for Agricultural and Food Ethics, Leuven, Belgium, 2–4 September, pp 38–42

Calder PC (2008) Joint Nutrition Society and Irish Nutrition and Dietetic Institute Symposium on 'Nutrition and autoimmune disease' PUFA, inflammatory processes and rheumatoid arthritis. Symposium on 'The challenge of translating nutrition research into public health nutrition'. Proc Nutr Soc 67:409–418

Calder PC, Jackson AA (2000) Undernutrition, infection and immune function. Nutr Res Rev 13:3–29

Calder PC, Newsholme P (2002) Glutamine and the immune system. In: Calder PC, Field CJ, Gill HS (eds) Nutrition and immune function. CABI, New York, NY, pp 109–132

Calder PC, Field CJ, Gill HS (2002) Nutrition and immune function. CABI, Wallingford

Carpentier YA (2001) Omega-3 fatty acids: from nutrition to pharmacological properties. Clin Nutr 20(Suppl 4):6–7

Coltherd JC, Babayan SA, Bünger L, Kyriazakis I, Allen JE, Houdijk JG (2011) Interactive effects of protein nutrition, genetic growth potential and *Heligmosomoides bakeri* infection pressure on resilience and resistance in mice. Parasitology 138:1305–1315

Conner EM, Grisham MB (1996) Inflammation, free radicals, and antioxidants. Nutrition 12:274–277

Coop RL, Holmes PH (1996) Nutrition and parasite interaction. Int J Parasitol 26:951–962

Coop RL, Kyriazakis I (1999) Nutrition-parasite interaction. Vet Parasitol 84:187–204

Coop RL, Huntley JF, Smith WD (1995) Effect of dietary protein supplementation on the development of immunity to *Ostertagia circumcincta* in growing lambs. Res Vet Sci 59:24–29

Dantzer R, O'Connor J, Freund GG, Johnson RW, Kelley KW (2008) From inflammation to sickness and depression: when the immune system subjugates the brain. Nature Reviews. Nat Rev Neurosci 9(1):46–56

de Vries M, Bokkers EAM, Dijkstr AT, van Schaik G, de Boer IJM (2011) Invited review: associations between variables of routine herd data and dairy cattle welfare indicators. J Dairy Sci 94:3213–3228

Dobson C, Bawden RJ (1974) Studies on the immunity of sheep to *Oesophagostomum columbianum*: effects of low-protein diet on resistance to infection and cellular reactions in the gut. Parasitology 69:23–255

Donaldson J, van Houtert MFJ, Sykes AR (1998) The effect of nutrition on the periparturient parasite status of mature ewes. Anim Sci 67:523–533

Drackley JK (1999) Biology of dairy cows during the transition period: the final frontier? J Dairy Sci 82:2259–2273

Drackley JK, Dann HM, Douglas GN, Janovick Guretzky NA, Litherland NB, Underwood JP, Loor JJ (2005) Physiological and pathological adaptations in dairy cows that may increase susceptibility to periparturient diseases and disorders. Ital J Anim Sci 4:323–344

Egger G, Dixon J (2009) Obesity and chronic disease: always offender or often just accomplice? Br J Nutr 102:1238–1242

Elsasser T, Ingvartsen KL, Khal S, Capuco AV (2006) Endocrine effects on immune function: the fining opportunities based on knowledge from growing calve and periparturient animal models. In: Sejrsen K, Hvelplund T, Nielsen MO (eds) Ruminant physiology. Wageningen Academic Publishers, Wageningen, pp 421–453

Finch JM, Turner RJ (1996) Effects of selenium and vitamin E on the immune responses of domestic animals. Res Vet Sci 60:97–106

Flores M (2005) A role of vitamin D in low-intensity chronic inflammation and insulin resistance in type 2 diabetes mellitus? Nutr Res Rev 18:175–182

Fossum C, Wattrang E, Fuxler L, Jensen KT, Wallgren P (1998) Evaluation of various cytokines (IL-6, IFN-alpha, IFN-gamma, TNF-alpha) as markers for acute bacterial infection in swine—a possible role for serum interleukin-6. Vet Immunol Immunopathol 64:161–172

Frape DL (1986) Equine nutrition and feeding. Longman Scientific and Technical, Harlow, Essex, pp 250–253

Fregonesi JA, Leaver JD (2001) Behavior, performance and health indicators of welfare for dairy cows hosed in strawyard or cubicle systems. Livest Prod Sci 68:205–216

Gershenzon J (2002) Secondary metabolites and plant defences. In: Taiz L, Zaiger E (eds) Plant physiology. Sinauder, Sunderland, pp 283–308

Goddeeris B (2010) Understanding nutritional immunomodulation: Th1 versus Th2. Dynamics in animal nutrition. In: Doppenberg J, van der Aar P (eds) Dynamics in animal nutrition. Wageningen Academic Publishers, Wageningen, pp 15–30

Grimble RF (1992) Dietary manipulation of the inflammatory response. Proc Nutr Soc 51:285–294

Grimble RF (1998) Modification of inflammatory aspects of immune function by nutrients. Nutr Res 18:1297–1317

Grimble RF (2001) Stress proteins in disease: metabolism on a knife edge. Clin Nutr 20 (6):469–476

Grimble RF (2002) Sulphur amino acids, glutathione and immune function. In: Calder PC, Field CJ, Gill HS (eds) Nutrition and immune function. CABI, New York, NY, pp 133–150

Hale MW, Spencer SJ, Conti B, Jasoni CL, Kent S, Radler ME, Reyes TM, Sominsky L (2015) Diet, behavior and immunity across the lifespan. Neurosci Biobehav Rev 58:46–62

Harrison GB, Pulford HD, Gatehouse TK, Shaw RJ, Pfeffer A, Shoemaker CB (1999) Studies on the role of mucus and mucosal hypersensitivity reactions during rejection of *Trichostrongylus colubriformis* from the intestine of immune sheep using an experimental challenge model. Int J Parasitol 29:459–468

Hewison M (2012) Vitamin D and immune function: an overview. Proc Nutr Soc 71:50–61. doi:10.1017/S0029665111001650

Horan B, Mee JF, O'Connor P, Rath M, Dillon P (2006) The effect of strain of Holstein-Friesian cow and feeding system on postpartum ovarian function, animal production and conception rate to first service. Theriogenology 63(3):950–971

Horrigan LA, Kelly JP, Connor TJ (2006) Immunomodulatory effects of caffeine: friend or foe? Pharmacol Ther 111:877–892

Houdijk JGM (2012) Differential effects of protein and energy scarcity on resistance to nematode parasites. Small Rumin Res 103:41–49

Houdijk JGM, Kyriazakis I, Jackson F, Huntley JF, Coop RL (2000) Can an increased metabolizable protein intake affect the periparturient relaxation in immunity against *Teladorsagia circumcincta* in sheep? Vet Parasitol 91:43–62

Houdijk JGM, Kyriazakis I, Jackson F, Coop RL (2001) The expression of immunity to *Teladorsagia circumcincta* and its relationship to protein nutrition depend on body protein reserves. Parasitology 122:661–672

Houdijk JGM, Kyriazakis I, Jackson F, Huntley JF, Coop RL (2003) Is the allocation of metabolisable protein prioritised to milk production rather than to immune functions in *Teladorsagia circumcincta* infected lactating ewes? Int J Parasitol 33:329–338

Houdijk JGM, Kyriazakis I, Jackson F, Huntley JF, Coop RL (2005) Effects of protein supply and reproductive status on local and systemic immune responses to *Teladorsagia circumcincta* in sheep. Vet Parasitol 129:105–117

Houdijk JGM, Jackson F, Kyriazakis I (2006) Rapid improvement of immunity to *Teladorsagia circumcincta* is achieved through a reduction in the demand for protein in lactating ewes. Int J Parasitol 36:219–227

Houdijk JGM, Jackson F, Kyriazakis I (2009) Nutritional sensitivity of resistance to *Trichostrongylus colubriformis* in lactating ewes. Vet Parasitol 160:258–266

Huntley JF, Newlands GF, Jackson F, Miller HR (1992) The influence of challenge dose, duration of immunity, or steroid treatment on mucosal mast cells and on the distribution of sheep mast cell proteinase in *Haemonchus*-infected sheep. Parasite Immunol 14:429–440

Huntley JF, Jackson F, Coop RL, Macaldowie C, Houdijk JGM, Familton AS, Xieh HL, Stankiewicz M, Sykes AR (2004) The sequential analysis of local inflammatory cells during abomasal nematode infection in periparturient sheep. Vet Immunol Immunopathol 97:163–176

Ing R, Su Z, Scott ME, Koski KG (2000) Suppressed T helper 2 immunity and prolonged survival of a nematode parasite in protein malnourished mice. Proc Natl Acad Sci 97:7078–7083

OIE (Organisation mondiale de la santé animale) (2015) Terrestrial animal health code, vol 1, section 7. Available from: http://www.oie.int/index.php?id=169&L=0&htmfile=chapitre_aw_introduction.htm. Last visit 25 Jan 2016

Israf DA, Coop RL, Stevenson LM, Jones DG, Jackson F, Jackson E, MacKellar A, Huntley JF (1996) Dietary protein influences upon immunity to *Nematodirus battus* infection in lambs. Vet Parasitol 61:273–286

Janovick Guretzky NA, Dann HM, Bionaz M, Trevisi E, Bertoni, G, Drackley JK (2007) Evaluation of acute phase reactants and indices of liver function in serum from dairy cows fed different levels on energy prepartum. J Dairy Sci 90(Suppl 1):408. In: Joint Annual Meeting ADSA, PSA, AMPA, ASAS, S. Antonio (Texas), 8–12 July 2007

Janovick NA, Drackley JK (2010) Prepartum dietary management of energy intake affects postpartum intake and lactation performance by primiparous and multiparous Holstein cows. J Dairy Sci 93:3086–3102

Johnson AK (2009) ASAS centennial paper: farm animal welfare science in the United States. J Anim Sci 87:2175–2179

Johnson RW, Finck BN (2001) TNFα and leptin: two players in an animal's metabolic and immunologic responses to infection. J Anim Sci 79(E. Suppl):E118–E127

Jones LA, Houdijk JGM, Knox DP, Kyriazakis I (2009) Immunomodulatory effects of dietary protein during *Nippostrongylus brasiliensis* infection in lactating rats. Parasite Immunol 31:412–421

Jones LA, Houdijk JGM, Sakkas P, Bruce AD, Mitchell M, Knox DP, Kyriazakis I (2011) Dissecting the impact of protein versus energy host nutrition on the expression of immunity to gastrointestinal parasites during lactation. Int J Parasitol 41:711–719

Kahn LP, Kyriazakis I, Jackson F, Coop RL (2000) Temporal effects of protein nutrition on the growth and immunity of lambs infected with *Trichostrongylus colubriformis*. Int J Parasitol 30:193–205

Kambara T, McFarlane RG (1996) Changes in T cell subpopulations of sheep due to age and dietary protein intake: association with protective immunity to *Trichostrongylus colubriformis*. Vet Immunol Immunopathol 51:127–135

Khafipour E, Krause DO, Plaizier JC (2009a) A grain-based subacute ruminal acidosis challenge causes translocation of lipopolysaccharide and triggers inflammation. J Dairy Sci 92:1060–1070

Khafipour E, Krause DO, Plaizier JC (2009b) Alfalfa pellet-induced subacute ruminal acidosis in dairy cows increases bacterial endotoxin in the rumen without causing inflammation. J Dairy Sci 92:1712–1724

Kimball SR, Jefferson LS (2006) New functions for amino acids: effects on gene transcription and translation. Am J Clin Nutr 83:500S–507S

Klasing KC (1988) Nutritional aspects of leukocyte cytokines. J Nutr 118:1436–1446

Koski KG, Scott ME (2001) Gastrointestinal nematodes, nutrition and immunity: breaking the negative spiral. Annu Rev Nutr 21:297–321

Koutsos EA, Klasing KC (2001) Interactions between the immune system, nutrition, and productivity of animals. In: Garnsworthy PC, Wiseman J (eds) Recent advances in animal nutrition. Nottingham University Press, Nottingham, pp 173–190

Kroner JC, Sommer A, Fabri M (2015) Vitamin D every day to keep the infection away? Nutrients 7(6):4170–4188

LeValley S (2010) Pregnancy toxemia (ketosis) in ewes and does. Fact Sheet No. 1.630. Colorado State University Extension sheep specialist, animal sciences. 8/2010

Lewis RA, Austen KF (1981) Mediation of local homeostasis and inflammation by leukotrienes and other mast cell-dependent compounds. Nature 293:103–108

MacRae JC (1993) Metabolic consequences of intestinal parasitism. Proc Nutr Soc 52:121–130

Maggini S, Wintergerst ES, Beveridge S, Hornig DH (2007) Selected vitamins and trace elements support immune function by strengthening epithelial barriers and cellular and humoral immune responses. Br J Nutr 98(Suppl 1):S29–S35

Maizels RM, Balic A, Gomez-Escobar N, Nair M, Taylor MD, Allen JE (2004) Helminth parasites-masters of regulation. Immunol Rev 201:89–116

Marley CL, Cook R, Barrett J, Keatinge R, Lampkin NH (2006) The effects of birdsfoot trefoil (*Lotus corniculatus*) and chicory (*Cichorium intybus*) when compared with perennial ryegrass (*Lolium perenne*) on ovine gastrointestinal parasite development, survival and migration. Vet Parasitol 138:280–290

Martin-Tereso J, Vestegen MWA (2011) A novel model to explain dietary factors affecting hypocalcaemia in dairy cattle. Nutr Res Rev 24:228–243
McClure SJ, McClure TJ, Emery DL (1999) Effects of molybdenum intake on primary infection and subsequent challenge by the nematode parasite *Trichostrongylus colubriformis* in weaned Merino lambs. Res Vet Sci 67:17–22
McKenzie RC, Arthur JR, Miller SM, Rafferty TS, Beckett GJ (2002) Selenium and the immune system. In: Calder PC, Field CJ, Gill HS (eds) Nutrition and immune function. CAB, Oxford, pp 229–250
Medzhitov R (2008) Original and physiological roles of inflammation. Nature 454:428–435
Michael E, Bundy DAP (1992a) Protein content of CBA/Ca mouse diet: relationship with host antibody responses and the population dynamics of *Trichuris muris* (Nematoda) in repeated infection. Parasitology 105:139–150
Michael E, Bundy DAP (1992b) Nutrition, immunity and helminth infection: effect of dietary protein on the dynamics of the primary antibody response to *Trichuris muris* (Nematoda) in CBA/Ca mice. Parasite Immunol 14:169–183
Min BR, Hart SP (2003) Tannins for suppression of internal parasites. J Anim Sci 81:E102–E109
Minuti A, Ahmed S, Trevisi E, Piccioli-Cappelli F, Bertoni G, Jahan N, Bani P (2014) Experimental acute rumen acidosis in sheep: consequences on clinical, rumen, gastrointestinal permeability conditions, and blood chemistry. J Anim Sci 92(9):3966–3977
Moore SE, Collinson AC, Tamba N'Gom P, Aspinall R, Prentice AM (2006) Early immunological development and mortality from infectious disease in later life. Proc Nutr Soc 65:311–318
Moraes LA, Piqueras L, Bishop-Baily D (2006) Peroxisome proliferator-activated receptors and inflammation. Pharmacol Ther 110:371–385
Morrow DA (1976) Fat cow syndrome. J Dairy Sci 59(9):1625–1629
Mueller-Harvey I (2006) Unravelling the conundrum of tannins in animal nutrition and health. J Sci Food Agric 86:2010–2037
Nagaraja TG, Chengappa MM (1998) Liver abscesses in feedlot cattle: a review. J Anim Sci 76:287–298
Napolitano F, Knierim U, Grasso F, De Rosa G (2009) Positive indicators of cattle welfare and their applicability to on-farm protocols. Ital J Anim Sci 8(Suppl 1):355–365
Neyestani TR (2008) Polyphenols and immunity. In: de Meester F, Watson RR (eds) Wild-type food in health promotion and disease prevention). Humana Press, Totowa, NJ, pp 431–434
Niezen JH, Charleston WAG, Robertson HA, Shelton D, Waghorn GC, Green R (2002) The effect of feeding sulla (*Hedysarum coronarium*) or lucerne (*Medicago sativa*) on lamb parasite burdens and development of immunity to gastrointestinal nematodes. Vet Parasitol 105:229–245
Nocek JE (1997) Bovine acidosis: implications on laminitis. J Dairy Sci 80:1005–1028
O'Boyle N, Corl CM, Gandy JC, Sordillo LM (2006) Relationship of body condition score and oxidant stress to tumor necrosis factor expression in dairy cattle. Vet Immunol Immunopathol 113:297–304
O'Sullivan BM, Donald AD (1973) Responses to infection with *Haemonchus contortus* and *Trichostrongylus colubriformis* in ewes of different reproductive status. Int J Parasitol 3:521–530
Osorio J, Trevisi E, Ji P, Drackley J, Bertoni G, Loor J, Luchini D (2014) Biomarkers of inflammation, metabolism, and oxidative stress in blood, liver, and milk reveal a better immunometabolic status in peripartal cows supplemented with Smartamine M or MetaSmart. J Dairy Sci 97(12):1–14
Ouchi N, Walsh K (2007) Adiponectin as an anti-inflammatory factor. Clin Chim Acta 380:24–30
Paolini V, Bergeaud JP, Grisez C, Prevot F, Dorchies P, Hoste H (2003a) Effects of condensed tannins on goats experimentally infected with *Haemonchus contortus*. Vet Parasitol 113:253–261

Paolini V, Frayssines A, De La Farge F, Dorchies P, Hoste H (2003b) Effects of condensed tannins on established populations and on incoming larvae of *Trichostrongylus colubriformis* and *Teladorsagia circumcincta* in goats. Vet Res 34:331–339

Paolini V, De La Farge F, Prevot F, Dorchies P, Hoste H (2005) Effects of the repeated distribution of sainfoin hay on the resistance and the resilience of goats naturally infected with gastrointestinal nematodes. Vet Parasitol 127:277–283

Pedersen S, Saeed I, Michaelsen KF, Friis H, Murrell KD (2002) Impact of protein energy malnutrition on *Trichuris suis* infection in pigs concomitantly infected with *Ascaris suum*. Parasitology 124:561–568

Potter A, Volker G, van Drunen S, van den Hurk L (2008) Veterinary vaccines: alternatives to antibiotics? Anim Health Res Rev 9:187–199

Rimbach G, Minihane AM, Majewics J, Fischer A, Pallauf J, Virgli F, Weinberg PD (2002) Regulation of cell signalling by vitamin E. Proc Nutr Soc 61:415–425

Ringseis R, Eder K (2009) Influence of conjugated linoleic acids on functional properties of vascular cells. Br J Nutr 102:1099–1116

Ríos-de Alvarez L, Greer AW, Jackson F, Athanasiadou S, Kyriazakis I, Huntley JF (2008) The effect of dietary sainfoin (*Onobrychis viciifolia*) on local cellular responses to *Trichostrongylus colubriformis* in sheep. Parasitology 135:1117–1124

Roche JR, Berry DP, Kolver ES (2006) Holstein-Friesian strain and feed effects on milk production, body weight and body condition score profiles in grazing dairy cows. J Dairy Sci 89:3532–3543

Rousing T, Wemelsfelder F (2006) Qualitative assessment of social behavior of dairy cows housed in loose housing systems. Appl Anim Behav Sci 101:40–53

Rushen J, de Passillé AM, von Keyserlingk MAG, Weary DM (2008) The welfare of cattle. Springer, Dordrecht

Sakkas P, Houdijk JGM, Athanasiadou S, Kyriazakis I (2012) Sensitivity of periparturient breakdown of immunity to parasites to dietary protein source. J Anim Sci 90:3954–3962. doi:10.2527/jas.2011-4829

Scrimshaw NS, Taylor CE, Gordon JE (1968) Interactions of nutrition and infection, 1st edn. World Health Organization Monograph Series no. 57. World Health Organization, Geneva

Sejian V, Lakritz J, Ezeji T, Lal R (2011) Assessment methods and indicators of animal welfare. Asia J Anim Vet Adv 6(4):301–315

Serhan CN, Chiag N, Van Dyke TE (2008) Resolving inflammation: dual anti-inflammatory and pro-resolution lipid mediators. Nat Rev 8:349–361

Shaw KL, Nolan JV, Lyneh JJ, Coverdole OR, Gill HS (1995) Effect of weaning supplementation and gender on acquired immunity to *Haemonchus contortus* in lambs. Int J Parasitol 25:381–387

Simopoulos AP (2002) Omega-3 fatty acids in inflammation and autoimmune diseases. J Am Coll Nutr 21(6):495–505

Spears JW, Weiss WP (2008) Role of antioxidants and trace elements in health and immunity of transition dairy cows. Vet J 176:70–76

Spinas E, Saggini A, Kritas SK, Cerulli G, Caraffa A, Antinolfi P, Pantalone A, Frydas A, Tei M, Speziali A, Saggini R, Pandolfi F, Conti P (2015) Can vitamin a mediate immunity and inflammation? J Biol Regul Homeost Agents 29(1):1–6

Steinbrenner H, Al-Quraishy S, Dkhil MA, Wunderlich F, Sies H (2015) Dietary selenium in adjuvant therapy of viral and bacterial infections. Adv Nutr 6(1):73–82

Stephen AI, Avenell A (2006) A systematic review of multivitamin and multimineral supplementation for infection. J Hum Nutr Diet 19:179–190

Strain SAJ, Stear MJ (2001) The influence of protein supplementation on the immune response to *Haemonchus contortus*. Parasite Immunol 23:527–531

Suchner U, Kuhn KS, Fürst P (2000) The scientific basis of immunonutrition. Proc Nutr Soc 59:553–563

Sykes AR, Xie HL, Stankiewicz M, Huntley JF, MacKellar A, Sedcole JR, McAnulty RW, Green R (2007) The effect of vaccinating infection during pregnancy and dietary protein supply on the periparturient immune response of sheep to infection with *Teladorsagia circumcincta* and *Trichostrongylus colubriformis* larvae. Animal 1:249–260

Tan J, Applegate TJ, Liu S, Guo Y, Eicher SD (2014) Supplemental dietary L-arginine attenuates intestinal mucosal disruption during a coccidial vaccine challenge in broiler chickens. Br J Nutr 112(7):1098–1109

Thompson PB (2010) Animal ethics and public expectations: the North American outlook. J Vet Med Educ 37(1):13

Thompson FN, Stuedemann JA (1993) Pathophysiology of fescue toxicosis. Agric Ecosyst Environ 44:263–281

Trevisi E, D'Angelo A, Gaviraghi A, Noé L, Bertoni G (2005) Blood inflammatory indices in goats around kidding. Ital J Anim Sci 4(Suppl 2):404

Trevisi E, Ferrari A, Piccioli-Cappelli F, Bertoni G (2008) Energy balance indexes and blood changes of dairy cows supplemented with rumen protected CLA in late pregnancy and early lactation. J Anim Sci 86(E-Suppl 2)/J Dairy Sci 91(E-Suppl 1):77

Trevisi E, Amadori M, Archetti I, Lacetera N, Bertoni G (2011a) Inflammatory response and acute phase proteins in the transition period of high-yielding dairy cows. In: Veas F (ed) Acute phase protein/book 2. InTech, Rijeka, pp 355–380

Trevisi E, Grossi P, Piccioli Cappelli F, Cogrossi S, Bertoni G (2011b) Attenuation of inflammatory response phenomena in periparturient dairy cows by the administration of an ω3 rumen protected supplement containing vitamin E. Ital J Anim Sci 10:277–285

Trevisi E, Jahan N, Ferrari A, Minuti A (2015) Pro-inflammatory cytokine profile in dairy cows: consequences for new lactation. Ital J Anim Sci 14:285–292

Tu T, Koski KG, Scott ME (2007a) Mechanisms underlying reduced expulsion of a murine nematode infection during protein deficiency. Parasitology 135:81–93

Tu T, Koski KG, Wykes LJ, Scott ME (2007b) Re-feeding rapidly restores protection against *Heligmosomoides bakeri* (Nematoda) in protein-deficient mice. Parasitology 134:899–909

Tzamaloukas O, Athanasiadou S, Kyriazakis I, Huntley JF, Jackson F (2006) The effect of chicory (*Cichorium intybus*) and sulla (*Hedysarum coronarium*) on larval development and mucosal cell responses of growing lambs challenged with *Teladorsagia circumcincta*. Parasitology 132:419–426

Uauy R (2007) Academic-industry partnerships in addressing nutrition- (infection-immunity-inflammation) interactions. Br J Nutr 98(Suppl 1):S17–S23

Underwood EJ, Suttle NF (1999) The mineral nutrition of livestock, 3rd revised edn. CABI, New York, NY

Van den Topp AM, Wensing T, Geelen MJH, Wentik GH, Van't Klooster AT, Beynen AC (1995) Time trends of plasma lipids and enzymes synthesizing hepatic triacylglycerol during post-partum development of fatty liver in dairy cows. J Dairy Sci 78:2208–2220

van Houtert MFJ, Barger IA, Steel JW, Windon RG, Emery DL (1995) Effects of dietary protein intake on responses of young sheep to infection with *Trichostrongylus colubriformis*. Vet Parasitol 56:163–180

Volman JJ, Ramakers JD, Plat J (2008) Dietary modulation of immune function by beta-glucans. Dietary modulation of immune function by beta-glucans. Physiol Behav 94:276–284

Von Keyserlingk MAG, Rushen J, de Passillè M, Weary DM (2009) Invited review: the welfare of dairy cattle—key concepts and the role of science. J Dairy Sci 92:4101–4111

Wannemuehler MJ (1995) Role of cytokines in intestinal health and disease. In: Myers MJ, Murtaugh MP (eds) Cytokines in animal health and disease. Marcel Dekker, New York, NY, pp 333–355

Wärnberg J, Cunningham K, Romeo J, Marcos A (2010) Role of physical activity on immune function. Physical activity, exercise and low grade systemic inflammation. Proc Nutr Soc 69:400–406

Watkins LR, Maier SF (2000) The pain of being sick: implications of immune-to-brain communication for understanding pain. Annu Rev Psychol 51:29–57

Watzl B, Girrbach S, Roller M (2005) Inulin, oligofructose and immunomodulation. Br J Nutr 93: S49–S55

Webster J (1994) Animal welfare. A cool eye towards Eden. Blackwell Science, Oxford

Williams CM (2000) Dietary fatty acids and human health. Ann Zootech 49:165–180

Wintergerst ES, Maggini S, Hornig DH (2007) Contribution of selected vitamins and trace elements to immune function. Ann Nutr Metab 51:301–323

Wullepit N, Raes K, Beerda B, Veerkamp RF, Fremaut D, De Smet S (2009) Influence of management and genetic merit for milk yield on the oxidative status of plasma in heifers. Livest Sci 123:276–282

Yang Q, Huang W, Jozwik C, Lin Y, Glasman M, Caohuy H, Srivastava M, Esposito D, Gillette W, Hartley J, Pollard HB (2005) Cardiac glycosides inhibit TNF-alpha/NF-kappaB signaling by blocking recruitment of TNF receptor-associated death domain to the TNF receptor. Proc Natl Acad Sci 102:9631–9636

Yates SG (1962) Toxicity of tall fescue forage: a review. Econ Bot 16:295–303

Chapter 6
Assessing Farm Animal Welfare from a Nutritional Perspective

Alex Bach, Marina A.G. von Keyserlingk, Tina M. Widowski, and Derek Haley

Abstract In many countries around the world genetic selection strategies have produced farm animals with a high productive (meat, milk, and eggs) potential. These high-producing animals need to be fed and housed under optimum conditions to ensure that the expression of their genetic potential does not compromise welfare. To strive for the highest level of welfare for food-producing animals, one must consider the animal's health and biological functioning and its affective state as well as ensure provision of environments where the animals can engage in behaviours they are highly motivated to perform. Animal-based models, including growth rate, body condition score, lameness, morbidity, and mortality, and environment-based indexes such as temperature, stocking density, and bedding quality are useful metrics to ensure a minimum level of welfare. However, their use on a particular farm can only be as good as the records kept on the farm. This chapter reviews the main nutritional challenges to welfare in farm animals and provides an overview of possible welfare indicators for cattle, swine, and poultry, with a special emphasis on those aspects related to nutrition.

A. Bach (✉)
ICREA, Barcelona, Spain
e-mail: alex.bach@icrea.cat

M.A.G. von Keyserlingk
Animal Welfare Program, Faculty of Land and Food Systems, University of British Columbia, Vancouver, BC, Canada

T.M. Widowski
Department of Animal Biosciences and Campbell Centre for the Study of Animal Welfare, University of Guelph, Guelph, ON, Canada

D. Haley
Department of Population Medicine and Campbell Centre for the Study of Animal Welfare, University of Guelph, Guelph, ON, Canada

C.J.C. Phillips (ed.), *Nutrition and the Welfare of Farm Animals*, Animal Welfare 16, DOI 10.1007/978-3-319-27356-3_6

6.1 Introduction

In food animal production systems, livestock and poultry are completely reliant on our ability and knowledge to provide them with sufficient food and water within an adequate environment that supports productivity and welfare. As major food sources for humans, milk, eggs, and meat products are produced under very different production systems around the world. Unfortunately, all too often husbandry conditions are far from optimum, and the animals suffer injuries, malnutrition, and other maladies and thus experience poor welfare. Equally unfortunate is the fact that there is often considerable disagreement as to what constitutes good or poor welfare.

Interest in on-farm assessment of welfare has grown in the past decade, in large part due to increased response to consumer demands for higher welfare standards (Fraser 2003). Welfare assessment research takes two broad approaches: resource-based measures and animal-based measures. Resource-based parameters describe the physical features of where the animal lives (housing) and the management of the farm, such as feeding facilities and associated management, air quality, and space allowance. In contrast, animal-based parameters record animals' reactions to their specific environments and include behaviour, health, and physiology (Johnsen et al. 2001). The advantage of using resource-based measures is that they are easy to record, can be measured objectively (making them highly reliable), and can serve as a basis for problem solving. However, resource-based measures are an indirect measure of welfare, and thus have been criticized (Whay et al. 2003). In contrast, animal-based parameters, which register the state of the animal itself, have been argued to be a much more direct measure of animal welfare (Whay et al. 2003). Although recording animal-based parameters can often be difficult and time-consuming, and in some cases the results may be complex to interpret (Johnsen et al. 2001), it has become increasingly clear that direct animal-based measures should be the preferred parameters in any farm animal welfare assessment.

On-farm assessment of animal welfare can be time-consuming and costly. A promising strategy to monitor animal welfare more efficiently is to first estimate the level of animal welfare on a farm, based on routine herd or flock data that are available at the farm level, and also through nation-wide databases. There have been several attempts to assess the welfare level of cattle under field conditions, including the JuBach index (Bach et al. 2004), Welfare Quality (2009b), and C.O.W.S. (2010; von Keyserlingk et al. 2012), among others. However, at present we lack clear relationships between objective data being collected at the herd level and the level of welfare of the individual animals within the herd.

This chapter reviews a few key examples associated with various livestock and poultry production systems and provides an overview of possible welfare indicators that can be used by stock people and farm managers to assess and improve the welfare of animals on their farms, with a special emphasis on those aspects related to nutrition.

6.2 Assessing the Welfare of the Milk-Fed Calf

Nutrition during the first weeks of life of any animal is particularly important. A low plane of nutrition for calves, for example, will contribute to both short- and long-term health and welfare consequences. Specifically, inadequate nutrition leading to weight loss can reduce immune function, and calves fed low volumes of milk often lose or fail to gain weight during the first week of life (Hammon et al. 2002; Jasper and Weary 2002), which in the case of artificially reared dairy calves, immediately precedes the period when calf morbidity and mortality is highest (USDA-NAHMS 2007). Traditionally, the main indicators of welfare for young calves have been (1) mortality rate, (2) morbidity rate, and (3) growth rate. To date, emphasis in some countries has been on the monitoring of mortality and morbidity rates in milk-fed heifer calves (USDA-NAHMS 2007). Although these routine measures are important, and we by no means advocate excluding them, relying exclusively on these measures does not ensure adequate welfare. For instance, farms that have successfully reduced the incidence of morbidity and mortality are able to rear a heifer more economically (Bach and Ahedo 2008), with an improved yielding ability (Bach 2012; Soberon et al. 2012) and longevity (Bach 2011), but we caution against the use of relying exclusively on mortality records, as death rates as the only indicator of welfare will probably have little traction with society. Moreover, these measures tell us little about whether the calf experienced hunger (De Paula Vieira et al. 2008) or was deprived of social contact (De Paula Vieira et al. 2010). Lastly, although morbidity does provide valuable information on the incidence of clinical disorders, this measure arguably grossly underestimates the number of calves that are subclinically ill.

There is considerable evidence of links between calf disease and feeding practices. The importance of an adequate intake of colostrum has long been known, but surveys continue to report that large numbers of dairy calves still receive either inadequate or marginal levels (USDA-NAHMS 2011; McGuirk and Collins 2004). The result of inadequate colostrum intake is a low concentration of circulating immunoglobulin (Ig) in the blood of the calf, a condition known as 'failure of passive transfer' (FPT; associated with a blood-serum concentration of IgG $\leq$ 10.0 g/L; McGuirk and Collins 2004) is a serious welfare concern. According to Wells et al. (1996), 31 % of calf deaths during the first 3 weeks of life could have been prevented if colostrum feeding had been adequate. Even where death is avoided, there can be long-term effects of an inadequate colostrum intake; calves with FPT have lower body weights 6 months later (Dewell et al. 2006).

The amount and quality of milk provided to the new-born calf is also important. For example, Godden et al. (2005) described that calves fed pasteurized waste milk experience lower morbidity and mortality than calves fed milk replacer, a difference likely to be due to the higher energy content of the whole-milk diet compared with the milk replacer (Khan et al. 2011). Some producers, especially those rearing calves for dairy purposes, have been reluctant to feed increased quantities of milk due to the common belief that this increases diarrhoea and rearing costs. However, a number of studies (Appleby et al. 2001; Jasper and Weary 2002; Khan et al. 2007) have demonstrated that calves fed more milk actually have the same or lower incidence of diarrhoea than limit-fed calves.

In response to the welfare concern that calves fed restricted milk amounts are hungry [see review by von Keyserlingk et al. (2009)], a growing body of research has focused on determining the effects of feeding more milk. Thus, calves provided with 20 % of their body weight (BW) equivalent as milk are able to maintain growth rates similar to those of calves suckling from their dam [see review by Khan et al. (2011)]. Furthermore, much work has explored alternative feeding systems that allow calves to express a more natural sucking behaviour (Khan et al. 2011), incorporate forage early on in life (Castells et al. 2012; Khan et al. 2012), and maintain growth rates of ca. 1 kg/day (Khan et al. 2011). Providing calves with more milk early in life creates additional challenges at weaning (Sweeney et al. 2010), but a substantial body of evidence now exists on proven management practices that minimize the negative effects associated with abruptly weaning calves from milk to a solid-based diet [see review by Khan et al. (2011)].

In summary, we encourage producers to track morbidity and mortality of all dairy calves born (including bull calves) and to routinely measure the number of calves that suffer from failure of passive transfer. Growths around 1 kg/day in *Bos taurus* calves are achievable and indicate adequate nutrition (De Paula Vieira et al. 2008). However, it is important to realize that growth rates are breed dependent and in tropical breeds a lower growth rate may still indicate adequate nutrition.

6.3 Assessing the Welfare of Dairy Cattle

There have been several objective parameters that have been proposed as indicators of welfare. Some efforts have focused on physiological measures; for instance plasma concentrations of growth hormone are reduced in cows that are deprived of lying, so low lying times could also affect milk production (Munksgaard and Lovendahl 1993) and lameness (Ito et al. 2010). Another objective parameter that has been proposed is blood haptoglobin (Hp). Huzzey et al. (2009) demonstrated that cows having blood Hp concentrations $\geq$1 g/L on day 3 after calving were 6.7 times more likely to develop severe or mild metritis than cows with basal levels of Hp. These results indicate that an acute phase inflammatory response precedes clinical metritis and that Hp screening may assist in the early detection of metritis, providing increased opportunities for early treatment and prevention. However, this predictive threshold had a sensitivity of 50 % and specificity of 87 %, which again makes it difficult to use as a sole indicator of health status or welfare.

Body condition of the animal is a reflection of their nutritional status and has been argued by some to be a useful indicator of welfare status of the animal [see review by Roche et al. (2009)], but insufficient by others [see review by von Keyserlingk et al. (2009)]. The high demand for nutrients at the onset of lactation seems to be a prime factor leading to diseases at this particular time. The majority of research on health issues in transition dairy cows has focused on nutrition, physiology, and metabolism. Despite decades of nutritional and epidemiological work, the incidence of disease around the calving period in lactating dairy cattle remains high, showing the need for better tools for identifying which cows are ill or are most likely to become ill after calving. Growing concerns regarding insulin resistance and loss of body condition and the failure to identify these cows from reductions in milk production or associated metabolic problems (Janovick et al. 2011) are alarming, as these cows may be feeling sick but go unnoticed. When the mobilization of body reserves is exaggerated (due to either excessive insulin resistance or inadequate nutrient supply) (Dann et al. 2006), the risk of cows succumbing to subclinical or clinical ketosis increases (Goldhawk et al. 2009; Vickers et al. 2013). However, equally concerning is the fact that subclinical ketosis can only be detected by sampling the blood for beta-hydroxybutyrate (b-OH-butyrate).

It is currently unclear whether dairy cows in early lactation have negative emotional responses to dietary inadequacies in the form of an unsatisfied hunger (Cooper et al. 2010). Research has indicated that cows with low feed intakes are at an increased risk of metabolic and infectious diseases during the transition period [see review by von Keyserlingk and Weary (2010)]. However, changes in feed intake must ultimately result from changes in feeding behaviour, and thus the latter may be useful for the prediction of disease in transition dairy cows (Huzzey et al. 2007).

Changes in the behavioural activity of farm animals are widely used as welfare indicators (Müller and Schrader 2003). Changes in feeding behaviour have long been used to help identify when animals become ill [reviewed by Weary et al. (2009)]. Sowell et al. (1999) reported that healthy feedlot steers spent 30 % more time at the feed bunk than morbid steers, and a greater percentage of healthy steers visited the feed bunk immediately following feed delivery. Recent research has shown that changes in feeding behaviour can also be useful in detecting illness in dairy cattle, especially during the transition period, when cows are most vulnerable to metabolic and infectious diseases. Patterns of feed intake have been shown to differ for healthy cows and cows diagnosed with metritis [see reviews by von Keyserlingk and Weary (2010) and Sepúlveda-Varas et al. (2013)], with the most dramatic differences occurring during times of highest bunk attendance between 06:00 and 18:00 h, when morbid cows clearly show a lower attendance. A number of studies have shown that feeding behaviour in the days before calving may play an important role in identifying cows at risk of disease in the early postpartum period. For example, Urton et al. (2005) showed that cows diagnosed with acute metritis after calving spent less time feeding during the prepartum period (days −12 to −2 prior to calving), perhaps because cows with lower intakes have poorer immune function (Hammon et al. 2006). Similarly, Huzzey et al. (2007) found that cows diagnosed with severe metritis 7–9 days postpartum consumed less feed and spent less time at the feed bunk during the 2-week period before calving, nearly 3 weeks before the first clinical signs of disease. The odds of having severe metritis increased by 1.72 for every 10-min decrease in feeding time during the week before calving, and the odds increased nearly threefold for every 1 kg decline in dry matter intake. More recent work has also shown that similar declines in feeding behaviour are observed in cows at risk of subclinical and clinical ketosis (González et al. 2008; Goldhawk et al. 2009). Furthermore, Bach et al. (2006) reported that the time devoted to eating decreased, and dry matter intake and milk production decreased, with increasing severity of lameness. Recently, Yunta et al. (2012) have shown that the time elapsed between fresh feed delivery time and attendance to the feedbunk could be a good proxy for identifying moderately lame cows.

A commonly proposed proxy for welfare of dairy cattle is culling rate (Langford and Stott 2012). Under intensive production systems, culling rates of dairy cattle tend to be relatively high, and these high rates have been perceived, in some instances, as a signal of poor welfare. However, it is not easy, and we argue not correct, to assess welfare of the animals based on culling incidence rate. Although we advocate for farms to implement collecting all information pertaining to culling, we believe that a high culling rate may sometimes co-exist with high levels of welfare at the herd level, as there are many situations in which cows are culled because producers have access to young stock with improved genetic and yielding potential. Farms with well-nourished and reproductively sound cows will have access to a relatively large number of high-genetic merit heifers that are eligible to replace older cows. Moreover, older well-nourished cows are arguably of much greater value for meat production and are more able to be transported and subsequently slaughtered than cows that are culled due to illness or lameness. If the number of cows in a herd were to be kept constant, under optimum nutrition and management conditions, $\geq$40 % of the cows could be culled to allow the heifers to enter the milking herd. However, under most practical situations, mortality of calves and heifers, as well as lactating cows, reduces this figure to much lower values. Nevertheless, caution should be applied when attempting to make inferences about welfare based on culling rates, as a high culling rate does not necessarily imply poor welfare.

Heat stress represents another aspect that can have a negative impact on the welfare of cattle. Failure to address heat stress has obvious detrimental impacts on animal welfare, as shown in 2006, when ~25,000 cows in California died as a result of a severe heat wave. Furthermore, heat stress diminishes productive and reproductive performance of cattle. Nutritional models used to feed cattle are based on the premise that the animal is in a thermoneutral environment, and when ambient temperatures rise beyond the thermoneutral zone, animals have an increased demand for nutrients for maintenance, but nevertheless tend to decrease intake in order to reduce the heat increment of digestion. The temperature–humidity index (THI) is an effective measure to evaluate the degree of exposure to heat stress. Heat stress in dairy cattle starts at a THI as low as 68 (Collier et al. 2011), which corresponds to 21 °C at 70 % humidity or 22 °C at 50 % humidity. In addition to decreasing ambient temperature (shade, ventilation, etc.), some nutritional interventions are needed to minimize discomfort due to heat stress. These include increasing dietary energy density (to compensate for the decrease in feed consumption) and providing a greater proportion of energy in the form of fat rather than carbohydrates.

Under extensive production systems, animals are exposed to prevailing climate conditions. For extensive production systems in hot climates, the most important issues become access to shade and water, whereas for cold climates the insufficient provision of nutrients to sustain performance and maintain body temperature is a common threat to welfare. Moreover, as stocking density increases, often coinciding with the rainy season, cows are exposed to mud. Also, the nutrition provided by pasture-based systems is not always in line with the requirements of cattle. Clearly, the welfare of cattle is undermined when they are sick (reviewed by Weary et al. 2009), and when mortality rates rise in extreme situations every effort should be made to rectify the situation. In general, comfort and survivability of animals under extreme cold conditions can be improved by achieving adequate body condition (or body energy reserves), which implies adequate stocking densities and feed supplementation before the cold weather arrives. Therefore, body condition score (BCS), in many instances, can be a useful metric to ensure a minimum level of welfare.

Lameness is widely recognized as one of the most serious welfare and production concerns in the dairy industry, with clinical lameness incidence rates that range in the UK and North America between 25 and 55 % (Espejo et al. 2006; Barker et al. 2010; von Keyserlingk et al. 2012). In addition, lameness is thought to be an important cause of economic losses due to treatment costs (Kossaibati and Esslemont 1997), milk production losses (Bach et al. 2006; Green et al. 2012), impaired fertility (Whay et al. 2003), and premature culling (Bicalho et al. 2007). There are several tests to classify cow lameness (Sprecher et al. 1997; Flower and Weary 2006), with most scoring systems differentiating between cows that are clinically lame and those that are severely lame. Given the magnitude of this problem, it is not surprising that lameness incidence has been a component of several welfare audits (Whay et al. 2003; Bach et al. 2004; Stull et al. 2004; Welfare Quality 2009b). Different audits, assurance programmes, and best practice documents vary in what they deem to be an acceptable level of lameness; for instance, the Canadian *Code of Practice on the Care and Handling of Dairy Cattle* states that dairy farmers must not have more than 10 % severe lameness in their cows (NFACC 2009).

6.4 Assessing the Welfare of Beef Feedlot Cattle

The digestive system of cattle evolved to enable efficient digestion of forage diets. However, in many beef feedlot production systems these diets do not support the growth potential and/or carcass classification (linked to profitability) of finishing cattle for slaughter. As a result, heavy grain feeding has become normal in feedlots [see review by Owens et al. (1998)]. For example, most North American feedlot finishing diets comprise 95 % grain, which has been shown to maximize cattle performance and ultimately profitability (Russell and Rychlik 2001). However, despite improved cattle performance and advantages in terms of profitability, the digestion of these concentrate-based diets creates distinct challenges in terms of cattle health and welfare (Nagaraja and Lechtenberg 2007a). These high-starch rations result in a rapid and extensive ruminal fermentation that produces high levels of volatile fatty acids (VFA) and low ruminal pH. In many cases, the unfortunate consequence of high grain feeding is ruminal acidosis, the predisposing factor to liver abscesses that arise when rumen bacteria enter the portal vein and travel to the liver (Nagaraja and Lechtenberg 2007b), and laminitis (Nocek 1997). Although there has been virtually no work to ascertain the welfare implications of feeding high-grain diets, there have been some behavioural indicators reported in the veterinary and nutritional literature, suggesting that the welfare of feedlot cattle is compromised to some extent when fed diets that cause acute or subacute ruminal acidosis. Rumen acidosis occurs when the pH of the rumen drops rapidly to between 5.0 and 5.5 (Penner et al. 2009), and this is thought to account for 25–30 % of all feedlot deaths (Gaylean and Rivera 2003). This, combined with the direct impact on liver abscesses (Nagaraja and Chengappa 1998), results in substantial economic losses to the feedlot industry. It is not uncommon for feedlot cattle to suffer chronic subclinical acidosis, with individual feed intakes fluctuating widely as animals reduce feed intake to recover from the low rumen pH [see reviews by González et al. (2012), and Schwartzkopf-Genswein et al. (2003)].

6.5 Assessing the Welfare of Pigs

In many countries around the world, genetic selection strategies have produced pigs with very high growth potential. Along with that characteristic, modern pigs have increased appetites and high motivation for feeding, which can lead to a number of nutrition-related welfare problems that should be monitored and assessed by specific means.

The welfare of suckling piglets will be severely compromised if they are not receiving enough milk. In the first instance this is a welfare problem, because the piglets experience the negative effect of feeling hungry. Over time, the lack of milk will additionally affect the piglets' normal biological functioning. With selection for large litter size, caretakers need to ensure that the sow has enough functional teats to feed all of her young. Behaviour during nursing can be used to assess whether piglets are receiving enough milk. Piglets that do not have access to a teat will fight vigorously and continuously during a nursing bout, as they try to gain access to a teat (de Passillé et al. 1988). This will be accompanied by distinct, high-pitched vocalizations (Jensen and Algers 1984). Caretakers can also assess piglets for facial lacerations, which may occur in litters where piglets have been fighting for teat access, particularly if the needle teeth of piglets have not been clipped. Piglets that spend more time massaging the udder in between nursing bouts generally have slower growth rates and lower weaning weights, and therefore extended periods of udder massage are a sign of nutritional need (Torrey and Widowski 2007). Piglets lacking sufficient milk will lose body condition and appear visibly thin. This can happen even in situations where creep feed is provided, because piglets do not have sufficient dentition to facilitate consumption of creep feed until they reach 19 days of age (Tucker and Widowski 2009).

The practice of cross-fostering piglets can help ensure each piglet has teat access. Studies show that cross-fostering early is best and that later cross-fostering is associated with increased pre-weaning piglet mortality (Straw et al. 1998). Selection for larger litter size has increased within-litter variability with respect to piglet birth weight, and larger piglets should be the ones fostered onto unfamiliar sows so that the smaller piglets have the advantage of staying with their own mother (Straw et al. 1998).

Weaning is another critical period for nutrition-related welfare problems because of the social, physical, and nutritional stress that most piglets experience when they are separated from the sow and switched from primarily sucking milk to eating solid food from a feeder. Many piglets do not adapt well to eating solid food and often develop behaviour problems such as ear biting, ear sucking, and belly nosing, which resembles udder massage but is directed at the abdomens of their pen-mates (Widowski et al. 2008). Piglets that belly nose spend less time at the feeder, spend more time at the drinker, and perform poorly. The recipients of belly nosing may also experience compromised welfare because they can develop lesions on their abdomen, or in severe cases, umbilical hernias. Sufficient access to water is also important for ensuring body weight uniformity and preventing ear biting (de Grau et al. 2005). Assessing the welfare of newly weaned piglets, therefore, should include observing their behaviour and body condition and scoring them for lesions on the abdomen and ears (Widowski et al. 2003; de Grau et al. 2005). On farms with weaning problems, management changes can be used to facilitate the development of feeding and drinking behaviour. These include weaning at a later age, changing from nipple to cup style drinkers, using feeding boards, or feeding wet mash for a few days (Widowski et al. 2008).

On large-scale intensive pig farms where high-quality, nutrient-dense feeds are used, growing (fattening) pigs are typically provided feed and water ad libitum. This should generally prevent the welfare problem of pigs feeling hungry or thirsty, but only if there is enough feeding and watering space to accommodate the number of pigs that are housed together. In addition to assessing the number of feeding spaces available, feeders must be adjusted to ensure flow of feed into the troughs (Myers et al. 2010). The first key measure that would indicate inadequate feeding or watering resources for growing pigs would be observing pigs fighting near the feeder, which can occur in an attempt to gain access to, or to defend, feeder access or feeding space. Alternatively, pigs can be observed for scratches or wounds on their bodies. These can also be caused by fighting after mixing new groups. However, beyond the early post-mixing period, the appearance of scratches and wounds warrants further investigation. Systems for scoring the severity of fighting injuries have been developed (Turner et al. 2006).

An important nutritionally related welfare problem seen on intensively raised pig farms is tail biting, which can cause injuries and lead to debilitating conditions among the pigs that are targeted. Therefore, inspection or scoring of bitten tails is important for assessing the welfare of fattening pigs. Tail biting is a very complex, multifactorial problem, but the oral nature of this behaviour may lead one to assume that this is a problem related to nutrition or feeding (Taylor et al. 2010). Indeed, nutrient deficiencies, competition for feeding space, and lack of foraging substrate can all lead to tail biting. Epidemiological research has uncovered a number of other risk factors, such as the use of partial or fully slatted flooring and a stocking density of grower pigs ≥ 110 kg/m^2. Researchers also determined that having straw available reduces the risk of tail biting (Moinard et al. 2003), probably by providing opportunities for more natural foraging behaviour patterns such as rooting and manipulating substrates.

In contrast to piglets and growing pigs, breeding sows and boars in intensive production systems are typically fed a restricted diet to prevent them from becoming overweight and to optimize their reproductive performance. While this practice decreases the probability that pigs will experience certain other welfare problems that stem from obesity and age (e.g. lameness), the main issue with the practice of restricted feeding is that the pigs are in a chronic state of hunger (D'Eath et al. 2009). Restricted-fed pigs exhibit distinct behaviour indicative of hunger. For example, it is common for gestating sows to stand up, vocalize, and be more active when the caretaker enters the room where they are being kept (D'Eath et al. 2009). These behaviour changes reflect the pigs anticipating being fed, which they learn to associate with the arrival of the caretaker. Even following feeding, pigs may still feel hungry and still be motivated to perform the appetitive and consummatory aspects of feeding behaviour. This can be assessed by how much time the pigs spend rooting the floor and their feeder, mouthing the bars of their stall, or engaging in sham chewing (D'Eath et al. 2009). A number of studies have been done examining ways to reduce the state of hunger in sows (e.g. by supplementing their diet with dense or bulky fillers; see D'Eath et al. 2009). Group-fed sows are highly aggressive when competing for feed, and subordinate sows may be unable to gain access to sufficient amounts of feed (Andersen et al. 1999). Therefore, assessing the welfare of breeding sows should include lesion and body condition scoring (Séguin et al. 2006).

For pigs of all ages, poor nutrition or the underfeeding of otherwise adequate nutrition can result in the welfare problem of pigs feeling hungry and losing weight (see earlier discussion on hunger). With respect to losing weight, a lack of body fat covering can be a welfare issue, especially for heavier pigs, because thin pigs lacking adequate fat reserves may be at a higher risk of developing pressure sores or other injuries from their boney protrusions being in contact with hard surfaces (e.g. concrete flooring, stall bars, etc.) (Velarde 2007). The standardized way of assessing whether pigs (and other livestock) are too thin is through the use of body condition scoring. This is a systematic method used on many livestock farms on an ongoing basis, to continually monitor animal performance. Different scoring systems exist (Charette et al. 1996; Fitzgerald et al. 2009), but all rely on evaluating the fat cover over specific control points on the pig's body, which sum together to contribute to an overall, individual, body condition score. Scoring and recording body condition on a regular basis allow farmers to track changes over time that might otherwise be missed as the condition of animals can gradually drift towards undesirable levels without being immediately noticed.

6.6 Assessing the Welfare of Broiler Chickens

Modern strains of meat chickens have been selected for large appetites, fast growth rates, and high feed efficiencies, making broiler meat the most economically efficient and the most widespread source of animal protein worldwide. The majority of broiler chickens are raised intensively, in relatively large flocks under loose-housing conditions on litter floors in either naturally or artificially ventilated barns (Weeks 2004). Welfare problems in broilers have to do with their fast rates of growth, the management of their environment during the rearing period, and, finally, conditions of handling and transport for slaughter. Key factors affecting broiler chicken welfare during rearing include genetics, air and litter quality, and stocking density.

Fast growth problems of modern strains of broilers occur because the rate of muscle deposition for meat outpaces the rates of development of the skeleton and cardiovascular system (Kestin et al. 2001; Julian 2005). Consequently, health problems such as lameness, ascites (water belly), and sudden death syndrome are common. Lameness increases with age of the bird and is most prevalent in heavier birds and in fast-growing genotypes, which affects their welfare through leg pain and impaired walking ability (Bradshaw et al. 2002; Nääs et al. 2009). The most practical method for assessing lameness is by gait scoring. Gait scores commonly range from 0 (normal, dexterous, and agile) to 5 (incapable of walking) (Garner et al. 2002; Welfare Quality 2009b). Assessment of cardiovascular problems mainly involves recording mortality and the prevalence of these conditions in flocks (Julian 2004, 2005). Birds suffering from ascites can be identified by their distended abdomen and laboured breathing. Reducing the voluntary nutrient intake of the birds either by altering diet form (e.g. mash versus crumble) or nutrient density or by increasing the dark period slows growth rates and has been shown to improve walking ability and alleviate some of the problems due to fast growth (e.g. Scott 2002; Brickett et al. 2007).

Broiler chickens spend the majority of time lying in direct contact with litter and droppings, which can result in contact dermatitis on the breast (breast blisters), hocks (hock burns), and feet (pododermatitis). These skin conditions are characterized by areas of discoloration and lesions that become crusted with exudate and litter, and they can be readily assessed by inspecting the feet, hocks, and areas of skin covering the ventral keel bone, either at the farm or at the slaughter plant (Berg 2004; Welfare Quality 2009a). Wet litter is the major contributing factor for skin and foot lesions. Therefore, assessment of bedding quality and moisture (e.g. ranging from dry and flaky to sticky or capped) is often used as an indirect measure of welfare for broilers (Welfare Quality 2009a). Factors that affect litter quality include type of litter material, ventilation management (i.e. keeping humidity within the recommended range), and drinker system (i.e. reducing spillage). Feed composition can also affect the risk of contact dermatitis because it affects the birds' water intake, consistency of their faeces, as well as nitrogen content of the faeces (Mayne 2005). Some epidemiological studies have shown significant associations between feed manufacturer and prevalence of footpad lesions (Haslam et al. 2007).

High stocking density is another major issue for broiler chicken welfare. Stocking birds at high rates increases the net profit of the barn even though the growth rates of individual birds are lower. As stocking rate increases, feed intakes and growth rates decline in a linear fashion, as birds have more difficulty accessing feeders and drinkers (e.g. McLean et al. 2002). Mortality is largely unaffected, but walking ability of birds is poorer in more crowded conditions, and resting birds get jostled more by other birds (Dawkins et al. 2004). High stocking density can also result in more carcass damage, which can be assessed by the amount of scratching and bruising at the slaughter plant (Hall 2001). Litter moisture, relative humidity, and air quality are all significantly affected by stocking rate, and ventilation management plays a major role in maintaining the health and welfare of the bird (Jones et al. 2005). Setting clear limits on what is considered to be an acceptable stocking rate for birds is difficult, and standards vary around the world (Estevez 2007). What is clear is that maintaining the birds' environment within recommended temperature and humidity ranges is more critical than the stocking density per se, and animal-based measures such as mortality, gait score, and skin lesions are best for determining the welfare of broiler chickens (Jones et al. 2005).

6.7 Assessing the Welfare of Broiler Breeders

The parent stock of the broiler chicken share the same high appetite and potential for growth as their offspring, but allowing them ad libitum feed intake results in obesity, leg problems, and reduced fertility as breeding adults (Mench 2002). Therefore, target body weights specified by breeding companies must be strictly adhered to ensure acceptable levels of health and productivity. This requires that broiler breeders be feed restricted throughout their entire lives. Feed restriction can be imposed either by providing a small allotment of food each day or by feeding a larger amount of ration on alternate days (a practice that is prohibited in some European countries) (Hocking 2004). Birds that are feed restricted show signs of chronic hunger and feeding frustration, such as prolonged pecking at the trough, increased activity, repetitive drinking, and spot pecking (Savory and Maros 1993; Savory and Lariviere 2000). Alternative diets that aim to reduce hunger through dilution with lower quality foodstuffs such as oat hulls to decrease the energy density of the diet (Zuidhof et al. 1995) or through the inclusion of appetite suppressants such as calcium propionate have been shown to reduce some, but not all, of the behavioural signs of hunger (Savory et al. 1996; Sandilands et al. 2006).

Assessing the welfare of broiler breeders involves measures similar to those of broilers, particularly aspects related to lameness, litter quality, and environmental control (Hocking 2004). However, because the birds are feed restricted, additional measures such as flock uniformity (e.g. coefficients of variation in body weight) and adequacy of feeding space are critical. Rates of mortality, cannibalism, and injuries due to aggression at the feeder or due to feeder design are also important for assessing the welfare of broiler breeders. Feather condition is also a good indicator of feather pecking in relation to feeding management (Morrisey et al. 2014).

6.8 Assessing the Welfare of Laying Hens

Except in the European Union, the majority of laying hens used for table egg production are still housed in cages (International Egg Commission 2011). Welfare concerns for caged laying hens include space restriction and lack of opportunities to perform natural behaviour such as nesting in an enclosed space, perching, and scratching and dust bathing in litter. Therefore, resource-based measures such as bird density, provision of nests, perches, and a litter substrate are sometimes used to assess the welfare of hens (Bartussek 1999; Rodenburg et al. 2008a). However, health and physical aspects of the birds' welfare are often compromised in non-cage as well as cage systems, and therefore animal-based measures are considered to be better measures of hen welfare because they reflect the effects of husbandry conditions on the animal regardless of the housing system used (Whay et al. 2003; Rodenburg et al. 2008a). Similarly, in any housing system, feed and water facilities together with body condition of the birds can be used to assess whether hens have access to adequate nutrition (LayWel 2006).

Feather condition and pecking injuries are widely used indicators of hen welfare, and a number of scoring systems have been developed to assess feather loss and damage from cannibalism in hens (e.g. LayWel 2006). Feather pecking and cannibalism are major welfare concerns in all types of layer housing systems. Feather removal and injury have been shown to be painful; feather loss reduces thermal comfort and predisposes birds to further injury, and feather-pecked birds have been shown to be more fearful and have higher stress responses. The cause of feather pecking is multifactorial; contributing factors include nutrition and feed form, early rearing experience with regard to foraging substrates, availability of foraging material in the laying environment, light intensity, and genetic predisposition (Rodenburg and Koene 2004). Reducing the energy content of the diet and increasing fibre content reduce feather pecking as birds may spend more time eating (van Krimpen et al. 2005). Similarly, hens fed pelleted diets are more likely to develop feather pecking than hens fed mash diets, possibly because the hens fed mash diets spend more time feeding or can forage in the mash (Savory et al. 1999). Providing foraging material such as straw (Aerni et al. 2000) or feeding higher levels of insoluble fibre or roughages such as silage or carrots (Steenfeldt et al. 2007) can reduce feather pecking and improve plumage condition.

Health measures are also used as indicators of hen welfare (Rodenburg et al. 2008b). These can include mortality records, health records, and records of parasite treatments (Whay et al. 2003). Methods for scoring wounds to hens' combs and vents and for scoring foot lesions similar to those used for broilers have been developed for assessing hen welfare (Tauson et al. 2005; Rodenburg et al. 2008b). Osteoporosis is a common problem in laying hens, which has resulted from selection for high rates of lay that deplete calcium stores and weaken bones. Keel bone deformities and keel bone breakage can be assessed by visual assessment of the ventral surface of the bird (Welfare Quality 2009a) or by palpation (Wilkins et al. 2004). Bone breakages in wing and leg bones can only be practically assessed by post mortem dissection (Rodenburg et al. 2008a).

6.9 Conclusions

In this chapter, we have provided a few illustrative examples where the failure to provide adequate nutrition can undermine the health and biological functioning of farm animals and thus compromise welfare. The use of animal-based indexes such as growth rate, body condition score, lameness, morbidity, and mortality are useful metrics to ensure a minimum level of welfare. However, the usefulness of these metrics on a particular farm can only be as good as the records kept on the farm; if producers do not routinely measure and record these metrics they will fail in their attempts to improve the welfare of the animals under their care. Lastly, to strive for the highest level of welfare for food-producing animals, one must consider the affective (emotional) state of the animal and provide environments where the animals can engage in behaviours they are highly motivated to perform.

References

Aerni V, El-Lethey H, Wechsler B (2000) Effect of foraging material and food form on feather pecking in laying hens. Br Poult Sci 41(1):16–21

Andersen IL, Bøe KE, Kristiansen AL (1999) The influence of different feeding arrangements and food type on competition at feeding in pregnant sows. Appl Anim Behav Sci 65(2):91–104

Appleby MC, Weary DM, Chua B (2001) Performance and feeding behaviour of calves on ad libitum milk from artificial teats. Appl Anim Behav Sci 74(3):191–201

Bach A (2011) Associations between several aspects of heifer development and dairy cow survivability to second lactation. J Dairy Sci 94(2):1052–1057

Bach A (2012) Ruminant nutrition symposium: optimizing performance of the offspring: nourishing and managing the dam and postnatal calf for optimal lactation, reproduction, and immunity. J Anim Sci 90(6):1835–1845

Bach A, Ahedo J (2008) Record keeping and economics of dairy heifers. Vet Clin N Am Food Anim Pract 24(1):117–138

Bach A, Juaristi JL, Rodríguez P (2004) Cow comfort. Servet Diseño y comunicación, Zaragoza
Bach A, Dinarés M, Devant M, Carré X (2006) Associations between lameness and production, feeding and milking attendance of Holstein cows milked with an automatic milking system. J Dairy Res 74(1):40–46
Barker ZE, Leach KA, Whay HR, Bell NJ, Main DC (2010) Assessment of lameness prevalence and associated risk factors in dairy herds in England and Wales. J Dairy Sci 93(3):932–941
Bartussek H (1999) A review of the animal needs index (ANI) for the assessment of animals' well-being in housing systems for Austrian proprietary products and legislation. Livest Prod Sci 61 (2–3):179–192
Berg C (2004) Pododermatitis and hock burn in broiler chickens. In: Weeks CA, Butterworth A (eds) Measuring and auditing broiler welfare. CABI, Wallingford, pp 37–49
Bicalho RC, Vokey F, Erb EN, Guard GM (2007) Visual locomotion scoring in the first seventy days in milk: impact on pregnancy and survival. J Dairy Sci 90(10):4586–4591
Bradshaw R, Kirkden R, Broom D (2002) A review of the aetiology and pathology of leg weakness in broiler in relation to welfare. Avian Poultry Biol Rev 13(2):45–103
Brickett KE, Dahiya JP, Classen HL, Annett CB, Gomis S (2007) Influence of dietary nutrient density, feed form, and photoperiod on walking ability and skeletal quality of broiler chickens. Poult Sci 86(10):2117–2125
C.O.W.S. (2010). The Novus C.O.W.S. Program. http://www.novusint.com/Services/COWS. Accessed 25 Jan 2016
Castells L, Bach A, Araujo G, Montoro C, Terré M (2012) Effect of different forage sources on performance and feeding behavior of Holstein calves. J Dairy Sci 95(1):286–293
Charette R, Bigras-Poulin M, Martineau GP (1996) Body condition evaluation in sows. Livest Prod Sci 46(2):107–115
Collier RJ, Zimbelman RB, Rhoads RP, Roads ML, Baumgard LH (2011) A re-evaluation of the impact of Temperature Humidity Index (THI) and Black Globe Humidity Index (BGHI) on milk production in high-producing dairy cows. In: Proceedings of the Western dairy management conference, Reno, NV, 9–11 March 2011, pp 113–125
Cooper MA, Arney DR, Phillips CJC (2010) The motivation of high and low yielding dairy cows to obtain supplementary concentrate feed. J Vet Behav Clin Appl Res 5(2):75–81
D'Eath RB, Tolkamp BJ, Kyriazakis I, Lawrence AB (2009) 'Freedom from hunger' and preventing obesity: the animal welfare implications of reducing food quantity or quality. Anim Behav 77(2):275–288
Dann HM, Litherland NB, Underwood JP, Bionaz M, D'Angelo A, McFadden JW, Drackley JK (2006) Diets during far-off and close-up dry periods affect periparturient metabolism and lactation in multiparous cows. J Dairy Sci 89(9):3563–3577
Dawkins MS, Donnelly CA, Jones TA (2004) Chicken welfare is influenced more by housing conditions than by stocking density. Nature 427(6972):342–344
de Grau AF, Dewey CE, Widowski TM, Friendship RM, deLange CFM, Millligana B (2005) Reducing weight variation and behaviour problems in nursery pigs on a commercial farm by improving water accessibility and providing environmental enrichment. J Anim Vet Adv 4 (1):51–57
de Passillé AMB, Rushen J, Harstock TG (1988) Ontogeny of teat fidelity in pigs and its relation to competition at suckling. Can J Anim Sci 68(2):325–338
De Paula Vieira A, Guesdon V, de Passillé AM, von Keyserlingk MAG, Weary DM (2008) Behavioural indicators of hunger in dairy calves. Appl Anim Behav Sci 109(2–4):180–189
De Paula Vieira A, von Keyserlingk MAG, Weary DM (2010) Effects of pair versus single housing on performance and behavior of dairy calves before and after weaning from milk. J Dairy Sci 93:3079–3085
Dewell RD, Hungerford LL, Keen JE, Laegreid WW, Griffin DD, Rupp GP, Grotelueschen DM (2006) Association of neonatal serum immunoglobulin G1 concentration with health and performance in beef calves. J Am Vet Med Assoc 228(6):914–921

Espejo LA, Endres MI, Salfer JA (2006) Prevalence of lameness in high-producing Holstein cows housed in freestall barns in Minnesota. J Dairy Sci 89:3052–3058

Estevez I (2007) Density allowances for broilers: where to set the limits? Poult Sci 86 (6):1265–1272

Fitzgerald RF, Stalder KJ, Dixon PM, Johnson AK, Karriker LA, Jones GF (2009) The accuracy and repeatability of sow body condition scoring. Prof Anim Sci 25(4):415–425

Flower FC, Weary DM (2006) Effect of hoof pathologies on subjective assessments of dairy cow gait. J Dairy Sci 89(1):139–146

Fraser D (2003) Assessing animal welfare at the farm and group level: the interplay of science and values. Anim Welf 12(4):433–443

Garner JP, Falcone C, Wakenell P, Martin M, Mench JA (2002) Reliability and validity of a modified gait scoring system and its use in assessing tibial dyschondroplasia in broilers. Br Poult Sci 43(3):355–363

Gaylean ML, Rivera JD (2003) Nutritionally related disorders affecting feedlot cattle. Can J Anim Sci 83(1):13–20

Godden SM, Fetrow JP, Feirtag JM, Green LR, Wells SJ (2005) Economic analysis of feeding pasteurized nonsaleable milk versus conventional milk replacer to dairy calves. J Am Vet Med Assoc 226(9):1547–1554

Goldhawk C, Chapinal N, Veira DM, Weary DM, von Keyserlingk MAG (2009) Prepartum feeding behavior is an early indicator of subclinical ketosis. J Dairy Sci 92(1):4971–4977

González LA, Tolkamp BJ, Coffey MP, Ferret A, Kyriazakis I (2008) Changes in feeding behavior as possible indicators for the automatic monitoring of health disorders in dairy cows. J Dairy Sci 91(3):1017–1028

González LA, Manteca X, Calsamiglia S, Schwartzkopf-Genswein KA, Ferret A (2012) Ruminal acidosis in feedlot cattle: interplay between feed ingredients, rumen function and feeding behavior—a review. Anim Feed Sci Technol 172(1):66–79

Green LE, Hedges VJ, Schukken YH, Blowey RW, Packington AJ (2012) The impact of clinical lameness on the milk yield of dairy cows. J Dairy Sci 85(9):2250–2256

Hall A (2001) The effect of stocking density on the welfare and behaviour of broiler chickens reared commercially. Anim Welf 10(1):23–40

Hammon HM, Schiessler G, Nussbaum A, Blum JW (2002) Feed intake patterns, growth performance, and metabolic and endocrine traits in calves fed unlimited amounts of colostrum and milk by automate, starting in the neonatal period. J Dairy Sci 85(12):3352–3362

Hammon DS, Evjen IM, Dhiman TR, Goff JP, Walters JL (2006) Neutrophil function and energy status in Holstein cows with uterine health disorders. Vet Immunol Immunopathol 113 (1–2):21–29

Haslam SM, Brown SN, Wilkins LJ, Kestin SC, Warriss PD (2007) Factors affecting the prevalence of foot pad dermatitis, hock burn and breast burn in broiler chicken. Br Poult Sci 48 (3):264–275

Hocking PM (2004) Measuring and auditing the welfare of broiler breeders. In: Weeks CA, Butterworth A (eds) Measuring and auditing broiler welfare. CABI, Wallingford, pp 19–35

Huzzey JM, Veira DM, Weary DM, von Keyserlingk MAG (2007) Prepartum behavior and dry matter intake identify dairy cows at risk for metritis. J Dairy Sci 90(7):3220–3233

Huzzey JM, Duffield TF, LeBlanc SJ, Veira DM, Weary DM, von Keyserlingk MAG (2009) Short communication: haptoglobin as an early indicator of metritis. J Dairy Sci 92(2):621–625

International Egg Commission (2011) Annual review 2011. IEC, London

Ito K, von Keyserlingk MAG, LeBlanc SJ, Weary DM (2010) Lying behavior as an indicator of lameness in dairy cows. J Dairy Sci 93(8):3553–3560

Janovick NA, Boisclair YR, Drackley JK (2011) Prepartum dietary energy intake affects metabolism and health during the periparturient period in primiparous and multiparous Holstein cows. J Dairy Sci 94(3):1385–1400

Jasper J, Weary DM (2002) Effects of ad libitum milk intake on dairy calves. J Dairy Sci 85 (11):3054–3058

Jensen P, Algers B (1984) An ethogram of piglet vocalizations during suckling. Appl Anim Ethol 11(3):237–248

Johnsen PF, Johannesson T, Sandoe P (2001) Assessment of farm animal welfare at herd level: many goals, many methods. Acta Agric Scand Sect A Anim Sci 51(Suppl 30):26–33

Jones TA, Donnelly CA, Dawkins MS (2005) Environmental and management factors affecting the welfare of chickens on commercial farms in the United Kingdom and Denmark stocked at five densities. Poult Sci 84(8):1155–1165

Julian RJ (2004) Evaluating the impact of metabolic disorders on the welfare of broilers. In: Weeks CA, Butterworth A (eds) Measuring and auditing broiler welfare. CABI, Wallingford, pp 51–60

Julian RJ (2005) Production and growth related disorders and other metabolic diseases of poultry—a review. Vet J 169(3):350–369

Kestin S, Gordon S, Su G, Sørensen P (2001) Relationships in broiler chickens between lameness, liveweight, growth rate and age. Vet Rec 148:195–197

Khan MA, Lee HJ, Lee WS, Kim HS, Ki KS, Hur T, Suh GH, Kang SJ, Choi YJ (2007) Structural growth, rumen development, and metabolic and immune responses of holstein male calves fed milk through step-down and conventional methods. J Dairy Sci 90(7):3376–3387

Khan MA, Weary DM, von Keyserlingk MAG (2011) Hay intake improves performance and rumen development of calves fed higher quantities of milk. J Dairy Sci 94(7):3547–3553

Khan MA, Weary DM, Veira DM, von Keyserlingk MAG (2012) Postweaning performance of heifers fed starter with and without hay during the milk-feeding period. J Dairy Sci 95 (7):3970–3976

Kossaibati M, Esslemont R (1997) The costs of production diseases in dairy herds in England. Vet J 154(1):41–51

Langford FM, Stott AW (2012) Culled early or culled late: economic decisions and risks to welfare in dairy cows. Anim Welf 21(S1):41–55

LayWel (2006) Photographic scoring system. www.LayWel.eu. Accessed 16 Nov 2013

Mayne RK (2005) A review of the aetiology and possible causative factors of foot pad dermatitis in growing turkeys and broilers. Worlds Poult Sci J 61(2):256–267

McGuirk SM, Collins M (2004) Managing the production, storage, and delivery of colostrum. Vet Clin N Am Food Anim Pract 20(3):593–603

McLean J, Savory C, Sparks N (2002) Welfare of male and female broiler chickens in relation to stocking density, as indicated by performance, health and behaviour. Anim Welf 11(1):55–73

Mench JA (2002) Broiler breeders: feed restriction and welfare. Worlds Poult Sci J 58(1):23–29

Moinard C, Mendl M, Nicol CJ, Green LE (2003) A case control study of on-farm risk factors for tail biting in pigs. Appl Anim Behav Sci 81(4):333–355

Morrisey KLH, Widowski TM, Leeson S, Sandilands V, Arnone A, Torrey S (2014) The effect of dietary alterations during rearing on feather condition in broiler breeder females. Poult Sci 93(1):1–8

Müller R, Schrader L (2003) A new method to measure behavioural activity levels in dairy cows. Appl Anim Behav Sci 83(4):247–258

Munksgaard L, Lovendahl P (1993) Effects of social and physical stressors on growth-hormone levels in dairy cows. Can J Anim Sci 73(4):847–853

Myers AJ, Goodband RD, Tokach MD, Dritz SS, Bergstrom JR, DeRouchey JM, Nelssen JL (2010) The effects of feeder space and adjustment on growth performance of finishing pigs. Kansas State University Swine Day 2010. Rep Prog 1038:172–177

Nääs IA, Paz ICLA, Baracho MS, Menezes AG, Bueno LGF, Almeida ICL, Moura DJ (2009) Impact of lameness on broiler well-being. J Appl Poult Res 18(3):432–439

Nagaraja TG, Chengappa MM (1998) Liver abscesses in feedlot cattle: a review. J Anim Sci 76 (1):287–298

Nagaraja TG, Lechtenberg KF (2007a) Acidosis in feedlot cattle. Vet Clin N Am Food Anim Pract 23:333–350

Nagaraja TG, Lechtenberg KF (2007b) Liver abscesses in feedlot cattle. Vet Clin N Am Food Anim Pract 23(2):351–369

National Farm Animal Care Council (NFACC) (2009) Code of practice for the care and handling of dairy cattle. National Farm Animal Care Council (NFACC), Lacombe, AB. ISBN 979-0-9812240-0-8

Nocek JE (1997) Bovine acidosis: implications on laminitis. J Dairy Sci 80(5):1005–1028

Owens FN, Secrist DS, Hill WJ, Gill DR (1998) Acidosis in cattle: a review. J Anim Sci 76 (1):275–286

Penner GB, Yu P, Christensen DA (2009) Effect of replacing forage or concentrate with wet or dry distillers' grains on the productivity and chewing activity of dairy cattle. Anim Feed Sci Technol 153(1–2):1–10

Roche JR, Friggens NC, Kay JK, Fisher MW, Stafford KJ, Berry DP (2009) Invited review: body condition score and its association with dairy cow productivity, health, and welfare. J Dairy Sci 92(12):5769–5801

Rodenburg TB, Koene P (2004) Feather pecking and feather loss. In: Perry GC (ed) Welfare of the laying hen. CABI, Wallingford, pp 227–238

Rodenburg TB, Tuyttens FAM, de Reu K, Herman L, Zoons J, Sonck B (2008a) Welfare assessment of laying hens in furnished cages and non-cage systems: assimilating expert opinion. Anim Welf 17(4):355–361

Rodenburg TB, Tuyttens FAM, de Reu K, Herman L, Zoons J, Sonck B (2008b) Welfare assessment of laying hens in furnished cages and non-cage systems: an on-farm comparison. Anim Welf 17(4):363–371

Russell JB, Rychlik JL (2001) Factors that alter rumen microbial ecology. Science 292 (5519):1119–1122

Sandilands V, Tolkamp BJ, Savory CJ, Kyriazakis I (2006) Behaviour and welfare of broiler breeders fed qualitatively restrictive diets during rearing: are there viable alternatives to quantitative restriction? Appl Anim Behav Sci 96(1–2):53–67

Savory CJ, Lariviere JM (2000) Effects of qualitative and quantitative food restriction treatments on feeding motivational state and general activity level of growing broiler breeders. Appl Anim Behav Sci 69(2):135–147

Savory CJ, Maros K (1993) Influence of degree of food restriction, age and time of day on behavior of broiler breeder chickens. Behav Process 29(3):179–190

Savory C, Hocking P, Mann J, Maxwell M (1996) Is broiler breeder welfare improved by using qualitative rather than quantitative food restriction to limit growth rate? Anim Welf 5 (2):105–127

Savory CJ, Mann JS, MacLeod MG (1999) Incidence of pecking damage in growing bantams in relation to food form, group size, stocking density, dietary tryptophan concentration and dietary protein source. Br Poult Sci 40(5):579–584

Schwartzkopf-Genswein KS, Beauchemin KA, Gibb DJ, Crews DH Jr, Hickman DD, Streeter M, McAllister TA (2003) Effect of bunk management on feeding behavior, ruminal acidosis and performance of feedlot cattle: a review. J Anim Sci 81(Suppl 2):E149–E158

Scott TA (2002) Evaluation of lighting programs, diet density, and short-term use of mash as compared with crumbled starter to reduce incidence of sudden death syndrome in broiler chicks to 35 days of age. Can J Anim Sci 82(3):375–383

Séguin M, Barney D, Widowski TM (2006) Assessment of a group-housing system for gestating sows: effects of space allowance and pen size on the incidence of superficial skin lesions, changes in body condition and farrowing performance. J Swine Health Prod 14(2):89–96

Sepúlveda-Varas P, Huzzey JM, Weary DM, von Keyserlingk MAG (2013) Behaviour, illness and management during the periparturient period in dairy cows. Anim Prod Sci 53(9):988–999

Soberon F, Raffrenato E, Everett RW, Van Amburgh ME (2012) Preweaning milk replacer intake and effects on long-term productivity of dairy calves. J Dairy Sci 95(2):783–793

Sowell BF, Branine ME, Bowman JGP, Hubbert ME, Sherwood HE, Quimby W (1999) Feeding and watering behavior of healthy and morbid steers in a commercial feedlot. J Anim Sci 77 (5):1105–1112

Sprecher DJ, Hostetler DE, Kanneene JB (1997) A lameness scoring system that uses posture and gait to predict dairy cattle reproductive performance. Theriogenology 47:1179–1187
Steenfeldt S, Kjaer JB, Engberg RM (2007) Effect of feeding silages or carrots as supplements to laying hens on production performance, nutrient digestibility, gut structure, gut microflora and feather pecking behaviour. Br Poult Sci 48(4):454–468
Straw BE, Bürgi EJ, Dewey CE, Duran CO (1998) Patterns of cross-fostering and piglet mortality on commercial US and Canadian swine farms. Prev Vet Med 33(1–4):83–89
Stull CL, Reed BA, Berry SL (2004) A comparison of three animal welfare assessment programs on California dairies. J Dairy Sci 88:1595–1600
Sweeney BC, Rushen JP, Weary DM, de Passille AMB (2010) Duration of weaning, starter intake, and weight gain of dairy calves fed large amounts of milk. J Dairy Sci 93(1):148–152
Tauson R, Kjaer J, Maria GA, Cepero R, Holm KE (2005) Applied scoring of integument and health in laying hens. In: Welfare implications of changes in production systems for laying hens. Laywel report SSPE-CT-2004-502315. http://www.laywel.eu/web/pdf/deliverables%2031-33%20health.pdf. Accessed 26 Dec 2013
Taylor NR, Main DC, Mendl M, Edwards SA (2010) Tail-biting: a new perspective. Vet J 186 (2):137–147
Torrey S, Widowski TM (2007) Relationship between growth and non-nutritive massage in suckling pigs. Appl Anim Behav Sci 107(1–2):32–44
Tucker AL, Widowski TM (2009) Normal profiles for deciduous dental eruption in domestic piglets: effect of sow, litter, and piglet characteristics. J Anim Sci 87(7):2274–2281
Turner SP, Farnworth MJ, White IMS, Brotherstone S, Mendl M, Knap P, Penny P, Lawrence AB (2006) The accumulation of skin lesions and their use as a predictor of individual aggressiveness in pigs. Appl Anim Behav Sci 96(3–4):245–259
Urton G, von Keyserlingk MAG, Weary DM (2005) Feeding behaviour identifies dairy cows at risk for metritis. J Dairy Sci 88(8):2843–2849
USDA-NAHMS [United States Department of Agriculture—National Animal Health Monitoring System] (2007) Dairy 2007, part I: reference of dairy cattle health and management practices in the United States. http://www.aphis.usda.gov/animal_health/nahms/dairy/#dairy2007. Accessed 13 Dec 2013
USDA-NAHMS (2011) Dairy 2011. Dairy Heifer Raiser, 2011. An overview of operations that specialize in raising dairy heifers. http://www.aphis.usda.gov/animal_health/nahms/dairy/downloads/dairyheifer11/HeiferRaiser.pdf. Accessed 15 Dec 2013
van Krimpen MM, Kwakkel RP, Reuvekamp BFJ, van der Peet-Schwering CMC, Den Hartog LA, Verstegen MWA (2005) Impact of feeding management on feather pecking in laying hens. Worlds Poult Sci J 61(4):663–685
Velarde A (2007) Skin lesions. In: Velarde A, Geers R (eds) On-farm monitoring of pig welfare. Wageningen Academic Publishers, Wageningen, pp 79–85
Vickers LA, Weary DM, Veira DM, von Keyserlingk MAG (2013) Feeding a higher forage diet prepartum decreases incidences of subclinical ketosis in transition dairy cows. J Anim Sci 91 (2):886–894
von Keyserlingk MAG, Weary DM (2010) Invited review: feeding behaviour of dairy cattle: measures and applications. Can J Anim Sci 90(3):303–309
von Keyserlingk MAG, Rushen J, de Passillé AMB, Weary DM (2009) Invited review: the welfare of dairy cattle—key concepts and the role of science. J Dairy Sci 92(9):4101–4111
von Keyserlingk MAG, Barrientos A, Ito K, Galo E, Weary DM (2012) Benchmarking cow comfort on North American freestall dairies: lameness, leg injuries, lying time, facility design, and management for high-producing Holstein dairy cows. J Dairy Sci 95(12):7399–7408
Weary DM, Huzzey JM, von Keyserlingk MAG (2009) Board-invited review. Using behavior to predict and identify ill health in animals. J Anim Sci 87(2):770–777
Weeks CA (2004) Introduction. In: Weeks CA, Butterworth A (eds) Measuring and auditing broiler welfare. CABI, Wallingford, pp xiii–xv
Welfare Quality (2009a) Assessment protocol for poultry. ASG Veehouderij BV, Lelystad

Welfare Quality (2009b) Welfare quality® assessment protocol for cattle. Welfare Quality® Consortium, Lelystad

Wells SJ, Dargatz DA, Ott SL (1996) Factors associated with mortality to 22 days of life in dairy herds in the United States. Prev Vet Med 29:9–19

Whay HR, Main DC, Green LE, Webster AJF (2003) Animal-based measures for the assessment of welfare state of dairy cattle, pigs and laying hens: consensus of expert opinion. Anim Welf 12(2):205–217

Widowski TM, Cottrell T, Dewey CE, Friendship RM (2003) Observations of piglet-directed behaviour patterns and skin lesions in eleven commercial swine herds. J Swine Health Prod 11 (4):181–185

Widowski TM, Torrey S, Bench CJ, Gonyou HW (2008) Development of ingestive behaviour and the relationship to belly nosing in early-weaned piglets. Appl Anim Behav Sci 110 (1–2):109–127

Wilkins LJ, Brown SN, Zimmerman P, Leeb C, Nicol CJ (2004) Investigation of palpation as a method for determining the prevalence of keel and furculum damage in laying hens. Vet Record 155(18):547–549

Yunta C, Guasch I, Bach A (2012) Short communication: lying behavior of lactating dairy cows is influenced by lameness especially around feeding time. J Dairy Sci 95(11):6546–6549

Zuidhof MJ, Robinson FE, Feddes JJR, Hardin RT, Wilson JL, McKay RI, Newcombe M (1995) The effects of nutrient dilution on the well-being and performance of female broiler breeders. Poult Sci 74(3):441–456

Chapter 7
Integrating Nutrition and Animal Welfare in Extensive Systems

Juan J. Villalba, Xavier Manteca, Philip E. Vercoe, Shane K. Maloney, and Dominique Blache

Abstract Extensive systems present herbivores multiple biotic and abiotic challenges such as temporal and spatial variation in the availability and quality of food and water, changes in the chemical and morphological defenses of plants, thermal stress, disease, predation, and competition. Due to variability in resource abundance and quality, food intake might not always satisfy their appetite or hunger. Animals can adapt by increasing their grazing time and/or by dispersing more widely. These changes in behaviour may improve the use of poor quality pastures, but may also have negative effects on productivity and welfare. For instance, poor pasture conditions lead to reductions in mating activity and reproductive performance, decrease the efficiency of behavioural strategies aimed at decreasing parasitic loads, and increase energetic costs. Thermoregulation has an energetic cost, which decreases animal production. However, a lack of exposure to thermal stress may not necessarily imply adequate animal welfare. Plant secondary compounds (PSC) can inhibit food digestion, increase metabolic costs, and potentially have toxic effects on animal tissues and metabolic processes, but at appropriate concentrations certain PSC may improve nutrition and immunity and alleviate some of the diseases that challenge herbivores in extensive systems. Predators have both a direct lethal impact on herbivores and indirect effects that challenge animal nutrition and welfare. They may restrict the use of high-quality habitats and increase the time invested in vigilance, which restricts foraging time. In contrast to this, facilitative interactions among animals with contrasting foraging strategies may

J.J. Villalba (✉)
Department of Wildland Resources, Utah State University, 84322 Logan, UT, USA
e-mail: Juan.villalba@usu.edu

X. Manteca
School of Veterinary Science, Universitat Autònoma de Barcelona, 08193 Barcelona, Spain

P.E. Vercoe • D. Blache
UWA Institute of Agriculture (Animal Production), The University of Western Australia, Crawley 6009, WA, Australia

S.K. Maloney
School of Anatomy, Physiology, and Human Biology (M311), The University of Western Australia, Crawley 6009, WA, Australia

C.J.C. Phillips (ed.), *Nutrition and the Welfare of Farm Animals*, Animal Welfare 16, DOI 10.1007/978-3-319-27356-3_7

positively impact on nutrition and welfare. In conclusion, some relationships between the variables described in this chapter (thermoregulation, PSC, animal–animal interactions) and welfare do not always vary in a uniform direction. Many of the challenges described in this chapter stem from variability and unpredictability of the environment, which is largely a function of natural—instead of managerial—influences. The challenge of managers is to provide the conditions and flexibility in their operations to allow animals to express their behaviour to cope with these challenges in a way that production and welfare are maximised within the constraints imposed by an ever-changing environment.

7.1 Introduction

Extensive production systems can be defined as those systems in which animals have little contact with humans, including infrequent handling (Hodgson 1990). In contrast, in intensive animal production systems there is frequent interaction between animals and their managers. Herbivore species evolved to graze and browse in a variety of ecosystems, such as rangeland, grassland, and open pasture or shrubby and woody landscapes, and a variety of herbivores have been domesticated (cattle, buffalo, sheep, goats, and camels) to be used for production purposes. From the time of their domestication, herbivores have been mainly kept in extensive production systems. Today, there are about three billion animals managed in grazing systems, which use 3.4 billion hectares of land. That means 26 % of the earth's ice-free land surface is used for animal production (FAO 2009). Developing countries use slightly more of their land mass (~30 %) than developed countries (20.5 %) for animal production. In principle, the welfare of herbivores kept in extensive grazing systems should be assisted by the fact that they have evolved to make the best use of such varied herbivorous environments. However, animals are not always kept where they evolved, and the unpredictability of environmental factors, coupled with the management of livestock by humans, is not always a match to the adaptive features of livestock. Consequently, these mismatches represent welfare problems for livestock kept in extensive systems.

7.2 Why Animal Welfare Is Compromised in Extensive Livestock Production Systems

A key aspect of animal welfare is the ability of an animal to cope with its environment (Broom 1986). Within this functional approach to animal welfare, it is essential to assess whether the biological systems of an animal are functioning in a satisfactory manner in response to environmental changes (Duncan and Fraser 1997). However, welfare not only depends on whether the animal can cope but also on whether coping attempts have negative consequences for the animal. Therefore, there is no simple relationship between challenge and response because multiple biological functions can be affected by a very specific challenge, such as variability in level or quality of nutrition (Blache et al. 2011). Consequently, a multidisciplinary assessment of animal welfare that measures behaviour, the level of productivity, physiology, health, and immunity is required (Gonyou 1986; Broom 1991; McGlone 2001). These multiple aspects of animal welfare are encompassed in the framework designed for animals kept in captive environments known as the 'Five Freedoms' (Farm Animal Welfare Council 1993), namely: (a) Freedom from thirst, hunger, and malnutrition, (b) Freedom from discomfort, (c) Freedom from pain, injury, and disease, (d) Freedom to express normal behaviour, and (e) Freedom from fear and distress. Initially the five freedoms framework might not be seen as directly applicable to livestock raised in extensive systems. As pointed out by Turner and Dwyer (2007), extensive systems should not compromise the expression of behaviour, compromise the health of the animals, or induce mental distress, since the animal is in a 'free range' or 'natural' environment. Thus, some hold the view that livestock in extensive production systems are not under our control and therefore we do not have a duty of care (Appleby 1996). Since these species have coevolved with the landscape to be able to cope with the degree of variability of the environment, how could the welfare of extensive livestock be compromised? The answer is twofold: (1) there is a potential mismatch between the coping capacity of the animals and the variability of the environment of extensive systems because, by controlling stocking density, the type of livestock, and landscape availability, managers control the animal distribution across time and space, and (2) animals kept in the environment they have evolved in can still face challenges from extreme variability in foods, habitats, and environmental conditions that exceed their capacity to cope. The second situation needs to be considered here because, although those factors are not 'controlled' by the manager, they represent a welfare challenge.

In the following sections, we will give examples of challenges faced by ruminant livestock either because of natural factors or managerial decisions such as stocking density, choice of species, breed, or timing of breeding. The direct and indirect consequences of each challenge on nutrition will be treated with particular attention since this is a focus of this book. In most codes of recommendations for the welfare of animals, adequate nutrition is one of the primary requirements to be satisfied (Farm Animal Welfare Council 1993; Kyriazakis and Savory 1997). For each challenge, the advantages and limitations of the adaptive behavioural and physiological responses of grazing animals will be presented. We then analyse the impact of these challenges and adaptations on animal welfare under the framework of the aforementioned Five Freedoms. The following challenges are most common and relevant to ruminants kept in extensive conditions: variability of the quality and availability of food and water, the impact of environmental toxicology, predator and fear effects on intake, and competition with conspecifics and non-livestock herbivores.

7.3 Challenge 1: Variability in the Availability and Quality of Food and Water

To fulfil the freedom from hunger and thirst, it is essential that livestock have access to feedstuffs of adequate quality and quantity to be able to meet their nutritional requirements. In extensive livestock production systems, both the quality and quantity of food and water available can be limited either by environmental or managerial factors.

Environmental factors, such as rainfall patterns, drive forage production especially in semiarid and arid systems (Campbell et al. 2006), and food resources can be both variable and unpredictable, and as a consequence, affect grazing productivity (McAllister 2012). Similarly, forage nutritive value can be affected by water availability, fertilisation, and disease because of their impact on plant development (Van Soest et al. 1978).

To optimise production, managers can adopt stocking rate strategies to compensate for the fluctuation in forage supply induced by changing environmental conditions, such as variable rainfall. They can opt for a conservative stocking strategy by maintaining a relatively constant, and low, stocking rate, which would be unlikely to exceed the carrying capacity even during dry years. This strategy aims to avoid vegetation degradation, livestock losses, and potential decreases in animal welfare caused by overstocking in dry years (Campbell et al. 2006). Low stocking rates (i.e. low number of animals on a given amount of land over a certain period of time) typically promote an increase in plant species diversity, although very low stocking rates may reduce plant diversity as a result of competitive exclusion (Grime 1979). At low stocking rates, grazing typically leads to the formation of patches of different forage quality and quantity. In these conditions, herbivores prefer to graze patches of high nutritive quality, which are generally those corresponding to the regrowth of previously grazed areas (i.e. patch grazing; Adler et al. 2001). This recurrent pattern of grazing contributes to the spatial heterogeneity of the vegetation generally observed in plant communities grazed at low stocking rates.

Alternatively to low stocking strategies, managers may adopt high and constant stocking rates, using supplemental feed to maintain high stocking rates during dry years (Campbell et al. 2006). The supplementation needs to be managed carefully because, when stocking rate exceeds the maximum number of animals that can be supported by a certain environment (i.e. carrying capacity), the biomass availability per animal declines and preferred plants (typically the most nutritive in a community) are subjected to a competitive disadvantage relative to non-preferred plants (typically less nutritive or more toxic). The biased grazing leads to changes in plant community structure and composition (Dumont et al. 2007) impacting on both the biomass and the nutritive quality of forage available for grazing (Walker 1995). Herbivores select plant and plant parts at different levels of stocking rates (Provenza and Villalba 2006). However, high animal densities increase competition for food resources and reduce selectivity (Bailey and Brown 2011). Additionally, the prevalence of non-preferred plants species in the context of high animal densities can increase the risk of consumption of poisonous plants (Pfister et al. 2002), which is a reduction in the freedom from disease.

In addition to the compromised freedom from hunger and disease, high animal densities may also induce social stress, meaning there is a decrease in freedom from stress by increased competition for resources. The increase in social stress leads to disturbed grazing patterns that can present some serious welfare issues since the grazing pattern and intensity ultimately affect growth rates in herbivores (Blanc and Theriez 1998).

Independent of climatic conditions, stocking density is one of the most important factors affecting forage availability and quality (Edwards 1980; Allison 1985; De Villiers et al. 1994). While high stocking rates can exacerbate the effects of drought on vegetation by increasing mortality of perennial grasses, negatively impacting on soils and thus contributing to range degradation (O'Connor 1995), conservative stocking rates lead to underuse of forage in wet years which may be inefficient, particularly as environmental conditions (e.g. rainfall across seasons) become more variable. Alternative to the extreme stocking densities listed above, an opportunistic strategy is to change stocking rates to match temporally variable forage supply, aiming to secure quality and quantity forage and therefore insure an adequate level of animal welfare. A variable stocking density strategy means that higher stocking rates are adopted in wet years and reduced stocking rates in dry years. This dynamic strategy aims to match grazing pressure to the existing vegetation in both wet and dry seasons and consequently avoid production losses, and potential declines in animal welfare in dry years (Campbell et al. 2006).

Ranchers now recognise that they not only need to manage to maximise animal production but to improve vegetation abundance and plant species composition and to enhance the ecological integrity of their operations (Grissom and Steffens 2013). Under this framework, they implement adaptive management strategies with variable timing, frequency and distribution of grazing with positive economic returns and benefits to animal welfare.

7.3.1 Adaptive Responses to Low Quantity and Quality Forage

To match nutrient intake to requirements, herbivores have developed adaptive responses to decreases in the amount and/or quality of food available. These adaptive strategies include modification of feeding behaviour such as diet selection, or changes in foraging time, or modification of social behaviour such as dispersion. However, in extreme cases the adoption of feeding strategies or changes in social behaviour can negatively impact on welfare.

7.3.2 Diet Selection

The selection of a feeding station by herbivores in a heterogeneous landscape has been described as a hierarchical decision process which takes place at different levels: the regional level, landscape, plant community, down to the individual plant and its parts (Senft et al. 1987). According to optimal foraging theory, when food availability changes in space and time, herbivores will focus on the most profitable foods if resources are abundant (MacArthur and Pianka 1966; Stephens and Krebs 1986), and broaden their diet when the availability of resources declines (e.g. dry season, drought, winter) (Owen-Smith and Novellie 1982; Stephens and Krebs 1986). In terms of habitat use, the selection process may lead to an ideal free distribution of grazing pressure among habitats of different qualities (Fretwell and Lucas 1970). Thus, optimality models suggest that as long as the stocking density is adequate, animals should not experience any form of malnutrition.

Most models used in optimal foraging theory use energy or protein as the currency to be maximised. Lambs and goats fed diets low in energy and protein prefer flavoured foods associated with intra-ruminal infusions of energy (starch, propionate, acetate) and nitrogen (urea, casein, gluten), respectively (Villalba and Provenza 1996, 1997a, b, c; Duncan and Young 2002). Thus, herbivores are able to associate different internal states with the ingestion, or lack thereof, of different nutrients (Egan 1980; Kyriazakis and Oldham 1993). Despite the importance of energy and nitrogen, foraging animals usually face a choice of a variety of habitats and they possess multiple nutrient requirements. Thus, models based on just one currency (e.g. energy) are poor predictors of habitat occupancy (Wallis de Vries and Schippers 1994). Complementary nutrients like phosphorous have improved those predictions in free-ranging cattle (Wallis de Vries and Schippers 1994). Consistent with this, lambs deficient in phosphorous and calcium increase preferences for flavours and supplements which provide those minerals (Villalba et al. 2006, 2008). Thus, a decrease in the diversity of foods and/or habitats may compromise animal welfare (Manteca et al. 2008; Villalba et al. 2011; Catanese et al. 2013). The inability to satisfy requirements for energy, protein, and minerals can lead to nutritionally unbalanced intake, health problems, and stress. In an unbalanced nutritional environment, animals may stop eating because their requirements for the nutrient in highest concentration are rapidly satisfied without reaching the requirements for nutrients in lower concentrations, a phenomenon known as incidental restriction (Raubenheimer 1992). Alternatively, the animal may continue foraging in order to satisfy the requirements for nutrients in lower concentrations, inevitably leading to overconsumption of the nutrient in highest concentration, a phenomenon known as incidental augmentation (Raubenheimer 1992). Excessive nitrogen concentrations can lead to ammonia build-up in the blood resulting in loss of appetite, infertility, and even death (Lobley and Milano 1997). In addition, the augmentation in nutrient-rich and homogeneous food environments will compromise animal welfare because excess nutrients can induce aversive behaviour (Provenza 1996). Lambs exposed to flavours associated with high doses of energy or nitrogen become averse to those flavours (Villalba and Provenza 1996, 1997a, b), suggesting that excesses of nutrients can be a source of stress. The provision of multiple alternatives regarding foods and habitats is necessary to reach the appropriate proportion of required nutrients (Bailey and Provenza 2008) and allow a ruminant to adjust its food selection to minimise the discomfort generated by several signals produced during the ingestive process, i.e. the supply of metabolisable energy, crude protein, and neutral-detergent fibre (Forbes 2007; Gregorini et al. 2015). In contrast, animals that cannot express their diet preferences may experience frustration (Rutter 2010) or negative postingestive feedback (Forbes 2007; Villalba et al. 2010b) and therefore have a compromised welfare.

7.3.3 Foraging Behaviour

Herbivores can vary greatly the time they spend feeding in response to pasture availability and quality. Ruminants adapt to poor forage conditions by modifying the biting rate (number of bites per unit time), bite weight (amount of food taken per bite), and total grazing time (Holmes 1989) to obtain the necessary daily herbage intake. On short overgrazed pasture, herbivores are not able to sustain the bite volume (Charcon and Stobbs 1976) and, because of the reduction in stratum bulk density (Burlison et al. 1991), the bite weight decreases relative to its optimal level (Arriaga-Jordan and Holmes 1986). Consequently, to maintain intake, grazing ruminants have to graze for longer periods as forage conditions deteriorate (Squires 1981; Arnold 1985a; Lu 1988). On good pastures, domestic ruminants graze between 4 and 9 h/day (Houpt 1991), whereas, on poor forage conditions, grazing times of up to 14 h have been recorded (Arnold and Dudzinski 1978). This increase in grazing time, which includes an increase in total jaw movements and walking, can lead to fatigue before nutritional requirements have been fulfilled (Williamson and Payne 1978; Birrell 1991). Consequently, daily energy requirements increase because animals walk more per unit of grazing time (Osuji 1974). For example, grazing ruminants require more energy than penned ruminants: 10–50 % for sheep (Holmes 1989) and 25 and 75 % for goats (NRC 1981).

The physical characteristics of the plant can also affect grazing time. The fibrousness of feed is positively correlated with total chewing time (Boever et al. 1990). It would be difficult to consider this parameter to assess the welfare of ruminants because chewing efficiency is also affected by age (older animals chew feed less efficiently than younger ones, Gill et al. 1966), body weight (larger animals eat faster than small ones, Deswysen 1986), pregnancy (pregnant animals need more chewing time per kg of dry matter than non-pregnant ones, Campling 1966), jaw force (Hooper and Welch 1983), number of chews per unit time (Grummer et al. 1987), percentage of true chewing (Deswysen et al. 1987), and tooth surface area (Welch and Smith 1970). On the other hand, animals can adapt to the physical characteristic of the forage by developing specific foraging skills (Flores et al. 1989a; Hodgson 1990). For example, lambs used to eating a particular kind of plant were more efficient at foraging from it than unfamiliar lambs (Flores et al. 1989b, c). It has to be noted that these specialised grazing techniques may be unsuitable for other types of plant communities (Arnold and Maller 1977) and could limit the feeding behaviour of animals moved between very different plant communities.

7.3.4 Dispersion

The degree of dispersion of grazing animals is measured as a function of the distance between individuals of the same group, mean group size, and the distance between groups (Arnold 1985b). Dispersion is a strategy that enables animals to find scattered pockets of forage and to avoid competition for food. As forage conditions deteriorate, animals tend to split into increasingly smaller subgroups, and the distance between these subgroups also increases (Dudzinski et al. 1978). Inter-individual distance depends mainly on the breed. In sheep, for example, the distance between nearest neighbours is much greater in hill breeds than in Merinos (Arnold 1985b). On poor pastures, Merino sheep remain as a single flock in most circumstances (Lynch 1974), whereas many other breeds form subgroups under almost all conditions (Arnold 1985b). Similar differences in dispersion strategy between breeds have been reported in cattle (Hall and Moore 1986). In breeds with limited dispersion the ability to feed on poor pastures becomes restricted (Kilgour et al. 1975). Dispersion is an effective adaptation to changes in pasture quality and availability; however, the welfare of gregarious animals could be compromised because when forage conditions deteriorate, dispersion may conflict with their natural social behaviour. Assessing the impact of dispersion on the social structure is not simple because some breeds can balance the need for social contact with the need to disperse. Hill sheep are highly social during winter, living in large subgroups influenced in their movements by other group members. In summer, however, there is a marked decrease in gregariousness, with a considerable reduction in subgroup size and reduced frequency of cohesive grazing behaviour. The summer behaviour corresponds with the growth of dispersed patches of grass (Hunter and Milner 1963; Lawrence and Wood-Gush 1987, 1988). In addition to breed, dispersion is also affected by other factors such as previous experience (Zimmerman 1980), maternal experience (Key and MacIver 1980), location of preferred feed such as specific plant communities (El Aich and Rittenhouse 1988) or feedblocks (Lawrence and Wood-Gush 1988), shade (Fowler 1984), and water source (Squires 1981).

7.3.5 *Water Intake*

Ruminants can go longer without food than they can without water, so their welfare is very much dependent on the ease of access to drinkable water. In intensive or semi-intensive animal production systems, livestock are either watered daily or allowed to drink whenever they like. In contrast, under pastoral conditions, livestock access watering points and develop rigorous watering regimes that vary from every 3 days in cattle and donkeys, through 5 days in sheep and goats, to 7–15 days in camels (Nicolson 1984). However, the frequency of drinking is affected by the level of humidity and ambient temperature. For example, in the Australian arid zone, Merinos drink every day and twice a day when ambient temperature exceeds 35 °C (Dawson et al. 1975).

Within species, breeds differ in their adaptation to water restriction. For example, Boran cattle have large intakes of water and the intake is faster than other breeds. Boran bulls can drink up to 105 L in 6 min without suffering water intoxication, whereas temperate cattle can only drink at a third of this rate safely (Nicolson 1984). In sheep, when the distance between food and water is short, Merinos and Border Leicesters drank twice per day, but when the distance exceeded 3.2 km, Merinos drank once per day while Border Leicesters continued drinking twice daily until the distance reached 4.8 km (Squires and Wilson 1971).

To access water, livestock may have to spend time walking long distances between watering points and grazing grounds (Squires 1981) that reduces time available for grazing (El Aich et al. 1991). The time spent travelling back and forth to the watering point depends on (1) the animal's need for water, which itself depends on species, breed, climatic conditions, water content of the forage, and water quality, (2) the distance to travel, and (3) the walking speed. All of these factors should be taken into consideration when assessing the welfare of livestock kept in extensive systems.

7.4 Challenge 2: Variability in the Toxicological Environment

In extensive grazing systems, animals encounter a diversity of plants and many of those contain plant secondary compounds (PSC), especially if the animals are grazing native 'unimproved' pasture (Estell 2010). Plants produce secondary compounds for many reasons, mainly as a defence mechanism in response to herbivory and other environmental stressors such as disease and competition (Cowan 1999; Mazid et al. 2011). More than 200,000 chemical structures have been identified as PSC, including flavonoids, tannins, saponins, alkaloids, non-protein amino acids, cyanogenic glycosides, glucosinolates, and terpenes (Cowan 1999; Bernhoft 2010). The chemical diversity of PSC is the basis of the diversity of their actions on animal's organs and systems, as well as on animal behaviour (Durmic and Blache

2012). A specific PSC can have both detrimental and/or beneficial effects on animal welfare depending on the form and the dose ingested, the duration of ingestion, and the species exposed (Greathead 2003; Bernhoft 2010). Plant secondary compounds can severely alter nutrient utilisation, digestive function, respiratory and cardiovascular function, immune function, as well as impacting on the nervous system and reproductive capacity (Vercoe et al. 2009).

Feed intake and feed efficiency can be reduced by PSC because they influence diet selection and grazing behaviour (Provenza and Villalba 2006). Plant secondary compounds can be bitter and often deter animals from eating the plant or interfere with the digestive and fermentation process and can directly affect host secondary metabolism. However, the dose and mix of PSC are critical. Ingestion of PSC can reduce feed conversion efficiency (Stienezen et al. 1996), while small quantities of some PSC can increase it (Carulla et al. 2005). Similarly, fibre digestion can be reduced (Patra and Saxena 2009), not affected (Wang et al. 2009), or stimulated (Hart et al. 2007) by PSC. The decrease in digestibility by PSCs is associated with the inhibition of ruminal protozoa, cellulolytic bacteria, and fungi (Ferme et al. 2004; Sivakumaran et al. 2004) and a general inhibition of microbial fermentation and microbial synthesis in the rumen (Oh et al. 1967, 1968; Nagy and Tengerdy 1968; Lu and Jorgensen 1987).

In addition to these specific impacts on intake and rumen function, PSC can affect all the components of the digestive system of herbivores, inducing gastroenteritis, ruminal atony (Aslani et al. 2004), lower viscosity of intestinal content (Mahgoub et al. 2008), and liver damage (Waghorn 2008), as well as bladder lesions (Sardon et al. 2005), renal disease, and kidney failure (see Durmic and Blache 2012).

It is not only the digestive system that is impacted by PSC as pointed out in a recent review by Durmic and Blache (2012). The respiratory system can also be dramatically affected after the ingestion of PSC, leading to serious health problems such as alveolar emphysema and pulmonary necrosis and edema in sheep and cattle (Wilson et al. 1977). Similarly, with the cardiovascular system, ingestion of PSC can lead to cardiac arrhythmia, haemorrhagic syndrome and cardiac necrosis, and haemolysis. Plant secondary compounds can act as foreign molecules and induce inflammation and reduced immune cell count and activity and cause anaemia, or act as anticoagulants. Photosensitisation and irritation can be common in animals consuming plants rich in some PSC, such as saponins and furanocoumarins (Revell and Revell 2007; Bernhoft 2010). The effects of PSC on the nervous system are often detrimental and can be lethal, for example, some alkaloids, cardiac glucosides, and cyanogenetic glucosides are powerful neurotoxics (Ingebrigtsen 2010).

The reproductive capacity of both males and females is also impacted by PSC, with some dramatic consequences on production and welfare [for review see Blache et al. (2008), Durmic and Blache (2012)]. PSC can stimulate the expression of male reproductive behaviours (Patel et al. 2011) but can also decrease sperm production and quality. In females, PSC that affect fertility can be grouped into plants with contraceptive, anti-implantation, and abortifacient activity (Kumar et al. 2012).

Although PSC, as described above, have a large range of negative impacts on health, welfare, and the productivity of livestock, ruminants have acquired adaptive features that allow them to cope with the presence of many PSC in pasture because of co-evolution. More importantly, managers of extensive livestock systems can also take advantage of the beneficial impacts of PSC on bodily function to improve both the quality of animal production and animal welfare without compromising productivity.

7.4.1 Herbivores Adaptations to Cope with PSC

The adaptive features that allow herbivores to cope with PSC include behavioural and metabolic adaptations (Foley et al. 1999). The main behavioural adaptation is the capacity of herbivores to develop food aversions when ingesting PSC, because some of these compounds induce nausea (Provenza 1996). These aversions restrict the intake of toxin-containing plants, an outcome that contributes to enhanced fitness of herbivores to a particular environment. The presence of PSC such as alkaloids (e.g., *Delphinium* spp.; Pfister et al. 1997), condensed tannins (e.g. *Coleogyne ramosissima* Torr.; Provenza et al. 1990), and terpenes (e.g., *Artemisia tridentata* Nutt.; Dziba et al. 2006; *Juniperus* spp. Utsumi et al. 2009) reduced feed intake in cattle, sheep, and goats. However, not all PSC cause food aversions (Pfister et al. 2010) and delayed toxic effects can limit the ability of herbivores to form food aversions. For instance, because horses do not develop a food aversion when illness is delayed they may be at risk for long-acting toxins like the alkaloids found in *Senecio* and *Equisetum* species (Houpt et al. 1990). Plant secondary compounds can also reduce forage digestibility and increase acidic burdens in consumers with negative impacts on intake and productivity (Foley et al. 1995, 1999).

The digestive system of herbivores has metabolic adaptations aimed at the detoxification of the ingested PSC (Freeland and Janzen 1974; Illius and Jessop 1995). However, detoxification comes with a substantial metabolic cost (Thomas et al. 1988; Sorensen et al. 2005) so adequate supplies of nutrients need to be available or provided by the livestock manager (Illius and Jessop 1995). The consumption of a diversity of PSC may reduce the overall toxic effect of the mix because (1) diluted mixtures are less toxic to herbivores as each individual PSC may be detoxified by a specific pathway (Freeland and Janzen 1974), thereby eliminating a risk of saturation of detoxification mechanisms, and (2) the formation of gastrointestinal complexes often reduces the absorption and activity of single compounds. For example, tannins and saponins chelate each other in the gastrointestinal tract. It has been shown that by appropriate diet selection, mice can choose a mix of feeds containing tannins and saponins in a way that nullifies the effect of each compound (Freeland et al. 1985). Likewise, goats have a greater feed intake when they are offered shrubs containing a combination of tannins and saponins, compared to when they are offered single shrubs (Rogosic et al. 2006).

7.4.2 PSC as Medicinal Compounds for Herbivores

As pointed out at the beginning of this section, plant secondary compounds can have positive effects on some bodily functions when they are ingested in the right quantity, for the right amount of time, or in the right combination. These medicinal effects of PSC have a great potential to improve both the health and welfare of ruminants kept in extensive systems, where the provision for animal health by managers is limited.

Internal parasites are one of the greatest disease problems in grazing livestock worldwide (Min and Hart 2003; Waller 2006). Failure to control gastrointestinal nematodes typically results in poor growth rates, ill-thrift, and death (Min et al. 2004). Considerable attention has been given recently to bioactive plants that affect internal parasite populations (Jackson and Miller 2006), such as plant-derived tannins, alkaloids, terpenes, saponins, and sesquiterpene lactones (Hoskin et al. 1999; Kayser et al. 2003; Hernández-Villegas et al. 2011). Several in vitro and in vivo studies have suggested that condensed tannins can have anthelmintic effects against ruminant nematode parasites. Livestock feeding on plants with tannins, such as sulla (*Hedysarum coronarium*) (Niezen et al. 1998, 2002), sainfoin (*Onobrychis viciifolia*; Paolini et al. 2003), Sericea lespedeza (*Lespedeza cuneata*; Shaik et al. 2006), heather (*Calluna vulgaris*; Osoro et al. 2007), a number of acacia species (for instance, *Acacia cyanophylla*; Akkari et al. 2008), and lentisk (*Pistacia lentiscus*; Landau et al. 2010), have lower faecal egg counts—an indirect measurement of parasitic burden—than those eating plants of similar quality, or the same rations without tannins. These actions result from several physiological mechanisms such as a direct anthelmintic effect (Athanasiadou et al. 2001), impaired larval establishment (Brunet et al. 2008), decreased parasitic nematodes fecundity in females (Martínez-Ortiz-de-Montellano et al. 2010), and increases in the supply of by-pass protein that enhances immune responses to intestinal parasites (Niezen et al. 2002; Min and Hart 2003). Recent results suggest that parasitised sheep and goats increase preferences for antiparasitic PSC when experiencing parasitic burdens relative to non-parasitised animals (Gradé et al. 2009; Osoro et al. 2007; Martínez-Ortiz-de-Montellano et al. 2010; Villalba et al. 2010b; Juhnke et al. 2012).

In addition to the control of gut pathogen load (Patra and Saxena 2009), PSC can facilitate rumination, prevent/cure bloating (Viegi et al. 2003), reduce acidosis (Hutton et al. 2010), reduce diarrhoea and constipation (McGaw and Eloff 2008), reduce kidney stones, and treat cystitis (Viegi et al. 2003). Moreover, some PSC can increase the protection of dietary proteins and reduce microbial proteolysis, peptidolysis, deamination, and degradation to ammonia, allowing nitrogen to escape to the duodenum and improve nitrogen utilisation and production (Waghorn

2008). The energy utilisation of feed can be increased by some PSC by reducing the amount of methane that is produced during fermentation (Bodas et al. 2008; Vercoe et al. 2009).

Some PSC have antitussive, emollient, and expectorant properties that improve lung function [for review see Durmic and Blache (2012)]. In the cardiovascular system, PSC can have positive impacts through hypotensive, cardioprotective, and vascular-stabilising actions (Svendsen et al. 2010). Consumption of PSC may also modify the immune system and blood composition. Some PSC stimulate the immune system because they enhance both innate and specific immunity via modulation of receptors, enzymes, and immune molecules and have anti-inflammatory properties (Provenza and Villalba 2010). At the skin level, PSC can help wound formation, reduce skin irritation, and be used to treat some skin diseases and ectoparasites. Some PSC have structural and functional similarity with benzodiazepines and have been exploited for their calmative and anxiolytic activity in livestock (Durmic and Blache 2012). While the mechanisms of action remain largely unknown, plants containing saponins, alkaloids, and tannins have been used in ethno-veterinary practices to improve the welfare of animals during, and around the time of, parturition by reducing dystocia, improving expulsion of the placenta, and recovery from difficult parturition (Viegi et al. 2003).

Overall, as long as the types and amounts of PSC can be managed, the medicinal properties of PSC described above hold exciting opportunities for the improvement of animal health and welfare and the productivity of large numbers of ruminants grazing poor quality (high fibre and low protein) diets while reducing ecological impact.

7.5 Challenge 3: Variability in the Thermal Environment

Animals are raised in extensive grazing systems located in very different climates, from temperate pastures (e.g. New Zealand), rangelands in extreme climatic zones (Texas, Patagonia, north-west Australia), Mediterranean grasslands (Western Australia, Israel, Uruguay), and tropical lowlands [Colombia; See part 3 of Hodgson and Illius (1996)]. Consequently, ruminant livestock can be exposed to temperatures varying from −20 to +50 °C. Ruminant species are generally well adapted to a range of ambient conditions; hence they can, through the process of thermoregulation, maintain homeostasis and the functioning of all physiological systems, over a broad range of temperatures. Thermostasis is important because most extensively managed livestock are mammals, and mammalian biochemical systems function optimally over a limited body temperature range. The range of ambient temperatures within which an endotherm can regulate its temperature without elevating its metabolic rate or evaporative water loss is called the thermoneutral zone (IUPS Thermal Commission 2001). The range of both the thermoneutral zone (the range of environmental temperatures where the cost of thermoregulation in minimal) and the thermoregulatory zone (the range of

Table 7.1 Core temperature and thermoneutral range in different species [data from Brody (1945), Hahn (1999), Robertshaw (2004)]

Species	Core temperature mean (range) (°C)	Thermoneutral zone (°C)
Sheep	39.1 (38.3–39.9)	21 to 25[a]
Beef cow	38.3 (36.7–39.1)	−18 to 23
Dairy cow	38.6 (38.0–39.3)	−15 to 26
Goat	39.1 (38.5–39.7)	13 to 21

Source: Data from Brody (1945), Ames (1980), Hahn (1999), Robertshaw (2004)
[a]Dependent on the degree of wool cover. The lower critical temperature can vary from −3 to 31 °C according to plane of nutrition and fleece characteristics

environmental temperatures where an animal has the capacity to maintain a stable core temperature) depends on the evolutionary history of a given species or breed (Table 7.1). The values for the thermoneutral range given in Table 7.1 are not absolute, as there are phenotypic and genetic differences within and between genotypes. For instance, there are differences in heat tolerance (e.g. estimated through panting score, tympanic temperature, respiration rate) between heat-tolerant, cold-intolerant *Bos indicus* cattle and heat-intolerant, cold-tolerant *Bos Taurus* (Hansen 2004; Gaughan et al. 2009). Livestock exposure to particular environments in high or low latitudes may lead to environmental adaptations that will also modify the thermoneutral ranges as animals adapt to local environments physiologically and behaviourally (Provenza 2008).

Extreme variations in temperature can compromise the welfare and the production of ruminants if the conditions exceed the limits of the breed/species to balance heat gain with heat loss to maintain a constant body temperature. Heat gain is the sum of metabolic heat produced by all exothermic biological reactions occurring in the body, including digestion and rumen fermentation, plus any environmental heat, such as solar radiant heat or convective heat gain when ambient temperature exceeds skin temperature. Heat loss from an animal occurs via four routes: conduction, convection, radiation, and evaporation (Jessen 2001).

7.5.1 Ruminants Coping with Thermal Stress

Since breeds and species of production animals are translocated to new grazing systems, such as highly productive dairy cattle transported to the tropics, the capacity of each breed to thermoregulate can be compromised when there is a mismatch between environmental conditions (humidity, maximum, and minimum temperature) and thermoneutral zone for the species or breed. Then animals are said to be heat or cold stressed when they are in an environment that is above or below the thermoneutral zone, respectively. Being heat or cold stressed does not mean that an animal's welfare is compromised because endothermic animals (mammals and birds) have evolved physiological mechanisms and capacities that defend body temperature in the face of thermal stress. The adaptations of ruminants to temperature stresses involve physiological and behavioural mechanisms aiming at controlling the production or loss of heat.

7.5.2 Adaptive Control of Heat Production

During cold stress, when the environmental temperature is below the thermoneutral zone, an animal must increase heat production. Cold stress thus induces an appropriate increase in heat production through increased metabolic rate, including shivering. The elevated metabolic rate then requires a matching increase in energy intake if energy balance is to be maintained. However, the required increase in food intake is possible only if forage is readily available and of high digestibility, which is not always the case. For example, both parameters decrease dramatically when the ambient temperature reaches freezing (Adams et al. 1986; Adams 1987). The capacity of individuals to sustain the elevated metabolic rate required to achieve heat balance in the cold depends on other factors such as health status and energy stores. For example, significant mortality rates have been seen in sheep exposed to cold weather and deprived of shelter in the first few days after shearing (Lynch et al. 1980). Even in the absence of mortality, the maintenance requirement increases during cold stress, so there is less energy available within the animal for production or growth. Shelter during cold weather may be provided by long dry grass to sheep at low stocking rates. Nevertheless, the lower quality of the food resource will lead to increased grazing time relative to the absence of shelter (Arnold and Dudzinski 1978).

Since heat gain can be increased by environmental heat, such as solar radiant heat, animal exposed to high temperature will seek shade. For instance, cattle in hot weather actively seek shade, which may reduce the radiant heat load by at least 30 % (Blackshaw and Blackshaw 1994). Thus, the provision of trees and shrubs is an important source of shade for hot weather and can improve the welfare and productivity of ruminants during periods of low environmental temperatures as well.

During heat stress, ruminants can decrease heat production by reducing food intake and rumination time, because digestion is an exothermic process. In sheep, heat stress induces decreases in food intake, feed efficiency, and utilisation, a disturbance of enzymatic reactions, hormonal secretions, and blood metabolites, all of which modify water, protein, energy, and mineral balances (Marai et al. 2007).

7.5.3 Adaptive Control of Heat Loss

Mammals use a range of physiological and behavioural mechanisms to increase heat loss during exposure to heat, including increases in evaporative heat loss, adoption of specific behaviours and body postures, and changes in daily patterns of activity and social group dynamics. In extreme heat, mammals rely on the most powerful form of heat loss, evaporative water loss, via panting or sweating. While the evaporation of water from the respiratory tract or general skin surface can help to maintain core body temperature, it also creates osmotic strain, and as such the welfare implications of heat stress are context dependent. Hot and humid conditions are more stressful than hot and dry conditions, because evaporation requires a vapour pressure (humidity) gradient. High absolute humidity reduces the amount of heat that an animal can dissipate by evaporative water loss. To increase evaporative water loss, the animal exposed to heat stress needs to have access to drinking water to avoid dehydration.

In hot conditions, grazing animals adopt behaviours that promote passive heat loss by convection and radiation, preferring, for example, areas with high wind speed and shaded areas where they can avoid direct solar radiation. Cattle exposed to heat stress will prefer to stand instead of lying down, and sheep spread their hind legs thereby exposing the well-vascularised skin under the flanks and promoting convective heat loss (Campbell et al. 1969). During prolonged exposure to high temperatures, cattle, bison, goats, and sheep change their daily activity pattern, shifting grazing activity to the coolest hours of the day (evening) and even switching to nocturnal grazing (Vallentine 2000). In cattle, breed has a strong effect on the impact of hot conditions. *Bos primigenius indicus* cattle are well adapted to heat, and breeds of indicus cattle and those derived from crosses with European breeds (Brahman and Tuli for example) spend less time in the shade and more time grazing than European breeds such as Angus (Forbes et al. 1998).

Heat stress may have a greater impact on welfare than cold stress because the 'safety margin' between normal core temperature and lethal hyperthermia is smaller than that for lethal hypothermia. In addition to the impact on welfare, exposure to heat and cold stress causes a decrease in production, because thermoregulation, directly or indirectly, costs energy. For example, it has been estimated that thermal stress could account for an increase of about 40 % in the maintenance energy requirements of range beef cows (Ames 1985). While thermoregulatory costs are reflected in decreased productivity (Webster 1991), the maintenance requirements of ruminants increase linearly during cold stress but exponentially during heat stress (Ames and Ray 1983). In cattle in the Southern USA, fertility is reduced from around 50 % in winter to less than 15 % in the summer (Thatcher and Collier 1986). At the extremes of climatic conditions it is not just productivity that decreases, but the animals themselves can be at risk.

7.6 Challenge 4: Pressure from Environmental Fauna

In extensive systems, livestock share the environment with other animals that either compete for the same resources, prey on them, or parasitise them. In all cases, herbivores have adaptive responses, but these responses are not always optimal and can compromise both their welfare and productivity.

7.6.1 Competition for Resources

According to niche theory, all species differ to some extent in the resources they use (Prins and Fritz 2008). In mammalian herbivore assemblages, different species of herbivores can be classified into grazers, browsers, and mixed feeders because of differences in morphophysiological adaptations and 'feeding types' (Hofmann 1984, 1989). Therefore, cohabitation between two species with morphophysiological differences will not translate to an overlap among their diets. In contrast, if two or more herbivore species have similar morphophysiological adaptations, they will share food resources and, in periods of short supply, the performance of at least one of the species will decrease (Mishra et al. 2004). The intensity of resource competition is greater in areas of low productivity or during periods of low resource availability (e.g. in seasonal environments during the dry season). As an animal species gets outcompeted from preferred resources, it is likely that members will increase the consumption of less suitable forage (Gordon and Illius 1989), which in turn will negatively impact their nutrition and welfare. However, the presence of competitors can also have a positive effect on diet quality of livestock if the resources are abundant, a phenomena called facilitation. For example, the presence of wild ungulates depressed the food intake and performance of cattle during the dry season (competition), but their presence enhanced diet quality and performance of cattle during the wet season (facilitation; Odadi et al. 2011). In the Serengeti, buffalo and zebra feed on and trample taller and coarser grasses, clearing the way for smaller herbivores (e.g. gazelle) to feed on exposed plant parts of higher nutritional quality (Bell 1971). In extensive systems, it is possible that the feeding behaviours of one species improve the feeding efficiency of individuals of one or more coexisting species. Thus, some wild and domestic species may improve their welfare and nutrition through facilitative interactions (Prins and Fritz 2008; duToit 2011).

7.6.2 Cost of Predation

In addition to the obvious impacts of predators directly killing herbivores (i.e. lethal effects), predators have significant nonlethal impacts on their prey (Edwards 1983; Hernández and Laundré 2005). They may scare their prey, not just when they are about to attack (Brown et al. 1999), but they establish a 'landscape of fear' for herbivores with a topography determined by the level of perceived predation risk (Laundré et al. 2001). The response of herbivores to predator presence impacts on their feeding behaviour, nutritional status, and social structure, all potentially leading to decreased welfare. In the presence or perception of predators, herbivores sacrifice feeding effort and invest more time in vigilance and shift from riskier to safer feeding stations (Edwards 1983; Hernández and Laundré 2005). For example, in elk, the reintroduction of wolves into Yellowstone National Park has led to an increase in vigilance and a decrease in feeding rates (Laundré et al. 2001). The estimated 20 % decline in foraging effort by female elk compromised survival rates and led to a decrease in the birth mass of calves during the spring (Laundré et al. 2001).

Predator presence, or the perception of predator pressure, impacts on the quality of the forage consumed by herbivores. In the absence of predation, the prey selects the highest quality habitats, but in response to predators, the prey moves to poorer quality habitat, resulting in decreased diet quality (Edwards 1983). For instance, elk in Yellowstone National Park now use forest edge areas more than open meadows, although elk preferred open meadow areas before the release of wolves in the park (Hernández and Laundré 2005). Together the 'lost opportunities' to feed in profitable patches, which are presumably perceived as risky and the increase in time and effort invested in vigilance and apprehension, lead to an increase in the energy cost of predation by reducing nutrient intake and increasing energy expenditure. Even small changes in habitat may lead to large increases in the cost of predation (Brown and Kotler 2004). When the predation pressure increases so much that the animals are not able to meet their nutritional requirements, both productivity and welfare can be dramatically compromised. To assess and manage the impact of predation, in addition to accounting for the killing of animals, it is possible to measure the times at which herbivores quit eating a preferred forage, activity levels in time and space, and habitat distributions with and without perceived predation risk (Brown and Kotler 2004).

7.6.3 Parasitic Insects

In extensive systems, it is not possible to control for insect infestation, such as flies. The effect of insects can be direct, such as the Australian sheep blowfly (*Lucilia cuprina*) that is responsible for initiating most blowfly strikes, or indirect, such as flies causing continuous irritation that led to a reduction in grazing time and a decrease in nutrient intake (Lefcourt and Schmidtmann 1989). In bad seasons, sheep may lose a large amount of grazing time due to harassment by *Oestrus ovis* flies (Blood et al. 1983). Ungulates have adopted individual and group behaviours to reduce fly irritation. By lying down, animals reduce the exposure of body surfaces, decrease the production of some fly attractants (e.g. sweat and carbon dioxide), and dislodge fewer flies from foliage (Espmark and Langvatn 1979). When forage conditions are poor, animals need to increase their grazing time, walk more while grazing, and have difficulty to lie down and avoid fly irritation. Social grouping is also used as a means to reduce fly harassment. As groups become larger, each individual attracts fewer flies (e.g. Duncan and Vigne 1979). Social grouping creates a clear conflict with the increased dispersion of animals associated with poor forage conditions.

7.7 Knowledge Gaps

Within the Five Freedom framework, it is clear that a lack of nutrients and restriction in feed availability and quality, induced by either management decisions, pasture quality, environmental factors such as climate or interactions with other species, is a source of metabolic stress that compromises welfare. At the same time, the impact on welfare of unbalanced nutrition because of lack of diversity or the impact of PSC is unknown. It is also not clear whether incidental restriction and/or augmentation induce stress in livestock grazing on rangelands. Studies to answer this question will need to be able to assess the level of stress of large numbers of animals dispersed in large areas.

In this chapter, we have often said that livestock can be moved to an environment different from that in which they have evolved, thus creating a mismatch between the adaptive capacity and the environmental pressures. However, over the last 15 years numerous studies have shown that experience early in life can cause epigenetic changes that influence foraging behaviour, habitat selection, and animal health (Provenza 2008). For example, lambs exposed to a high-sodium-containing shrub during uterine life grow faster and handle a salt load better than lambs from mothers on pasture (Chadwick et al. 2009). Epigenetic effects suggest that future generations of livestock could be better adapted to the environment than their parents, and that experience has the potential to improve nutrition and welfare in the same habitat. To take full advantage of these rapid adaptive processes, we need to understand how important early life experiences are in the adaptation of an animal to low-quality diets and/or harsh environments. Should we change the environment in order to increase animal productivity and welfare, or shape an animal (i.e. through experiences in utero and early in life) to better fit its environment?

Livestock that graze in extensive systems are generalist herbivores, meaning that they evolved consuming diverse diets as opposed to monotonous pastures. Single foods generate orosensorial and postingestive signals that cause animals to satiate (Rolls 1986), and satiety may be aversive (Provenza 1996; Provenza and Villalba 2006). In addition, diverse diets increase the likelihood of ingesting beneficial chemicals that enhance the health and welfare of animals (Villalba et al. 2010a). However, the effects of exposure to low- vs. high-diversity diets on stress, health, and animal welfare are unknown.

The interactions between dispersion, increased grazing time, and the expression of other behaviours need to be explored to understand how grazing in extensive systems can decrease the richness and increase the metabolic cost of natural behaviours. For instance, courtship in free-ranging cattle involves a great deal of interaction between animals (Blockey 1976) that requires time and energy and, therefore, may increase stress and reduce production.

The impact of PSC on animal health, welfare, and productivity is receiving an increasing interest. However, only a few compounds have been studied and we do not know much about the optimal dosage, the efficiency of mixture, or even the amount that is consumed and how PSC are metabolised (Provenza and Villalba 2006). Research on the impact of detoxification pathways on diet selection and energetic costs conducted on possums (e.g. Marsh et al. 2006; Nersesian et al. 2011) and woodrats (e.g. Sorensen et al. 2005) is needed on ungulates grazing on rangelands. New studies should explore the effects of specific compounds, doses, and interactions, particularly in animals grazing in extensive systems where PSC are abundant and diverse. Then, we will be able to minimise the health risks and possibly improve the welfare and productivity of animals grazing in rangelands through better preparation and management of animals. To do so, we need to test and monitor novel plants and new plant varieties using bioassays and with animals under experimental conditions prior to them being introduced into grazing systems (Revell and Revell 2007). In addition to these direct effects of PSC on animal health, we need to quantify the metabolic cost associated with the neutralisation/detoxification of PSC by herbivores as well as to explore their ability to select diets that meet their nutritional requirements, while avoiding intoxication by PSC (Jensen et al. 2015).

While ambient conditions above and below the thermoneutral zone are described as heat and cold stress, the relationship between imposed thermal stress and animal welfare is more complex than simply assuming that thermal stress causes decreased welfare. The complexity of the adaptive response of herbivores to extreme temperature under extensive system conditions and the importance of breed on the long-term adaptation to climate change are poorly understood.

The long-term effects of a particular management regime need to be evaluated in order to better predict the nutrition and welfare of animals in the long run. For instance, supplementation programs at high stocking densities may be beneficial for animals in the short run. However, the negative long-term effects on rangelands due to overgrazing will certainly impact the nutrition and welfare of animals in the future.

7.8 Conclusions

Many of the challenges described in this chapter are rooted in the variability and unpredictability of the environment. Foods, habitats, and environmental conditions are not 'provided' by the manager, but they occur naturally. The challenge for managers is to provide the conditions and flexibility in their operations that allow animals to express their behaviour and cope with challenges in a way that production and welfare are maximised given the constraints imposed by the ever-changing environmental conditions.

The environmental challenges that animals experience in extensive conditions may represent a source of 'positive stress' if animals possess the experience and resources to effectively solve the problems with which they are presented (Meehan and Mench 2007). Thus, some of the challenges that herbivores experience in extensive conditions may become opportunities for managers to create innovative programs that provide animals the training and resources to rectify their states of unbalance, which will ultimately enrich the environment where animals live.

References

Adams DC (1987) Influence of winter weather on range livestock. In: Proceedings of the Grazing Livestock Nutrition Conference, University of Wyoming, Laramie, WY, pp 1037–1042

Adams DC, Nelsen TC, Reynolds WL, Knapp BW (1986) Winter grazing activity and forage intake of range cows in the northern great plains. J Anim Sci 62:1240–1246

Adler PB, Raff DA, Lauenroth WK (2001) The effect of grazing on the spatial heterogeneity of vegetation. Oecologia 128:465–479

Akkari H, Ben Salem H, Gharbi M, Abidi S, Darghouth MA (2008) Feeding *Acacia cyanophylla* Lindl. foliage to Barbarine lambs with or without PEG: effect on the excretion of gastro-intestinal nematode eggs. Anim Feed Sci Technol 147:182–192

Allison CD (1985) Factors affecting forage intake by range by range ruminants: a review. J Range Manage 38:305–311

Ames DR (1980) Thermal environment affects production efficiency of livestock. Bioscience 30 (7):457–460

Ames DR (1985) Managing cows during winter. In: Proceedings of The Range Beef Cow Symposium IX, Chadron, NE, pp 11–16

Ames DR, Ray DE (1983) Environmental manipulation to improve animal productivity. J Anim Sci 57:209–220

Appleby MC (1996) Can we extrapolate from intensive to extensive conditions? Appl Anim Behav Sci 49:23–27

Arnold GW (1985a) Ingestive behaviour. In: Fraser AF (ed) Ethology of farm animals. Elsevier, Amsterdam, 500 pp

Arnold GW (1985b) Territoriality. In: Fraser AF (ed) Ethology of farm animals. Elsevier, Amsterdam, 500 pp

Arnold GW, Dudzinski ML (1978) Ethology of free ranging domestic animals. Elsevier, Amsterdam, 198 pp

Arnold GW, Maller RA (1977) Effects of nutritional experience in early and adult life on the performance and dietary habits of sheep. Appl Anim Ethol 3:5–26

Arriaga-Jordan CM, Holmes W (1986) The effect of concentrate supplementation on high-yielding dairy cows under two systems of grazing. J Agric Sci (Camb) 107:453–461

Aslani MR, Movassaghi AR, Mohri M, Abbasian A, Zarehpour M (2004) Clinical and pathological aspects of experimental oleander (*Nerium oleander*) toxicosis in sheep. Vet Res Commun 28:609–616

Athanasiadou S, Kyriazakis I, Jackson F, Coop RL (2001) Direct anthelmintic effects of condensed tannins towards different gastrointestinal species: in vitro and in vivo studies. Vet Parasitol 99:205–219

Bailey DW, Brown JR (2011) Rotational grazing systems and livestock grazing behavior in shrub-dominated semi-arid and arid rangelands. Rangel Ecol Manage 64:1–9

Bailey DW, Provenza FD (2008) Mechanisms determining large-herbivore distribution. In: Prins HHT, van Langevelde F (eds) Resource ecology: spatial and temporal dynamics of foraging. Springer, Dordrecht, pp 7–28

Bell RHV (1971) A grazing ecosystem in the Serengeti. Sci Am 225:86–93

Bernhoft A (2010) A brief review. In: Bernhoft A (ed) Bioactive compounds in plants—benefits and risks for man and animals. Nor Acad Sci Lett, Oslo, pp 11–17

Birrell HA (1991) The effect of stocking rate on the grazing behaviour of Corriedale sheep. Appl Anim Behav Sci 28:321–331

Blache D, Maloney SK, Revell D (2008) Use and limitations of alternative feed resources to sustain and improve reproductive performance in sheep and goats. Anim Feed Sci Technol 147:140–157

Blache D, Maloney SK, Terlouw EMC (2011) Physiology. In: Appleby MC, Hughes BO, Mench JA, Olsson A (eds) Animal welfare. CAB International, Wallingford, pp 155–182

Blackshaw JK, Blackshaw AW (1994) Heat stress in cattle and the effect of shade on production and behaviour: a review. Aust J Exp Agric 34:285–295

Blanc F, Theriez M (1998) Effects of stocking density on the behaviour and growth of farmed red deer hinds. Appl Anim Behav Sci 56:297–307

Blockey MA d B (1976) Sexual behaviour of bulls at pasture: a review. Theriogenology 6:387–392

Blood DC, Radostis OM, Henderson JA (1983) Veterinary medicine. Bailliere Tindall, Eastbourne

Bodas R, López S, Fernandez M, García-González R, Rodríguez AB, Wallace RJ, González JS (2008) In vitro screening of the potential of numerous plant species as antimethanogenic feed additives for ruminants. Anim Feed Sci Technol 145:245–258

Brody S (1945) Bioenergetics and growth. Reinhold, New York, NY, 786 pp

Broom DM (1986) Indicators of poor welfare. Br Vet J 142:524–526

Broom DM (1991) Animal welfare: concepts and measurement. J Anim Sci 69:4167–4175

Brown JS, Kotler BP (2004) Hazardous duty pay and the foraging cost of predation. Ecol Lett 7:999–2014

Brown JS, Laundré JW, Gurung M (1999) The ecology of fear: optimal foraging, game theory, and trophic interactions. J Mammal 80:385–399

Brunet S, Martinez-Ortiz de Montellano C, Torres-Acosta JFJ, Sandoval-Castro CA, Aguilar-Caballero AJ, Capetillo-Leal C, Hoste H (2008) Effect of the consumption of Lysiloma latisiliquum on the larval establishment of gastrointestinal nematodes in goats. Vet Parasitol 157:81–88

Burlison AJ, Hodgson J, Illius AW (1991) Sward canopy structure and the bite dimensions and bite weight of grazing sheep. Grass Forage Sci 46:29–38

Campbell JB, Stringham E, Gervais P (1969) Pasture activities of cattle and sheep. In: Campbell JB (ed) Experimental methods for evaluating herbage. Canada Department of Agriculture, Publication 1315

Campbell BM, Gordon IJ, Luckert MK, Petheram L, Vetter S (2006) In search of optimal stocking regimes in semi-arid grazing lands: one size does not fit all. Ecol Econ 60:75–85

Campling RC (1966) A preliminary study of the effect of pregnancy and of lactation on the voluntary intake of food by cows. Br J Nutr 20:25–39

Carulla JE, Kreuzer M, Machmuller A, Hess HD (2005) Supplementation of *Acacia mearnsii* tannins decrease methanogenesis and urinary nitrogen in forage-fed sheep. Aust J Agric Res 56:961–970

Catanese F, Obelar M, Villalba JJ, Distel RA (2013) The importance of diet choice on stress-related responses by lambs. Appl Anim Behav Sci 148:37–45

Chadwick MA, Vercoe PV, Williams IH, Revell DK (2009) Programming sheep production on saltbush: adaptations of offspring from ewes that consumed high amounts of salt during pregnancy and early lactation. Anim Prod Sci 49:311–317

Charcon E, Stobbs TH (1976) Influence of progressive defoliation of a grass sward on the eating behaviour of cattle. Aust J Agric Res 27:709–727

Cowan MM (1999) Plant products as antimicrobial agents. Clin Microbiol Rev 12:564–582

Dawson TJ, Denny MJS, Russell EM, Ellis B (1975) Water usage and diet preferences of free ranging kangaroos, sheep and feral goats in the Australian arid zone during summer. J Zool 177:1–23

de Boever JL, Andries JI, de Brabander DL, Cottyn BG, Buysse FX (1990) Chewing activity of ruminants as a measure of physical structure. A review of factors affecting it. Anim Feed Sci Technol 27:281–291

De Villiers JF, Botha WA, Wandrag JJ (1994) The performance of lambs on kikuyu as influenced by stocking rate and grazing system. S Afr J Anim Sci 24:133–139

Deswysen AG (1986) Physiologie digestive comparee et valeur alimentaire des ensilages. Effect des acides acetique et lactique. In: Sanofi Animal Health (ed) L'ensilage: nouveaux aspects biologiques, Paris, 231 pp
Deswysen AG, Ellis WC, Pond KR (1987) Interrelationships among voluntary intake, eating and ruminating behaviour and ruminal motility of heifers fed corn silage. J Anim Sci 64:835–841
Dudzinski ML, Schuh HJ, Wilcox D, Gardiner G, Morrissey T (1978) Statistical and probabilistic estimations of forage conditions from grazing behaviour of Merino sheep in a semi-arid environment. Appl Anim Ethol 4:357–368
Dumont B, Garel JP, Ginane C, Decuq F, Farruggia A, Pradel P, Rigolot C, Petit M (2007) Effect of cattle grazing a species-rich mountain pasture under different stocking rates on the dynamics of diet selection and sward structure. Animal 1:1042–1052
Duncan IJH, Fraser D (1997) Understanding animal welfare. In: Appleby MC, Hughes BO (eds) Animal welfare. CAB International, Wallingford, pp 19–31
Duncan P, Vigne N (1979) The effect of group size in horses on the rate of attacks by blood-sucking flies. Anim Behav 27:623–625
Duncan AJ, Young SA (2002) Can goats learn about foods through conditioned food aversions and preferences when multiple food options are simultaneously available? J Anim Sci 80:2091–2098
Durmic Z, Blache D (2012) Bioactive plants and plant products: effects on animal function, health and welfare. Anim Feed Sci Technol 176:150–162
DuToit JT (2011) Coexisting with cattle. Science 333:1710–1711
Dziba LE, Hall JO, Provenza FD (2006) Feeding behavior of lambs in relation to kinetics of 1,8-cineole dosed intravenously or into the rumen. J Chem Ecol 32:391–408
Edwards PJ (1980) The use of stocking rate/animal performance models in research and extension. Proc Grassld Soc S Afr 15:73–77
Edwards J (1983) Diet shifts in moose due to predator avoidance. Oecologia 60:185–189
Egan AR (1980) Host animal-rumen relationships. Proc Nutr Soc 39:79–87
El Aich A, Rittenhouse LR (1988) Use of habitats by free-grazing sheep. Appl Anim Behav Sci 21:223–231
El Aich A, El Asraoui M, Rittenhouse LR (1991) Effect of trailing to water on temporal behaviour and ingestion of herded sheep in Morocco. Appl Anim Behav Sci 31:251–257
Espmark Y, Langvatn R (1979) Lying down as a means of reducing fly harassment in red deer (*Cervus elaphus*). Behav Ecol Sociobiol 5:51–54
Estell RE (2010) Coping with shrub secondary metabolites by ruminants. Small Rumin Res 94:1–9
FAO (2009) The state of food and agriculture 2009. Livestock in the balance. FAO, Rome
Farm Animal Welfare Council (1993) Second report on priorities for research and development in farm animal welfare. MAFF, Tolworth
Ferme D, Banjac M, Calsamiglia S, Busquet M, Kamel C, Avgustin G (2004) The effects of plant extracts on microbial community structure in a rumen-simulating continuous-culture system as revealed by molecular profiling. Folia Microbiol (Praha) 49:151–155
Flores ER, Provenza FD, Balph DF (1989a) Role of experience in the development of foraging skills of lambs browsing the shrub servicerry. Appl Anim Behav Sci 23:271–278
Flores ER, Provenza FD, Balph DF (1989b) Relationship between plant maturity and foraging experience of lambs grazing hycrest crested wheatgrass. Appl Anim Behav Sci 23:279–284
Flores ER, Provenza FD, Balph DF (1989c) The effect of experience on the foraging skills of lambs: importance of plant form. Appl Anim Behav Sci 23:285–291
Foley WJ, McLean S, Cork SJ (1995) Consequences of biotransformation of plant secondary metabolites on acid-base metabolism in mammals—a final common pathway? J Chem Ecol 21:721–743
Foley WJ, Iason GR, McArthur C (1999) Role of plant secondary metabolites in the nutritional ecology of mammalian herbivores-how far have we come in 25 years? In: Jung HJG, Fahey GC Jr (eds) Nutritional ecology of herbivores, 5th International Symposium on the Nutrition of Herbivores. American Society of Animal Science, Savoy, IL, pp 130–209

Forbes JM (2007) A personal view of how ruminant animals control their intake and choice of food: minimal total discomfort. Nutr Res Rev 20:132–146

Forbes TD, Rouquette FM, Holloway JW (1998) Comparisons among Tuli-, Brahman-, and Angus-sired heifers: intake, digesta kinetics, and grazing behavior. J Anim Sci 76:220–227

Fowler DG (1984) Reproductive behaviour of rams. In: Lindsay DR, Pearce DT (eds) Reproduction in sheep. Cambridge University Press, Cambridge, pp 39–46, 414 pp

Freeland WJ, Janzen DH (1974) Strategies in herbivory by mammals: the role of plant secondary compounds. Am Nat 108:269–286

Freeland WJ, Calcott PH, Anderson LR (1985) Tannins and saponin: interaction in herbivore diets. Biochem Syst Ecol 13:189–193

Fretwell SD, Lucas HL Jr (1970) On territorial behavior and other factors influencing habitat distribution in birds. I. Theoretical development. Acta Biotheor 19:16–36

Gaughan JB, Mader TL, Holt SM, Sullivan ML, Hahn GL (2009) Assessing the heat tolerance of beef cattle genotypes. Int J Biometeorol 54(6):617–627

Gill J, Campling RC, Westgarth DR (1966) A study of chewing during eating in the cow. Br J Nutr 20:13–23

Gonyou HW (1986) Assessment of comfort and well-being in farm animals. J Anim Sci 62:1769–1775

Gordon IJ, Illius AW (1989) Resource partitioning by ungulates on the Isle of Rhum. Oecologia 79:383–389

Gradé JT, Tabuti JRS, Van Damme P (2009) Four footed pharmacists: indications of self-medicating livestock in Karamoja, Uganda. Econ Bot 63:29–42

Greathead H (2003) Plants and plant extracts for improving animal productivity. Proc Nutr Soc 62:279–290

Gregorini P, Villalba JJ, Provenza FD, Beukes PC, Forbes JM (2015) Modelling preference and diet selection patterns by grazing ruminants: a development in a mechanistic model of a grazing dairy cow, MINDY. Anim Prod Sci 55:360–375

Grime JP (1979) Plant strategies and vegetation processes. Wiley, Chichester, 456 pp

Grissom G, Steffens T (2013) Adaptive grazing management at Rancho Largo Cattle Company. Rangelands 35:35–44

Grummer RR, Jacob AL, Woodford JA (1987) Factors associated with variation in milk fat depression resulting from high grain diets fed to dairy cows. J Dairy Sci 70:613–619

Hahn G (1999) Dynamic responses of cattle to thermal heat loads. J Anim Sci 77(Suppl 2):10–20

Hall SJG, Moore GF (1986) Feral cattle of Swona, Orkney Islands. Mammal Rev 16:9–96

Hansen PJ (2004) Physiological and cellular adaptations of zebu cattle to thermal stress. Anim Reprod Sci 82:349–360

Hart KJ, Yanez-Ruiz DR, Duval SM, McEwan NR (2007) Plant extracts to manipulate rumen fermentation. Anim Feed Sci Technol 147:8–35

Hernández L, Laundré JW (2005) Foraging in the 'landscape of fear' and its implications for habitat use and diet quality of elk *Cervus elaphus* and bison *Bison bison*. Wildl Biol 11:215–220

Hernández-Villegas MM, Borges-Argáez R, Rodriguez-Vivas RI, Torres-Acosta JFJ, Méndez-Gonzalez M, Cáceres-Farfan M (2011) Ovicidal and larvicidal activity of the crude extracts from *Phytolacca icosandra* against *Haemonchus contortus*. Vet Parasitol 179:100–106

Hodgson J (1990) Grazing management: science into practice. Longman, Essex, 203 pp

Hodgson J, Illius AW (1996) The ecology and management of grazing systems. CAB International, Oxford, 466 pp

Hofmann RR (1984) Comparative anatomical studies imply adaptive variations of ruminant digestive physiology. Can J Anim Sci 64(Suppl):203–205

Hofmann RR (1989) Evolutionary steps of ecophysiological adaptation and diversification of ruminants. A comparative view of their digestive system. Oecologia 78:443–457

Holmes W (1989) The utilization of pasture. In: Jarrige R (ed) Ruminant nutrition: recommended allowances and feed tables. INRA, Paris, 389 pp

Hooper AP, Welch JG (1983) Chewing efficiency and body size of kid goats. J Dairy Sci 66:2551–2556

Hoskin SO, Barry TN, Wilson PR, Charleston WAG, Hodgson J (1999) Effects of reducing anthelmintic input upon growth and faecal egg and larval counts in young farmed deer grazing chicory (*Cichorium intybus*) and perennial ryegrass (*Lolium perenne*)/white clover (*Trifolium repens*) pasture. J Agric Sci (Camb) 132:335–345

Houpt KA (1991) Domestic animal behavior for veterinarians and animal scientists, 2nd edn. Iowa State University Press, Ames, IA, 416 pp

Houpt KA, Zahorik DM, Swartzman-Andert JA (1990) Taste aversion learning in horses. J Anim Sci 68:2340–2344

Hunter RF, Milner C (1963) The behaviour of individual, related and groups of South Country Cheviot hill sheep. Anim Behav 11:507–513

Hutton PG, Durmic Z, Vercoe PE (2010) Investigating Eremophila glabra as a bioactive agent for preventing lactic acidosis in sheep. Anim Prod Sci 50:449–453

Illius AW, Jessop NS (1995) Modeling metabolic costs of allelochemical ingestion by foraging herbivores. J Chem Ecol 21:693–719

Ingebrigtsen K (2010) Main plant poisonings in livestock in the Nordic countries. In: Bernhoft A (ed) Bioactive compounds in plants—benefits and risks for man and animals. Norwegian Academy of Science and Letters, Oslo, pp 30–43

Iups Thermal Commission (2001) Glossary of terms for thermal physiology: third edition. Jpn J Physiol 51:245–280

Jackson F, Miller J (2006) Alternative approaches to control—Quo vadit? Vet Parasitol 139:371–384

Jensen LM, Wallis IR, Foley WJ (2015) The relative concentrations of nutrients and toxins dictate feeding by a vertebrate browser, the greater glider *Petauroides volans*. PLoS One 10:e0121584

Jessen C (2001) Temperature regulation in humans and other mammals. Springer, Berlin, 193 pp

Juhnke J, Miller J, Hall JO, Provenza FD, Villalba JJ (2012) Preference for condensed tannins by sheep in response to challenge infection with Haemonchus contortus. Vet Parasitol 188:104–114

Kayser O, Kiderlen AF, Croft SL (2003) Natural products as antiparasitic drugs. Parasitol Res 90: S55–S62

Key C, MacIver RM (1980) The effects of maternal influences on sheep: breed differences in grazing, resting and courtship behavior. Appl Anim Ethol 6:33–48

Kilgour R, Pearson AJ, de Langen H (1975) Sheep dispersal patterns on hill country: techniques for study and analysis. Proc N Z Soc Anim Prod 35:191–197

Kumar D, Kumar A, Prakash O (2012) Potential antifertility agents from plants: a comprehensive review. J Ethnopharmacol 140:1–32

Kyriazakis I, Oldham JD (1993) Diet selection in sheep: the ability of growing lambs to select a diet that meets their crude protein (nitrogen × 6.25) requirements. Br J Nutr 69:617–629

Kyriazakis I, Savory CJ (1997) Hunger and thirst. In: Appleby MC, Hughes BO (eds) Animal welfare. CAB International, Wallingford, pp 49–62

Landau S, Azaizeh H, Muklada H, Glasser T, Ungar ED, Baram H, Abbas N, Markovics A (2010) Anthelmintic activity of Pistacia lentiscus foliage in two Middle Eastern breeds of goats differing in their propensity to consume tannin-rich browse. Vet Parasitol 173:280–286

Laundré JW, Hernández L, Altendorf KB (2001) Wolves, elk, and bison: reestablishing the "landscape of fear" in Yellowstone National Park, U.S.A. Can J Zool 79:1401–1409

Lawrence AB, Wood-Gush DGM (1987) Social behaviour of hill sheep; more to it than meets the eye. Appl Anim Behav Sci 17:382

Lawrence AB, Wood-Gush DGM (1988) Influence of social behaviour on utilization of supplemental feedblocks by Scottish hill sheep. Anim Prod 46:203–212

Lefcourt AM, Schmidtmann ET (1989) Body temperature of dry cows on pasture: environmental and behavioural effects. J Dairy Sci 72:3040–3049

Lobley GE, Milano GD (1997) Regulation of hepatic nitrogen metabolism in ruminants. Proc Nutr Soc 57:547–563

Lu CD, Jorgensen NA (1987) Alfalfa saponins affect site and extent of nutrient digestion in ruminants. J Nutr 117:919–927

Lu CD (1988) Grazing behaviour and diet selection of goats. Small Rumin Res 1:205–216

Lynch JJ (1974) Merino sheep: some factors affecting their distribution in very large paddocks. In: Geist V, Walther F (eds) The behaviour of ungulates and its relation to management, New series. No. 24. IUCN, Morges, pp 679–707

Lynch JJ, Mottershead BE, Alexander G (1980) Sheltering behaviour and lamb mortality amongst shorn Merino ewes lambing in paddocks with a restricted area of shelter or no shelter. Appl Anim Ethol 6:163–174

MacArthur RH, Pianka ER (1966) On optimal use of a patchy environment. Am Nat 100:603–609

Mahgoub O, Kadim IT, Tageldin MH, Al-Marzooqi WS, Khalaf SQ, Ambu Ali A (2008) Clinical profile of sheep fed non-conventional feeds containing phenols and condensed tannins. Small Rumin Res 78:115–122

Manteca X, Villalba JJ, Atwood SB, Dziba L, Provenza FD (2008) Is dietary choice important to animal welfare? J Vet Behav 3:229–239

Marai IFM, El-Darawany AA, Fadiel A, Abdel-Hafez MAM (2007) Physiological traits as affected by heat stress in sheep—a review. Small Rumin Res 71:1–12

Marsh KJ, Wallis IR, McLean S, Sorensen JS, Foley WJ (2006) Conflicting demands on detoxification pathways influence how common brushtail possums choose their diets. Ecology 87:2103–2112

Martínez-Ortiz-de-Montellano C, Vargas-Magaña JJ, Canul-Ku HL, Miranda-Soberanis R, Capetillo-Leal C, Sandoval-Castro CA, Hoste H, Torres-Acosta JFJ (2010) Effect of a tropical tannin-rich plant *Lysiloma latisiliquum* on adult populations of *Haemonchus contortus* in sheep. Vet Parasitol 172:283–290

Mazid M, Khan TA, Mohamad F (2011) Role of secondary metabolites in defence mechanism in plants. Biol Med 3:232–249

McAllister RRJ (2012) Livestock mobility in arid and semiarid Australia: escaping variability in space. Rangel J 34:139–147

McGaw LJ, Eloff JN (2008) Ethnoveterinary use of southern African plants and scientific evaluation of their medicinal properties. J Ethnopharmacol 119:559–574

McGlone JJ (2001) Farm animal welfare in the context of other society issues: toward sustainable systems. Livest Prod Sci 72:75–81

Meehan CL, Mench JA (2007) The challenge of challenge: can problem solving opportunities enhance animal welfare? Appl Anim Behav Sci 102:246–261

Min BR, Hart SP (2003) Tannins for suppression of internal parasites. J Anim Sci 81:E102–E109

Min BR, Pomroy WE, Hart SP, Sahlu T (2004) The effect of short-term consumption of a forage containing condensed tannins on gastro-intestinal nematode parasite infections in grazing wether goats. Small Rumin Res 51:279–283

Mishra C, vanWieren SE, Ketner P, Heitkonig MA, Prins HHT (2004) Competition between domestic livestock and wild bharal *Pseudois nayaur* in the Indian Trans-Himalaya. J Appl Ecol 41:344–354

Nagy JG, Tengerdy RP (1968) Antibacterial action of essential oils of Artemisia as an ecological factor. II. Antibacterial action of the volatile oils of Artemisia tridentata (big sagebrush) on bacteria from the rumen of mule deer. Appl Microbiol 16:441–444

Nersesian CL, Banks PB, McArthur C (2011) Titrating the cost of plant toxins against predators: determining the tipping point for foraging herbivores. J Anim Ecol 80:753–760

Nicolson L (1984) Boran livestock in heavy drinking bouts. ILCA Newsletter (International Livestock Centre for Africa) 3:1–4. Addis Abeba, Ethiopia

Niezen JH, Robertson HA, Waghorn GC, Charleston WAG (1998) Production, faecal egg counts and worm burdens of ewe lambs which grazed six contrasting forages. Vet Parasitol 80:15–27

Niezen JH, Charleston WA, Robertson HA, Shelton D, Waghorn GC, Green R (2002) The effect of feeding sulla (Hedysarum coronarium) or lucerne (Medicago sativa) on lamb parasite burdens and development of immunity to gastrointestinal nematodes. Vet Parasitol 105:229–245

NRC (1981) Nutrient requirements of goats: angora, dairy and meat goats in temperate and tropical countries. National Academy Press, Washington, DC

O'Connor TG (1995) Transformation of a savanna grassland by drought and grazing. Afr J Range Forage Sci 12:53–60
Odadi WO, Karachi MK, Abdulrazak SA, Young TP (2011) African wild ungulates compete with or facilitate cattle depending on season. Science 333:1753–1755
Oh HK, Sakai T, Jones MB, Longhurst WM (1967) The effect of various essential oils isolated from Douglas-fir needles upon sheep and deer rumen microbial activity. Appl Microbiol 15:777–784
Oh HK, Jones MB, Longhurst WM (1968) Comparison of rumen microbial inhibition resulting from various essential oils isolated from relatively unpalatable plant species. Appl Microbiol 16:39–44
Osoro K, Mateos-Sanz A, Frutos P, Garcia U, Ortega-Mora LM, Ferreira LMM, Celaya R, Ferre I (2007) Anthelmintic and nutritional effects of heather supplementation on Cashmere goats grazing perennial ryegrass-white clover pastures. J Anim Sci 85:861–870
Osuji PO (1974) The physiology of eating and the expenditure of the ruminant at pasture. J Range Manage 27:437–447
Owen-Smith N, Novellie P (1982) What should a clever ungulate eat? Am Nat 119:151–178
Paolini V, Dorchies P, Hoste H (2003) Effects of sainfoin hay on gastrointestinal nematode infections in goats. Vet Rec 152:600–601
Patel DK, Kumar R, Prasad SK, Hemalatha S (2011) Pharmacologically screened aphrodisiac plant-A review of current scientific literature. Asian Pac J Trop Biomed 1:S131–S138
Patra AK, Saxena J (2009) Dietary phytochemicals as rumen modifiers: a review of the effects on microbial populations. Anton Leeuw 96:363–375
Pfister JA, Provenza FD, Manners GD, Gardner DR, Ralphs MH (1997) Tall larkspur ingestion: can cattle regulate intake below toxic levels? J Chem Ecol 23:759–777
Pfister JA, Provenza FD, Panter KE, Stegelmeier BL, Launchbaugh KL (2002) Risk management to reduce livestock losses from toxic plants. J Range Manage 55:291–300
Pfister JA, Gardner DR, Cheney CC, Panter KE, Hall JO (2010) The capability of several toxic plants to condition taste aversions in sheep. Small Rumin Res 90:114–119
Prins HH, Fritz H (2008) Species diversity of browsing and grazing ungulates: consequences for the structure and abundance of secondary production. In: Gordon IJ, Prins HH (eds) The ecology of browsing and grazing. Springer, Berlin, pp 179–200
Provenza FD (1996) Acquired aversions as the basis for varied diets of ruminants foraging on rangelands. J Anim Sci 74:2010–2020
Provenza FD (2008) What does it mean to be locally adapted and who cares anyway? J Anim Sci 86:E271–E284
Provenza FD, Villalba JJ (2006) Foraging in domestic vertebrates: linking the internal and external milieu. In: Bels VL (ed) Feeding in domestic vertebrates: from structure to function. CABI, Oxfordshire, pp 210–240
Provenza FD, Villalba JJ (2010) The role of natural plant products in modulating the immune system: an adaptable approach for combating disease in grazing animals. Small Rumin Res 89:131–139
Provenza FD, Burritt EA, Clausen TP, Bryant JP, Reichardt PB, Distel RA (1990) Conditioned flavor aversion: a mechanism for goats to avoid condensed tannins in blackbrush. Am Nat 136:810–828
Raubenheimer D (1992) Tannic acid, protein, and digestible carbohydrate dietary imbalance and nutritional compensation in locusts. Ecology 73:1012–1027
Revell CK, Revell DK (2007) Meeting 'duty of care' obligations when developing new pasture species. Field Crop Res 104:95–102
Robertshaw D (2004) Temperature regulation and the thermal environment. In: Reece W (ed) Dukes' physiology of domestic animals. Cornell University Press, Ithaca
Rogosic J, Estell RE, Skobic D, Martinovic A, Maric S (2006) Role of species diversity and secondary compound complementarity on diet selection of Mediterranean shrubs by goats. J Chem Ecol 32:1279–1287
Rolls BJ (1986) Sensory-specific satiety. Nutr Rev 44:93–101
Rutter SM (2010) Review: grazing preferences of sheep and cattle: implications for production, the environment and animal welfare. Can J Anim Sci 90:285–293

Sardon D, de la Fuente I, Calonge E, Perez-Alenza MD, Castano M, Dunner S, Pena L (2005) H-ras immunohistochemical expression and molecular analysis of urinary bladder lesions in grazing adult cattle exposed to bracken fern. J Comp Pathol 132:195–201

Senft RL, Coughenour MB, Bailey DW, Rittenhouse LR, Sala OE, Swift DM (1987) Large herbivore foraging and ecological hierarchies. Bioscience 37:789–799

Shaik SA, Terrill TH, Miller JE, Kouakou B, Kannan G, Kaplan RM, Burke JM, Mosjidis JA (2006) Sericea lespedeza hay as a natural deworming agent against gastrointestinal nematode infection in goats. Vet Parasitol 139:150–157

Sivakumaran S, Molan AL, Meagher LP, Kolb B, Foo LY, Lane GA, Attwood GA, Fraser K, Tavendale M (2004) Variation in antimicrobial action of proanthocyanidins from Dorycnium rectum against rumen bacteria. Phytochemistry 65:2485–2497

Sorensen JS, McLister JD, Dearing MD (2005) Novel plant secondary metabolites impact dietary specialists more than generalists (*Neotoma* spp.). Ecology 86:140–154

Squires V (1981) Livestock management in the arid zone. Inkata Press, Melbourne

Squires VR, Wilson AD (1971) Distance between food and water supply and its effects on drinking frequency, and food and water intake of Merino and Border Leicester sheep. Aust J Agric Res 22:283–290

Stephens DW, Krebs JR (1986) Foraging theory. Princeton University Press, Princeton, NJ, 262 pp

Stienezen GC, Waghorn GC, Douglas GB (1996) Digestibility and effects of condensed tannins on digestion of sulla (*Hedysarum coronarium*) when fed to sheep. N Z J Agric Res 39:215–221

Svendsen M, Ellingsen I, Tonstad S (2010) Compounds that may be responsible for reduced risk for cardiovascular diseases related to consumption of vegetables, fruit and berries. In: Bernhoft A (ed) Bioactive compounds in plants—benefits and risks for man and animals. Norwegian Academy of Science and Letters, Oslo, pp 71–77

Thatcher WW, Collier RJ (1986) Effects of climate on bovine reproduction. In: Morrow DA (ed) Current therapy in theriogenology, 2nd edn. W.B. Saunders, Philadelphia, PA, pp 301–309

Thomas DW, Samson C, Bergeron J (1988) Metabolic costs associated with the ingestion of plant phenolics by Microtus pennsylvanicus. J Mammal 69:512–515

Turner SP, Dwyer CM (2007) Welfare assessment in extensive animal production systems: challenges and opportunities. Anim Welf 16:189–192

Utsumi SA, Cibils AF, Estell RE, Soto-Navarro SA, Van Leeuwen D (2009) Seasonal changes in one seed juniper intake by sheep and goats in relation to protein and plant secondary metabolites. Small Rumin Res 81:152–162

Vallentine JF (2000) Grazing management, 2nd edn. Academic, San Diego, CA, 533 pp

Van Soest PJ, Mertens DR, Deinum B (1978) Preharvest factors influencing quality of conserved forage. J Anim Sci 47:712–720

Vercoe PE, Durmic Z, Revell DK (2009) Rumen microbial ecology: helping to change landscapes. Options Méditérr 85:225–236

Viegi L, Pieroni A, Guarrera PM, Vangelisti R (2003) A review of plants used in folk veterinary medicine in Italy as basis for a databank. J Ethnopharmacol 89:221–244

Villalba JJ, Provenza FD (1996) Preference for flavored wheat straw by lambs conditioned with intraruminal administrations of sodium propionate. J Anim Sci 74:2362–2368

Villalba JJ, Provenza FD (1997a) Preference for wheat straw by lambs conditioned with intraruminal infusions of starch. Br J Nutr 77:287–297

Villalba JJ, Provenza FD (1997b) Preference for flavored foods by lambs conditioned with intraruminal administrations of nitrogen. Br J Nutr 78:545–561

Villalba JJ, Provenza FD (1997c) Preference for flavored wheat straw by lambs conditioned with intraruminal infusions of acetate and propionate. J Anim Sci 75:2905–2914

Villalba JJ, Provenza FD, Hall JO, Peterson C (2006) Phosphorus appetite in sheep: dissociating taste from postingestive effects. J Anim Sci 84:2213–2223

Villalba JJ, Provenza FD, Hall JO (2008) Learned appetites for calcium, phosphorus and sodium in sheep. J Anim Sci 86:738–747

Villalba JJ, Provenza FD, Manteca X (2010a) Links between ruminants' food preference and their welfare. Animal 4:1240–1247

Villalba JJ, Provenza FD, Hall JO, Lisonbee LD (2010b) Selection of tannins by sheep in response to gastrointestinal nematode infection. J Anim Sci 88:2189–2198

Villalba JJ, Catanese F, Provenza FD, Distel RA (2011) Relationships between early experience to dietary diversity, acceptance of novel flavors, and open field behavior in sheep. Physiol Behav 105:181–187

Waghorn GC (2008) Beneficial and detrimental effects of dietary condensed tannins for sustainable sheep and goat production—progress and challenges. Anim Feed Sci Technol 147:116–139

Walker JW (1995) Viewpoint: grazing management and research now and in the next millennium. J Range Manage 48:350–357

Waller PJ (2006) Sustainable nematode parasite control strategies for ruminant livestock by grazing management and biological control. Anim Feed Sci Technol 126:277–289

Wallis de Vries MF, Schippers P (1994) Foraging in a landscape mosaic: selection for energy and minerals in free-ranging cattle. Oecologia 100:107–117

Wang CJ, Wang SP, Zhou H (2009) Influences of flavomycin, ropadiar, and saponin on nutrient digestibility, rumen fermentation, and methane emission from sheep. Anim Feed Sci Technol 148:157–166

Webster A (1991) Metabolic responses of farm animals to high temperature. In: Ronchi B, Nardone A, Boyazoglu J (eds) Animal husbandry in warm climates. Pudoc Wageningen, Viterbo, pp 15–22

Welch JG, Smith AM (1970) Forage quality and rumination time in cattle. J Dairy Sci 53:797–800

Williamson G, Payne WJA (1978) An introduction to animal husbandry in the tropics, 3rd edn. Longman, Essex, 881 pp

Wilson BJ, Garst JE, Linnabary RD, Channell RB (1977) Perilla ketone: a potent lung toxin from the mint plant, *Perilla frutescens* Britton. Science 197:573–574

Zimmerman EA (1980) Desert ranching in central Nevada. Rangelands 2:184–186

Chapter 8
Feeding Cattle for Improved Productivity, Health, and Welfare in Modern Farming Enterprises

David E. Beever and Alex Bach

Abstract The increasing world demand for animal protein, the increasing costs of energy, the need to reduce greenhouse gas emissions, and the declining availability of cereals for feeding to farm animals are all placing pressure on systems of milk and meat production from ruminant livestock. In many parts, producers have responded by increasing herd sizes whilst continuing to introduce genetically superior stock (in terms of performance), both seen as ways to improve economic returns. However, despite progressive increases in herd sizes, management practices have not always improved. The combination of high productivity and inadequate nutrition and management presents challenges to the health and welfare of ruminant livestock. This chapter describes the most important points to consider to ensure adequate nutrient supply to cattle to sustain productivity, as well as health and welfare.

8.1 Introduction

The last 30–40 years have seen many changes in the way cattle and sheep are kept for the production of meat and milk. One of the most common changes across the globe has been a relatively steady increase in farm size. This trend, started in the poultry industry, was closely followed by the pig industry and is now rapidly progressing in the cattle industry. However, increases in farm sizes have not always been followed by improvements in management.

David E. Beever was deceased at the time of publication.

D.E. Beever (✉)
Formerly of the University of Reading, Reading, UK

A. Bach
Department of Ruminant Production, Institut de Recerca i Tecnologia Agroalimentaries, Catalonia, Spain

Institució Catalana de Recerca i Estudis Avançats, Catalonia, Spain

C.J.C. Phillips (ed.), *Nutrition and the Welfare of Farm Animals*, Animal Welfare 16, DOI 10.1007/978-3-319-27356-3_8

Over the same period, the way in which beef is produced has also changed. Larger units have been established, with beef feedlots commonplace in a number of important beef producing countries. These include North America, South America, and Australia, where large numbers of cattle (>20,000) are often held in modest accommodation from weaning through slaughter. But feedlots have not gained universal acceptance and a significant number of smaller beef units still exist in many countries. In dairying, most countries have seen significant consolidation as numbers of dairy farmers have declined and herd sizes have increased, although herds of more than 500 cows are not commonplace and herds of more than 1000 cows are still relatively rare. At the same time, milk yields per cow have increased and total cow populations have declined or remained static. To maintain profitability, farmers have needed to seek ways to continue to increase milk yields per cow, with an increasing number of farms now providing purpose-built accommodation suitable for year-round housing. There is, however, a significant number of large herds (>400 cows) maintained exclusively on pasture and not housed, as frequently seen in New Zealand, South Africa, Ireland, and other regions where good year-round water supplies are more reliable.

At the same time, significant progress in both animal breeding and animal genetics has occurred. Many native breeds capable of modest levels of production have been replaced by more productive breeds. A good example is the extensive introduction of the Holstein breed for increased milk yields being one such example. In pigs and poultry, structured cross-breeding programmes have resulted in new, more highly productive breeds, whilst in all sectors of animal production, genetic selection has been relentless. In most cases, the focus has been towards increased levels of meat, milk, or egg production. More recently, however, many breeding initiatives have widened their selection criteria to counter specific production issues that have become more evident. Leg weaknesses in laying hens and compromised fertility in dairy cows, as well as increased lameness, provide suitable examples. The need to provide animals capable of high levels of production and able to cope with the types of production system in which they are maintained is being increasingly recognized by those involved in livestock production. In the last months, and certainly, in the near future, efforts are and will be made to select for improved efficiency rather than just increased production.

The increasing human population is demanding more food and a more secure supply chain. With greater affluence being enjoyed by more people, the demand for animal products is increasing. Countering this, however, is the human's thirst for fuel, which is growing at an even faster rate. Most governments are pursuing policies of reduced fossil fuel use and increased reliance on renewables, the production of biofuels from farmed crops being one such example. The last decade has witnessed an exponential increase in biofuel production, and such trends are likely to continue. Currently, starch in cereal grains is the preferred substrate, with the use of lignocellulosic materials (forages) still only in the development stage. Inevitably, this has led to an increased proportion of the annual production of cereals being diverted to bioethanol production from direct or indirect (by farmed livestock) use for human food. This is most evident in North and South America.

The two outcomes are increased grain prices and increased scarcity of cereals for feeding to farm livestock. Higher grain prices are affecting all systems of animal production, and higher levels of efficiency of conversion of feed into animal product and optimal animal health and welfare will be required for such systems to remain financially viable. Feeding large amounts of grains was an affordable, simple, and reliable way to improve ruminant animal productivity at competitive cost. However, this is likely to change with increasing feed prices (or market volatility) and compromised grain supply, and over the next decade increased use of forages, together with an increasing stream of co-product feeds arising from the biofuels industry, will merit consideration. The question then remains: Will the current ruminant genotypes capable of high levels of milk and meat production be fit for purpose or will further changes in genetics, breeds, as well as systems of production be required?

This chapter considers these aspects in more detail by examining a number of defined production systems for ruminant livestock. It reviews progress in these systems, identifying where improvements can be made, and provides a forward-looking view as to how systems of animal production are likely to change over the next 1–2 decades, by taking account of the often conflicting demands of society for secure food supplies (including product wholesomeness), high animal health and welfare standards, minimization of environmental impact, and the changing nature of the animal feed supply chain.

8.2 Calves

In the management of dairy animals, most of the attention has focused on lactating and transition cows. However, dairy replacement heifers not only define the long-term future of the dairy enterprise but also represent close to 10 % of the costs associated with milk production, being the third largest after taking account of feed for the milking herd and labour costs. It follows that optimizing heifer growth (and welfare) is likely to have an important impact on herd profitability through cost rationalization and improved performance of the future herd.

It is now evident that nutrient supply and hormonal signals at specific windows during development (both prenatal and early postnatal) can exert permanent changes on metabolic processes in humans (Fall 2011) and equally affect the performance, body composition, and metabolic function of the offspring of livestock (Wu et al. 2006) through processes generically referred to as foetal programming and metabolic imprinting. It is highly possible that the present-day cow, with high milk yield potential but significant reproductive and metabolic challenges, is not only a consequence of genetic selection, but the result of the way her dam was fed at the time of her pregnancy, as well as the way in which the dam was fed during her early life (Bach 2012).

The first weeks of life appear to have long-lasting consequences on the physiological function of neonates. The pioneering work of McCance (1962) illustrated that limit-feeding rats during the first 21 days of life resulted in a lifetime programming of growth pattern that was reduced compared with rats fed ad libitum. Interestingly, when the same dietary restriction was applied for 21 days but at a more advanced age, the intervention had no lasting effect, as the underfed rats were able to demonstrate compensatory growth gains when re-fed. Studies in cattle have confirmed that a nutritional restriction up to weaning age limits their capacity for compensatory growth (Café et al. 2006; Greenwood et al. 2006), whilst others have shown that increased average daily gain (ADG) during the first 2 months of life results in greater body weight (BW) at 24 months of age (Robelin and Chilliard 1989; Moallem et al. 2010).

Evidence from the literature (Hoffman and Funk 1992; Krpálková et al. 2014) and our own field data suggests that age at first calving has little correlation with milk production during the first lactation, provided calving does not occur before 22 months of age. In contrast, BW has a much larger effect on milk production. Bach and Ahedo (2008) reported that each additional 70 kg of BW at calving equated to an additional 1000 kg milk yield during the first 305 days of the first lactation. Furthermore, several studies (Bach and Ahedo 2008; Bach 2012; Soberon et al. 2012) have shown a positive correlation between ADG from birth to 2 months of age with first lactation milk yield, possibly by as much as an additional 225 kg milk for each additional 0.1 kg ADG (Bach 2012). Therefore, providing adequate nutrition to sustain high growth rates (>750 g/day) during the first 2 months should result in systems of heifer rearing that are not only economically more efficient but also more effective in terms of increased milk production. Despite indisputable evidence that feeding more milk replacer will improve growth rates, many farmers still limit feed heifer calves to not more than two daily feeds of milk replacer each of 2 L. This level of feeding will inevitably compromise growth and most probably future performance and at the same time can significantly increase the risk of disease. To achieve the desired results, monitoring of concentrate starter feed intake is also essential, especially if early gains from additional milk feeding are not to be lost after weaning. A high-quality milk replacer with 27 % crude protein (CP) and 15–17 % fat is recommended, with Hill et al. (2009) suggesting a requirement of 56 g of crude protein/Mcal of ME for optimal performance. In addition, the starter feed must comprise palatable feed ingredients. Nevertheless, if high milk volumes of milk are to be fed (e.g., 8 L/day), it must be kept in mind that milk feedings should be spread out throughout the day in small meals (~3 L maximum). Bach et al. (2013a) reported that offering 4 L of milk (or milk replacer) in two meals led to increased insulin resistance in young calves.

Overall, successful management of young calves requires a focus on (1) optimization of growth; (2) minimization of health issues; and (3) an efficient and stress-free transition from liquid to solid feed. Shortly after birth, high-quality colostrum should be provided to all calves to ensure an adequate passive immune transfer. This reduces the risk of disease at a time when scouring and influenza frequently challenge the welfare of calves. Management and cleanliness, if not of a high standard, may add to any negative effects of poor colostrum feeding practices. Most dairy replacement calves are housed in individual hutches during the first weeks of life to minimize disease transfer and facilitate control of starter feed intake. A reduction in morbidity can be achieved by reducing the amount of stress calves experience at this time, as this is known to induce immune-suppression. Good hygiene is critical to reduce environmental microbial load throughout the whole rearing period. To minimize stress, it has been typically recommended to keep calves individually housed for an additional 1–2 weeks after weaning, as this helps separate the adverse effects of weaning and moving into groups. However, when Ziegler et al. (2008) compared the performance of calves that were immediately placed in groups of 10 animals at weaning with those retained in individual stalls for an additional 14 days, no differences in performance during the following 112 days were noted. In contrast, a study by Bach et al. (2010) involving 320 calves concluded that animals grouped immediately after weaning reached target BW 6 days earlier and experienced fewer respiratory disorders than calves grouped at 6 days post-weaning. In a second study, Bach et al. (2010) used 240 female calves to assess the impact of grouping before weaning on ADG and health. Calves grouped at 49 days of age (when milk replacer feeding was reduced by half) had a greater ADG and BW at 56 days of age as a result of increased solid feed consumption compared with those grouped at 56 days of age and with fewer cases of respiratory disease. Furthermore, De Paula Vieira et al. (2010) reported that calves raised in pairs performed better right after weaning, and more recently (Jensen et al. 2015) have reported that pair housing also improves growth performance before weaning. Commingling calves before weaning is especially important when feeding large amounts of milk or milk replacer, which may delay the consumption of dry feed and normal development of the rumen (Davis and Drackley 1998). In a recent study, Terré et al. (2007) found duodenal microbial flow of calves following an enhanced-growth feeding programme involving increased milk feeding was lower than that of calves fed conventionally, suggesting the existence of a reduced rumen microflora, which may have negatively affected both solid feed intake and overall feed digestibility after weaning. Social facilitation to foster intake may be effective to ease the weaning process, especially if enhanced-feeding programmes are being used. In this regard, offering a poor quality (nutritionally) forage source, such as mature ryegrass hay or oats straw, has been shown to improve rumen health, solid feed consumption, and performance (Castells et al. 2012). Thus, it is recommended to start offering chopped forages to calves at around 45 days of life (Bach et al. 2013b).

During the transition from milk to solid feeding in calves, an increased incidence of respiratory diseases can compromise performance. Ensuring good air quality is important and can be improved by increased air flow, noting the welfare of the calf is more important than that of the stockman. Frequent provision of clean bedding is important, but should always be provided by hand, as mechanical straw blowers that are often used as a labour-saving device can increase dust levels, leading to increased irritation of the calves' respiratory tract. Consideration of calves' previous health history may help reduce the overall incidence of respiratory problems, including pneumonia (Bach et al. 2011). This not only improves calf welfare and performance but also extends the productive life of the adult cow (Bach 2011).

8.3 Management and Nutrition During the Growing Phase

Once weaned, calves are typically moved to groups that progressively increase in size, but inevitably with less attention being devoted to such animals, thus increasing their vulnerability to welfare problems. Feed represents the single greatest expense associated with heifer rearing, and at this time small improvements in feed conversion efficiency (FCE) can result in relatively large cost savings. For this reason, it is important to continually monitor actual intakes to allow the quality of the ration or the amount offered (even when calves are at pasture) to be adjusted if production targets are not being met. Together with routine observation and weighing of heifers, it should be possible to achieve optimal growth, at the same time ensuring adequate welfare and minimal disease, whilst controlling levels of feeding to avoid the animals becoming over-conditioned.

More recently, the popularity of limit-feeding heifers to improve feed efficiency and avoid over-conditioning has been proposed. Hoffman et al. (2007) suggested that limit feeding a diet of increased density whilst ensuring provision of all required nutrients could be an alternative approach to control over-conditioning and improve feed efficiency. However, when Zanton and Heinrichs (2007) limit fed young (130 kg) Holstein heifers for 35 weeks on a diet containing 25 % forage and compared these with similar heifers fed a greater allocation of a diet containing 75 % forage, no differences in ADG or skeletal growth between the two groups were noted. Furthermore, increasing ration nutrient density invariably increases the overall costs of the ration. Thus, limit feeding will only be economic if the increase in ration unit cost is offset by the reduced intake of feed. One concern with limit feeding, however, is possible effects on animal welfare, where animals have no access to feed for several hours each day. This can rapidly lead to stress and adverse behavioural patterns. An alternative approach now being proposed is to ad libitum feed a ration of reduced energy density by the inclusion of mature forages or cereal straw. Provided the ration is well mixed to avoid ration sorting, hunger issues will not occur and satisfactory weight gains can be achieved, at the same time avoiding over-conditioning of the animals. The provision of additional physical fibre should improve rumen function, which in turn may lead to improvements in FCE, and ration cost is likely to be significantly reduced.

8.4 Managing and Feeding the Transition Cow

Most of the health problems in dairy cows, both metabolic and infectious, occur during early lactation, frequently related to low-energy intake after calving, with the associated negative energy balance resulting in conditions such as ketosis and fatty liver syndrome. It is now recognized that most of these problems originate from the way in which the cow is fed and managed during the periparturient period. Traditional practice was to feed high-energy rations in the immediate pre-calving period to minimize body fat mobilization, ketosis, and fatty liver after calving, whilst adapting the rumen microflora to more nutrient-dense rations intended for the following lactation. However, overfeeding to the point where cows become over-conditioned can increase the incidence of calving difficulties and postpartum metabolic problems. Despite such issues, NRC (2001) recommended a relatively high energy-dense ration (1.60 Mcal of NE_L/kg) before calving, and this practice has been followed by many farmers. However, several studies (e.g. Rabelo et al. 2003; Loor et al. 2006) have shown that when cows were fed to NRC (2001) recommendations with ad libitum access to high-energy diets during the dry period, a greater reduction in feed intake before parturition occurred along with a slower recovery of intake postpartum. Further studies (Dann et al. 2005, 2006; Douglas et al. 2006; Loor et al. 2006) have shown similar outcomes, with cows moderately overfed during the dry period, even without becoming overtly obese, considered to be at increased risk of developing periparturient health problems. The current view is that the recommended high-energy feeding during the dry period, especially as calving approaches, may be detrimental to cow health. An average prepartum cow requires about 15 Mcal of net energy for lactation (NE_L) per day, and feeding a ration energy density of 1.60 Mcal of NE_L/kg ration would readily provide more than 19 Mcal of NE_L per day. Consequently, lower energy rations (approx. 1.32 Mcal of NE_L/kg) are now being recommended as sufficient to meet the energy requirements of dry cows. Cereal straw provides an excellent source of fibre in such rations, as well as an important energy diluent, provided the ration is well mixed to avoid ingredient selection by the cows. With respect to dietary protein, feeding rations of approximately 13 % crude protein are recommended and possibly marginally greater when significant numbers of first calvers are being fed.

8.5 Low Input/Low Output Systems of Milk Production

In countries like New Zealand and Ireland capable of high annual yields of forage from grazed pasture, systems of milk production based on extensive use of fresh grass as the principal feed for milk production prevail. The lower feed cost compared with conserved or purchased feeds represents a competitive advantage. Cows are bred to calve to coincide with the expected re-commencement of pasture growth and then milked until pasture supply or quality declines or fails with advancing season. Supplementary feeding is generally limited to the first few weeks of lactation, when higher milk yields are expected.

One major issue, however, is the amount of grass dry matter (DM) cows voluntarily consume each day. Even when pasture amount and quality are optimal, it is rare for lactating cows to consume more than 18 kg DM/day, with 14–15 kg of grass DM/day being a more realistic figure (Beever et al. 2000). Only rarely does grass DM exceed 20 % and on many occasions it is significantly less. To achieve a daily DM intake of 20 kg, a cow must therefore consume more than 100 kg, maybe as much as 120 kg of fresh forage per day, equal to 20–25 % of her BW. This restriction in feed intake, along with pasture-fed cows generally having shorter lactations as pasture availability declines with advancing season, results in low average annual lactation yields, rarely exceeding 5000 kg without supplementation. Thus low input/low output systems tend to favour increased herd sizes in order to maximize annual milk sales. This is evident in New Zealand, whilst gaining support in parts of Ireland, the only difference being the more challenging Northern Hemisphere winters, which frequently require the cows to be housed at this time.

Systems where the cow harvests the grass at minimal cost to the producer are popular to many producers, yet there are a number of underlying issues that many advocating the merits of such systems often overlook. Of most concern is the control of body condition (BC). When cows are dried off early due to failing pasture quality and supply, the risk of these cows becoming over-conditioned prior to calving increases and over-conditioned cows at calving are more likely to be predisposed to health issues, including assisted calvings, ketosis, and fatty liver. After calving, all cows are geared towards milk production, and a sustained period when total energy intake fails to meet total energy output can be expected. The extent of this shortfall will be exacerbated when cows are over-conditioned at calving as this will affect post-calving appetite. Inevitably, cows lose BC, which can often be both extensive and prolonged. Furthermore, over-feeding of protein in early lactation may stimulate milk production, often with an associated increased BC loss. When pasture-fed cows are consuming early season growth at a time of maximum lactation potential, overfeeding of protein may be unavoidable due to the high protein levels in the grass at this time. This can often result in a shorter lactation with a high peak, further challenging the cow's metabolism, welfare, and if energy is in short supply, fertility.

Significant BC loss after calving frequently leads to cows failing to conceive. Albeit the study did not include pasture-fed cows, Butler and Smith (1989) reported cows losing between 0.5 and 1.0 BC score units between calving and first service had a mean pregnancy rate of 53 % to that service, compared with only 17 % for cows losing over 1.0 BC score at this time. Subsequently, Beam and Butler (1999) found that increased negative energy balance reduced the pulse frequency of luteinizing hormone, which has a direct impact on the subsequent fate of the developing follicle.

In a large study with pasture-fed cows conducted in Ireland, Mee (2004) noted that, over 7 years, average first service conception rate declined by 1 % unit per year (60–54 %), with a corresponding 10-day increase in mean calving interval. The study also found the number of cows experiencing abnormal reproductive cycles doubled (13–26 %), all of these changes occurring without any marked increase in annual milk production over the study period. Mee (2004) also reported many cows had less overt oestrus behaviour and concluded that

> ... strategies are required to improve or halt the decline in reproductive performance (and that) these must include feeding systems to reduce negative energy balance and maintain body condition.

This was in agreement with the earlier summary of Buckley et al. (2003), who, also with pasture-fed cows in Ireland, concluded that

> reproductive performance, especially the probability of conception, may be negatively associated with the magnitude and duration of negative energy balance in early lactation.

Compromised fertility affects both cow welfare and farm profitability in low input/low output systems. Cows with delayed conception calve after the first flush of spring grass, reducing the opportunity to exploit a high quality feed, whilst repeat breeding adds to overall costs associated with feed supplements. Where rigorous culling policies are practised, delayed conception increases the risk of cows being prematurely removed from the herd, to be replaced by heifers or cows with more timely calving patterns. The most obvious strategy to reduce BC loss in early lactation is to extend supplementary feeding. Many producers, however, only see this as a short-term issue to overcome any early season shortfall in pasture supply, and judge the benefits solely in terms of daily milk volumes. Where the milk yield response to supplementary feeding appears marginal (additional milk income less supplement costs), many producers quickly cease this practice, without consideration of possible positive outcomes on BC score, fertility, and cow longevity.

It is important to recognize the nutritional limitations of grazed pasture, as well as its potential. Grass is not a consistent feedstock in terms of its nutritional composition, changing between seasons and days, and even within days. Spring grass has high levels of soluble carbohydrates, often with elevated levels of protein, but reduced amounts of fibre. With advancing season, fibre levels increase at the expense of soluble carbohydrates, whilst protein levels generally decline, often with a noted increase in nonprotein nitrogen content, which is nutritionally inferior to true protein (Beever and Siddons 1986). Supplementary feeds need to account for such changes, providing energy rather than protein during early season and more readily available energy and possibly some good quality protein as the season advances. But almost all pasture for milking cows in temperate grazing systems lacks adequate levels of physically effective fibre, due to forage type and grazing practices, which aim to optimize the annual growth and consumption of pasture. Many of these pasture-fed cows have compromised rumination activity, especially during spring and early summer, which inevitably has negative effects on rumen function. Added to this, high intakes of soluble carbohydrates promote rapid and extensive acid production in the rumen, with many pasture-fed cows operating for long periods each day at suboptimal rumen pH. Williams et al. (2005), working with cows grazing irrigated pasture in Australia, reported rumen pH was below 5.5, considered suboptimal for fibre digestion, for an average of 17 h/day, whilst O'Grady et al. (2008) in Ireland indicated low rumen pH values approaching 5.0. Such levels affect not only fibre digestion but overall FCE. Grass-induced rumen acidosis differs from grain-induced acidosis as it is not due to lactic acid accumulation, but nonetheless, the need for avoidance is just as great. Feeding supplements with high starch content, especially in large discrete meals at milking, is not recommended for the cows' welfare, and supplements containing more slowly degrading energy sources with significant amounts of physically effective fibre are preferred. Feeding mixed rations optimized for both chemical and physical parameters, and ideally before each milking, is likely to have less adverse effect on rumen acidity by promoting rumination, improving rumen function and FCE, whilst reducing forage substitution rate.

For low input/low output systems to thrive due to the low feed costs, the issues of excessive BC loss, compromised fertility, increased health issues at calving, and suboptimal rumen function need to be addressed, and consideration given to a more proactive system of supplementary feeding.

8.6 High Input/High Output Systems of Milk Production

Contrary to low input systems, cows in intensive dairy production systems are typically supplied with high-quality rations according to their immediate needs. However, two major nutrition threats or challenges can occur shortly after calving. First, due to the high genetic potential of the cows involved in such production systems, milk yield during early lactation is supported, in part, through mobilization of body reserves, which can lead to the development of ketosis. The second challenge relates to rumen health, as in attempting to provide sufficient energy to support high levels of milk production, the proportion of non-fibrous carbohydrates in the diet is frequently increased at the expense of fibre or forage content. Feeding such rations frequently leads to a more vigorous fermentation in the rumen, which in turn increases rumen acid levels, with a concomitant fall in rumen pH.

Ketosis is a common affliction in dairy cattle during early lactation, closely related to liver lipidosis. It is the result of an inadequate supply of energy and glucose precursors from feed to support the level of milk output being achieved.

Ruminal acidosis commonly affects more than 20 % of early- and mid-lactation cows (Garrett et al. 1997; Oetzel et al. 1999). It occurs as a result of an accumulation of fermentation end-products, mainly volatile fatty acids (VFA) and lactic acid, in the rumen, due to either an excessively vigorous fermentation that produces a large amount of VFA, or an inadequate removal of these VFAs from the rumen, principally via absorption through the rumen wall.

Rumen acidosis is usually associated with erratic feed intake. This is mostly due to the protection mechanisms to control rumen pH. Basically, when rumen pH decreases, some rumen bacteria sensitive to pH lyse and their contents are liberated into the rumen fluid. These cell contents from lysed bacteria, in combination with the accumulation of fermentation end-products responsible for the decrease in rumen pH, cause an increase in the osmolality of rumen fluid. Carter and Grovum (1990) reported that appetite of the dairy cattle declines as rumen osmolality increases. Furthermore, low rumen pH may damage the rumen epithelium, inducing a local inflammatory response that may also affect appetite. Once the animal ceases eating due to subacute rumen acidosis (SARA), the rumen environment progressively returns to its normal conditions, pH increases, osmolality and local inflammation decrease, and appetite resumes. SARA is not easily observed in commercial, group-fed operations with loose-housed animals because not all animals are affected at the same time.

An important welfare consequence of rumen acidosis is lameness. Lameness, in conjunction with secondary reproductive failure and low milk production, is commonly one of the most important causes of premature and involuntary culling. Ruminal acidosis is probably the most common digestive upset of dairy cattle, as more than 60 % the hoof lesions are associated with laminitis in some form (Nocek 1997), and rumen acidosis is implicated in most laminitis. Laminitis not only causes pain to cows, but it also compromises milk production (Warnick et al. 2001; Green et al. 2002; Bach et al. 2007).

Ruminal acidosis can be prevented and controlled through good ration formulation, adequate particle size of the ration, and feed bunk management. Others suggest the inclusion of ionophores, organic acids, probiotics, and buffers and alkalinizers in the ration, but the results of such interventions can be quite variable. The most common recommendation to prevent rumen acidosis is to provide an adequate amount of physically effective fibre in the ration.

A better recognition of the differences in physical form of neutral detergent fibre (NDF) could improve the control of rumen pH. Physically effective fibre can be defined as the proportion of dietary NDF contained in particles above 1.8 mm according to Mertens (1997) and between 8 and 19 mm according to Lammers et al. (1996). Physically effective fibre affects chewing times, saliva production, and, ultimately, rumen buffering capacity. Numerous studies have been conducted to evaluate the effects of physically effective fibre, but results have not been conclusive. Some authors (Soita et al. 2000; Krause et al. 2002; Beauchemin et al. 2003; Calberry et al. 2003) reported a positive relationship between peNDF and rumen pH, whereas others (Yang et al. 2001; Kononoff et al. 2003; Kononoff and Heinrichs 2003; Einarson et al. 2004) have failed to do so. Because of interactions between concentrate inclusion level, rate of fermentation, forage-to-concentrate ratio, and total DMI, it is difficult to recommend an optimum physically effective fibre value for preventing and controlling SARA. Beauchemin and Yang (2005) proposed the use of peNDF and dietary fermentable organic matter to predict rumen pH and control SARA. Meanwhile, Zebeli et al. (2006) presented a model that accounted for physically effective fibre (>1.8 mm) and the amount of fermentable organic matter from forages and rumen degradable starch in the ration, which improved the prediction of rumen pH compared with using only estimated peNDF. More recently, Rustomo et al. (2006) evaluated the effect of ration acidogenic value and forage particle length on rumen pH. As the acidogenic value of the concentrate fed increased, rumen pH decreased, and time below pH 5.6 increased. However, the effect of forage particle size on rumen pH was only evident when long forage was offered in conjunction with a highly acidogenic concentrate. Clark and Armentano (2002) reported significant linear increases in chewing and rumination times as particle size of alfalfa hay in the TMR increased, although there were no differences in milk fat percentage across particle sizes. Similarly, Beauchemin et al. (2003) reported a good correlation between effective NDF in the TMR and ruminal pH and chewing activity, but no effects were found on milk fat percentages. Feeding forages of small particle size has been correlated with decreased chewing activity and saliva secretion, low rumen pH, and low milk fat percentages (Cassida and Stokes 1986). It is important to remember that feed selection estimates made with individually fed cows are likely to underestimate feed selection in free-stall barns (Leonardi and Armentano 2007). Intake of the longest particles expressed as a percentage of the predicted intake was 73.2 % in a tie-stall barn and 63.3 % in a free-stall barn. Thus, in loose-housed cows, the occurrence of SARA could be more frequent, because the diet will be different depending on the time any given cow reaches the feed bunk (reflecting the level of particle selection exerted on that ration by the other cows).

In summary, it appears that just balancing for total NDF in the ration, independently of particle size, may explain only about half of the variation in milk fat content. Manipulating the particle size of forages does not consistently result in increased ruminal pH or reduced risk of SARA, especially if sorting by the cows against large particles takes place.

Beyond ensuring adequate rumen function and avoiding SARA, it is important the cow receives the exact amount of nutrients she needs. For example, an excessive supply of protein not only will carry an economic loss and toll for the environment, but it will also force the cow to divert energy resources to metabolize the surplus of amino acids into urea and finally excrete nitrogen in the urine. To avoid imbalance in supplies of nutrients, it is pivotal to formulate rations accounting for actual DM intake of the cows and then modify nutrient density of the ration accordingly.

8.7 Beef Production

Most of the world's beef comes from beef breeds reared specifically for that purpose. Some countries produce a proportion of their annual output from beef breeds used as terminal sires on those dairy cows not required or considered suitable to produce dairy heifer replacements, whilst culled dairy cows are also a useful source of beef (Phillips 2010). This section will focus on cow–calf or suckler beef operations with recognized beef breeds.

Following puberty, suckler cows are expected to produce one live calf each year, with most herds either spring calving, with significant reliance on pasture whilst the calf is with the cow, or autumn calving, when cows and calves are generally housed and fed conserved forage and supplements during the first 6 months of the calf's life. Weaning usually takes place when the calves are 6–8 months old at target weights of 300–350 kg for males and 275–300 kg for females, according to breed and genetics. Before weaning the cow should have been re-bred, whilst after drying off only a modest plane of nutrition through to the next calving will be required.

Weaning can be traumatic for the calf, both socially and in terms of performance (intake and weight gain) and health. Providing stress-free conditions at this time is important, as well as a plentiful supply of a well-balanced ration to avoid or minimize any weaning check. The calves should be fed to promote skeletal and muscle growth at high levels of feed efficiency, whilst avoiding digestive upsets and exaggerated fat deposition. The growing phase provides the main opportunity to optimize financial returns and feeds need to be well presented, palatable, and of average nutrient density. Significant amounts of forage, both grazed and conserved, can be included, as rations containing high energy and high starch are not required at this time. In contrast, the finishing phase for beef cattle represents the period of maximum fat deposition to achieve the required carcass fat cover. High-energy rations with increased levels of starch are required, and such rations are generally more expensive. Also, as fat deposition is energetically more costly than protein (lean) deposition, feed efficiency inevitably declines (Owens et al. 1995). Consequently, the fattening phase should be as short as possible to avoid losing many of the earlier financial gains captured during the growing phase.

From this brief analysis of cow–calf operations, it would appear such systems are relatively easy to implement, yet there is considerable evidence of this not being so on many farms. Maximizing the number of calves weaned each year should be the first priority, with 95 weaned calves produced per 100 cows bred considered acceptable in top managed herds (Phillips 2010). Yet, survey data indicate many farms in extensive rangelands achieving much lower levels (Bortolussi et al. 2005). Secondly, herds should aim for tight calving patterns with a maximum calving spread of 9 weeks, yet, many farms report calving spread of 16 weeks or more. This presents an obvious management issue, disadvantageous to both spring-calving herds, where a significant number of freshly calved cows would miss the first grass flush, and autumn-calving herds having an increased number of underweight calves at the end of winter (Phillips 2010). There would be knock-on effects for the next expected calving, and ultimately the ideal of a spring- or autumn-calving herd is lost. Farms that operate both calving periods generally do so due to fertility failures and not as a deliberate policy decision.

On average, a suckler cow will annually consume around 4 t of feed DM, a significant overhead cost to be carried by the weaned offspring. In low-performing herds this cost is exaggerated, where achieving only 80 weaned calves per 100 cows bred increases cow feed cost by almost 20 % per weaned calf compared with top-performing herds. Compared with dairy cows, the nutrient demands of the suckler cow are modest, rarely exceeding 12–13 kg DM/day at peak lactation, whilst during the dry period as little as 8 kg DM/day of a lower quality ration should suffice. Thus, unlike dairy cows, excessive BC loss in early lactation is not a major issue, rather the propensity of suckler cows to become over-conditioned needs to be addressed. Over-fatness must be avoided at the time when cows are being bred and also as they approach calving, making management of BC probably the most important issue with suckler cows. Over-fat cows during the breeding period can be avoided with careful nutrition management, provided cows do not calve with excessive body condition. Subsequently, as the cow enters the second trimester of pregnancy with milk yield declining, some BC gain can be allowed, possibly through to the first part of the third trimester. This provides an excellent opportunity to exploit late summer regrowths for autumn-calvers and lower quality conserved forages for spring-calvers. Prior to calving, however, all cows should be fed to achieve an acceptable BC, which inevitably requires significant nutritional intervention in some herds.

Spring-born calves generally will be housed soon after weaning and fed moderate energy rations with adequate amounts of protein for continued skeletal development and lean tissue deposition. Meanwhile, most autumn-born calves after weaning should be of suitable age and weight to be turned on to well-managed pasture with minimal supplementation until grass quality and availability becomes an issue. Average daily weight gains for these cattle should be between 1.2 and 1.4 kg/day according to breed and gender, but will be influenced by both level of feed consumption and overall FCE. As noted earlier, FCE declines as cattle mature, but a target feed efficiency during the growing phase of 170 g weight gain/kg of feed DMI with a daily intake of 7.5 kg should be sufficient to secure target weight gains.

The start of the finishing phase is affected by animal weight and breed, with early-maturing breeds often requiring only a minimal period of higher energy feeding compared with later-maturing breeds. Maximum achievable FCE will have dropped further to around 120–140 g/kg of DM, requiring a daily feed DMI approaching 11 kg to achieve a daily gain of ca. 1.4–1.5 kg.

However, as with dairy cows, achieving satisfactory feed efficiencies is more than simply supplying a ration of adequate energy density. It is crucial that this feed energy becomes fully available to the animal through normal digestive processes, which requires optimal rumen function. There are many examples where high grain rations fail to deliver the expected outcome in terms of daily weight gain and in most of these the principal cause is suboptimal digestion. In such situations, high acid loads in the rumen can result in subacute or even acute rumen acidosis. Unlike high acid loads noted with grass-fed cattle, lactic acid is the principal acid accumulating in the rumen of grain-fed cattle, and being a stronger acid than rumen volatile fatty acids its effects are more difficult to reverse. Feeding mixed rations, either as the total feed or as a supplement for grazing animals, containing an adequate amount of physically effective fibre, will negate possible adverse effects of high grain feeding on rumen function, whilst replacing some of the faster degrading grain, such as rolled wheat or barley, with more slowly degrading starch sources such as maize will further improve rumen function and overall feed efficiency, to achieve daily weight gains closer to expectations.

8.8 Conclusions

Compromised animal health and welfare lead to suboptimal animal performance, which in turn reduces the yield of animal product per unit of feed supplied. This is not acceptable at a time when the world continues to strive to produce more food for the expanding population and undoubtedly adds an unnecessary additional environmental burden. The use of feed for milk production illustrates the magnitude of the problem, where low-input herds with poor fertility, leading to extended dry periods, and high annual herd replacements were estimated to be using over 30 % of the annual feed use for non-milk producing stock.

All livestock industries must be committed to optimizing animal health, welfare, and feed conversion efficiency. At the same time, significant changes are happening in respect to the nature of the feed base for future meat and milk production, as the surge in biofuels production gathers momentum. This position may change if current initiatives to utilize lignocellulosic materials for biofuel production prove financially viable, but until that time, all livestock industries will face increased costs and reduced availability of grains. Ruminant production systems will have to cope with both higher grain prices and lower availability. Those countries capable of producing high-quality forages will be able to exploit such feeds, and provided these are suitably fed to optimize rumen function, will likely have a competitive advantage. Nevertheless, the key to ensure adequate productivity while avoiding health disorders is to ensure the provision of the right amount of nutrients to the animal. For that, it is key to know how much feed is being consumed and then balance the ration accordingly based on daily amounts rather than percentage of nutrients in the diet.

References

Bach A (2011) Associations between several aspects of heifer development and dairy cow survivability to second lactation. J Dairy Sci 94(2):1052–1057

Bach A (2012) Ruminant nutrition symposium: optimizing performance of the offspring: nourishing and managing the dam and postnatal calf for optimal lactation, reproduction, and immunity. J Anim Sci 90(6):1835–1845

Bach A, Ahedo J (2008) Record keeping and economics of dairy heifers. Vet Clin N Am Food Anim Pract 24(1):117–138

Bach A, Iglesias C, Devant M (2007) Daily rumen pH pattern of loose-housed dairy cattle as affected by feeding pattern and live yeast supplementation. Anim Feed Sci Technol 131 (1–2):146–153

Bach A, Ahedo J, Ferrer A (2010) Optimizing weaning strategies of dairy replacement calves. J Dairy Sci 93(1):413–419

Bach A, Tejero C, Ahedo J (2011) Effects of group composition on the incidence of respiratory afflictions in group-housed calves after weaning. J Dairy Sci 94(4):2001–2006

Bach A, Domingo L, Montoro C, Terré M (2013a) Short communication. J Dairy Sci 96 (7):4634–4637

Bach A, Terré M, Pinto A (2013b) Performance and health responses of dairy calves offered different milk replacer allowances. J Dairy Sci 96(12):7790–7797

Beam SW, Butler WR (1999) Effects of energy balance on follicular development and first ovulation in post-partum dairy cows. J Reprod Fertil 54(Suppl):411–424

Beauchemin KA, Yang WZ (2005) Effects of physically effective fiber on intake, chewing activity, and ruminal acidosis for dairy cows fed diets based on corn silage. J Dairy Sci 88 (6):2117–2129

Beauchemin KA, Yang WZ, Rode LM (2003) Effects of particle size of alfalfa-based dairy cow diets on chewing activity, ruminal fermentation, and milk production. J Dairy Sci 86 (2):630–643

Beever DE, Siddons RC (1986) Digestion and metabolism in the grazing ruminant. In: Milligan LP, Grovum WL, Dobson A (eds) Control of digestion and metabolism in ruminants. Prentice-Hall, Englewood Cliffs, NJ, pp 479–497

Beever DE, Offer N, Gill M (2000) The feeding value of grass and grass products. In: Hopkins A (ed) Grass: its production and utilisation. Blackwell Scientific, Oxford, pp 140–195

Bortolussi G, McIvor JG, Hodgkinson JJ, Coffey SG, Holmes CR (2005) The northern Australian beef industry, a snapshot. 2. Breeding herd performance and management. Aust J Exp Agric 45 (9):1075–1091

Buckley F, O'Sullivan K, Mee JF, Evans RD, Dillon P (2003) Relationships among milk yield, body condition, cow weight and reproduction in spring-calved Holstein-Friesians. J Dairy Sci 86(7):2308–2319

Butler WR, Smith RD (1989) Interrelationships between energy balance and postpartum reproductive function in dairy cattle. J Dairy Sci 72(3):767–783

Café LM, Hearnshaw H, Hennessy DW, Greenwood PL (2006) Growth and carcass characteristics at heavy market weights of Wagyu-sired steers following slow or rapid growth to weaning. Aust J Exp Agric 46(6–7):951–955

Calberry JM, Plaizier JC, Einarson MS, Mcbride BW (2003) Effects of replacing chopped alfalfa hay with alfalfa silage in a total mixed ration on production and rumen conditions of lactating dairy cows. J Dairy Sci 86:3611–3619

Carter RR, Grovum WL (1990) A review of the physiological significance of hypertonic body fluids on feed intake and ruminal function: salivation, motility and microbes. J Anim Sci 68 (9):2811–2832

Cassida KA, Stokes MR (1986) Eating and resting salivation in early lactation dairy cows. J Dairy Sci 69(5):1282–1291

Castells L, Bach A, Araujo G, Montoro C, Terré M (2012) Effect of different forage sources on performance and feeding behavior of Holstein calves. J Dairy Sci 95(1):286–293

Clark PW, Armentano LE (2002) Influence of particle size on the effectiveness of the fiber in alfalfa silage. J Dairy Sci 85(11):3000–3007

Dann HM, Morin DE, Bollero GA, Murphy MR, Drackley JK (2005) Prepartum intake, postpartum induction of ketosis, and periparturient disorders affect the metabolic status of dairy cows. J Dairy Sci 88(9):3249–3264

Dann HM, Litherland NB, Underwood JP, Bionaz M, D'Angelo A, McFadden JW, Drackley JK (2006) Diets during far-off and close-up dry periods affect periparturient metabolism and lactation in multiparous cows. J Dairy Sci 89(9):3563–3577

Davis CL, Drackley JK (1998) The development, nutrition, and management of the young calf. Iowa State University Press, Ames, IA

De Paula Vieira A, von Keyserlingk MAG, Weary DM (2010) Effects of pair versus single housing on performance and behavior of dairy calves before and after weaning from milk. J Dairy Sci 93(7):3079–3085

Douglas GN, Overton TR, Bateman HG, Dann HM, Drackley JK (2006) Prepartal plane of nutrition, regardless of dietary energy source, affects periparturient metabolism and dry matter intake in Holstein cows. J Dairy Sci 89(6):2141–2157

Einarson MS, Plaizier JC, Wittenberg KM (2004) Effects of barley silage chop length on productivity and rumen conditions of lactating dairy cows fed total mixed rations. J Dairy Sci 87(9):2987–2996

Fall CHD (2011) Evidence for the intra-uterine programming of adiposity in later life. Ann Hum Biol 38(4):410–428

Garrett EF, Nordlund KV, Goodger WJ, Oetzel GR (1997) A cross-sectional field study investigating the effect of periparturient dietary management on ruminal pH in early lactation dairy cows. J Dairy Sci 80(Suppl 1):169

Green LE, Hedges VJ, Schukken YH, Blowey RW, Packington AJ (2002) The impact of clinical lameness on the milk yield of dairy cows. J Dairy Sci 85(9):2250–2256

Greenwood PL, Café LM, Hearnshaw H, Hennessy DW, Thompson JM, Morris SG (2006) Long-term consequences of birth weight and growth to weaning for carcass, yield and beef quality characteristics of Piedmontese- and Wagyu-sired cattle. Aust J Exp Agric 46(2):257–269

Hill TM, Bateman HG II, Aldrich JM, Schlotterbeck RL (2009) Optimizing nutrient ratios in milk replacers for calves less than five weeks of age. J Dairy Sci 92(7):3281–3291

Hoffman PC, Funk DA (1992) Applied dynamics of dairy replacement growth and management. J Dairy Sci 75(9):2504–2516

Hoffman PC, Simson CR, Wattiaux M (2007) Limit feeding of gravid Holstein heifers: effect on growth, manure nutrient excretion, and subsequent early lactation performance. J Dairy Sci 90 (2):946–954

Jensen MB, Duve LR, Weary DM (2015) Pair housing and enhanced milk allowance increase play behavior and improve performance in dairy calves. J Dairy Sci 98(4):2568–2575

Kononoff PJ, Heinrichs AJ (2003) The effect of corn silage particle size and cottonseed hulls on cows in early lactation. J Dairy Sci 86(7):2438–2451

Kononoff PJ, Heinrichs AJ, Lehman HA (2003) The effect of corn silage particle size on eating behavior, chewing activities, and rumen fermentation in lactating dairy cows. J Dairy Sci 86 (10):3343–3353

Krause KM, Combs DK, Beauchemin KA (2002) Effects of forage particle size and grain fermentability in midlactation cows. II. Ruminal pH and chewing activity. J Dairy Sci 85 (8):1947–1957

Krpálková L, Cabrera VE, Vacek M, Štípková M, Stádník L, Crump P (2014) Effect of prepubertal and postpubertal growth and age at first calving on production and reproduction traits during the first 3 lactations in Holstein dairy cattle. J Dairy Sci 97(5):3017–3027

Lammers BP, Buckmaster DR, Heinrichs AJ (1996) A simple method for the analysis of particle sizes of forage and total mixed ration. J Dairy Sci 79(5):922–928

Leonardi C, Armentano LE (2007) Short communication: feed selection by dairy cows fed individually in a tie-stall or as a group in a free-stall barn. J Dairy Sci 90(5):2386–2389

Loor JJ, Dann HM, Guretzky NAJ, Everts RE, Oliveira R, Green CA, Litherland NB, Rodriguez-Zas SL, Lewin HA, Drackley JK (2006) Plane of nutrition prepartum alters hepatic gene expression and function in dairy cows as assessed by longitudinal transcript and metabolic profiling. Physiol Genomics 27(1):29–41

McCance RA (1962) Food, growth, and time. Lancet 280(7258):671–676

Mee JF (2004) Temporal trends in reproductive performance in Irish dairy herds and associated risk factors. Ir Vet J 57(3):158–166

Mertens DR (1997) Creating a system for meeting the fiber requirements of dairy cows. J Dairy Sci 80(7):1463–1481

Moallem U, Werner D, Lehrer H, Zachut M, Livshitz L, Yakoby S, Shamay A (2010) Long-term effects of ad libitum whole milk prior to weaning and prepubertal protein supplementation on skeletal growth rate and first-lactation milk production. J Dairy Sci 93(6):2639–2650

Nocek JE (1997) Bovine acidosis: implications on laminitis. J Dairy Sci 80(5):1005–1028

NRC (2001) Nutrient requirements of dairy cattle, 7th edn. National Academy of Sciences, Washington, DC

O'Grady L, Doherty ML, Mulligan FJ (2008) Subacute ruminal acidosis (SARA) in grazing Irish dairy cows. Vet J 176(1):44–49

Oetzel GR, Nordlund KV, Garrett EF (1999) Effect of ruminal pH and stage of lactation on ruminal lactate concentration in dairy cows. J Dairy Sci 82(Suppl 1):35

Owens FN, Gill DR, Secrist DR, Coleman SW (1995) Review of some aspects of growth and development of feedlot cattle. J Anim Sci 73(10):3152–3172

Phillips CJC (2010) Principles of Cattle Production, 2nd edn. CABI, Wallingford

Rabelo E, Rezende RL, Bertics SJ, Grummer RR (2003) Effects of pre- and postfresh transition diets varying in dietary energy density on metabolic status of periparturient dairy cows. J Dairy Sci 88(12):4375–4383

Robelin J, Chilliard Y (1989) Short-term and long-term effects of early nutritional deprivation on adipose tissue growth and metabolism in calves. J Dairy Sci 72(2):505–513

Rustomo B, Al-Zahal O, Odongo NE, Duffield TF, Mcbride BW (2006) Effects of rumen acid load from feed and forage particle size on ruminal pH and dry matter intake in the lactating dairy cow. J Dairy Sci 89(12):4758–4768

Soberon F, Raffrenato E, Everett RW, Van Amburgh ME (2012) Preweaning milk replacer intake and effects on long-term productivity of dairy calves. J Dairy Sci 95(2):783–793

Soita HW, Christensen DA, McKinnon JJ (2000) Influence of particle size on the effectiveness of the fiber in barley silage. J Dairy Sci 83(1):2295–2300

Terré M, Devant M, Bach A (2007) Effect of level of milk replacer fed to Holstein calves on performance during the preweaning period and starter digestibility at weaning. Livest Sci 110 (1–2):82–88

Warnick LD, Janssen D, Guard CL, Gröhn YT (2001) The effect of lameness on milk production in dairy cows. J Dairy Sci 84(9):1988–1997

Williams YJ, Doyle PT, Egan AR, Stockdale CR (2005) Increasing the intake of highly digestible Persian clover herbage reduces rumen fluid pH and the rate of degradation of neutral detergent fibre in grazing dairy cows. Aust J Exp Agric 45(12):1529–1537

Wu G, Bazer FW, Wallace M, Spencer TE (2006) Intrauterine growth retardation: implications for the animal sciences. Board-invited review. J Anim Sci 84(9):2316–2337

Yang WZ, Beauchemin KA, Rode LM (2001) Effects of grain processing, forage to concentrate ratio, and forage particle size on rumen pH and digestion by dairy cows. J Dairy Sci 84 (10):2203–2216

Zanton GI, Heinrichs AJ (2007) The effects of controlled feeding of a high-forage or high-concentrate ration on heifer growth and first-lactation milk production. J Dairy Sci 90 (7):3388–3396

Zebeli Q, Tafaj M, Steingass H, Metzler B, Drochner W (2006) Effects of physically effective fiber on digestive processes and milk fat content in early lactating dairy cows fed total mixed rations. J Dairy Sci 89:651–668

Ziegler D, Ziegler B, Raeth-Knight M, Larson R, Golombeski G, Linn J, Chester-Jones H (2008) Performance of post-weaned Holstein heifer calves transitioned to group housing using different management strategies while fed a common diet. J Dairy Sci 91:465 (Abstr. TH191)

Chapter 9
Water and the Welfare of Farm Animals

Marina A.G. von Keyserlingk, Clive J.C. Phillips, and Birte L. Nielsen

Abstract Provision of adequate water supplies is essential for the welfare of farm animals. Water forms the largest component of an animal's body and is an essential nutrient required for all biological functions, including temperature regulation, digestion, foetal development, and production. This essential nutrient can only be restricted for short periods of time. Water deprivation results in substantial welfare concerns, as it can hinder biological functioning, and has been associated with morbidity and, in cases of extreme deprivation, mortality. It is likely associated with a highly negative emotional state in farm animals (e.g. in humans referred to as thirst). Both quality and quantity of water may limit water intake, and management factors such as high stocking density at the water source can also negatively affect water intake. Providing examples from the primary production species, cattle, pigs, and poultry, we describe the importance of water quality and quantity in food animal production.

9.1 Introduction

The importance of water in the maintenance of life becomes clear when one compares it to that of solid food. It has long been known that animals deprived of water live for a much shorter period of time than their counterparts deprived of food (French 1956). Fitzsimmons (1979a, b) recognized that thirst, the subjective sensation resulting from a lack of water, is only surpassed for its impact on welfare by

M.A.G. von Keyserlingk (✉)
Animal Welfare Program, Faculty of Land and Food Systems, University of British Columbia, Vancouver, BC V6T 1Z6, Canada
e-mail: nina@mail.ubc.ca

C.J.C. Phillips
School of Veterinary Science, The University of Queensland, Building # 8114, Gatton, QLD 4343, Australia

B.L. Nielsen
INRA, Neurobiologie de l'Olfaction, Université Paris-Saclay, Jouy-en-Josas 78350, France

C.J.C. Phillips (ed.), *Nutrition and the Welfare of Farm Animals*, Animal Welfare 16, DOI 10.1007/978-3-319-27356-3_9

severe pain and gasping for air in the hierarchy of physiological drives, highlighting the significance of water for homeostatic functioning and animal welfare.

Water forms the largest component of an animal's body and is an essential nutrient required for all biological functions, including temperature regulation, digestion, foetal development, and milk production. Thus, if water intake declines due to restricted access or inferior quality, it has a negative impact on feed consumption, production, health, and the welfare of the animal. In this chapter, we provide illustrative examples involving what is arguably the most frequently ignored nutrient. We address the importance of both water quality and quantity in food animal production systems involving the primary food production species of cattle, pigs, and poultry.

9.2 Cattle

It has been known for many decades that when cattle have restricted access to water, their feed intake is lowered (Balch et al. 1953), demonstrating that consumption of adequate feed and water is essential for maintaining healthy rumen function. Moreover, animals fed diets high in salt may suffer from intoxication unless they have sufficient water available to flush the excess salt through their digestive system (Riggs et al. 1953). Water provides the means to maintain osmolality with differing intakes of osmotically active compounds, such as sodium and protein. Increases in the intake of these nutrients require increased water intake, in part because they are excreted in urine. For example, when the sodium (Na) content in the diet of beef cattle was increased from 4.0 up to 7.9 g/kg DM, an extra 0.18 L was drunk per 1 g additional Na intake (Chiy et al. 1993).

In cases of extreme dehydration, the concentration of water in body tissues is decreased, leading to haemoconcentration and less efficient circulation and oxygen transport (Sykes 1955). Reduced skin turgidity and sunken eyes can provide a useful indicator of water inadequacy in such circumstances. Brain lesions (Padovan 1980) and hypernatraemia, the clinical manifestation of high serum Na concentrations, have also been noted in cases of severe dehydration (Knowles 1956). Other effects of reduced water consumption include depression of thermoregulation, which is particularly pertinent for cattle grazing in hot climates (Ittner et al. 1951).

Adult and growing cattle require a large volume of water to maintain homeostasis; growing beef cattle, for example, drink ~27–66 L per day (Winchester and Morris 1956; NRC 2001). Dairy cows producing over 30 kg of milk per day require a significant supply of freshwater as they can consume from 80 to 100 kg of water or more per day when healthy (although this declines when they are ill) (Huzzey et al. 2007; Fig. 9.1). In the case of lactating dairy cattle, if the water supply is interrupted for more than 12 h an alternative watering method should be used. Current recommendations on the depth of water required to facilitate normal drinking behaviour are scarce, but the *Code of Practice for the Care and Handling of Dairy Cattle* in Canada recommends a depth of at least 10 cm in water troughs

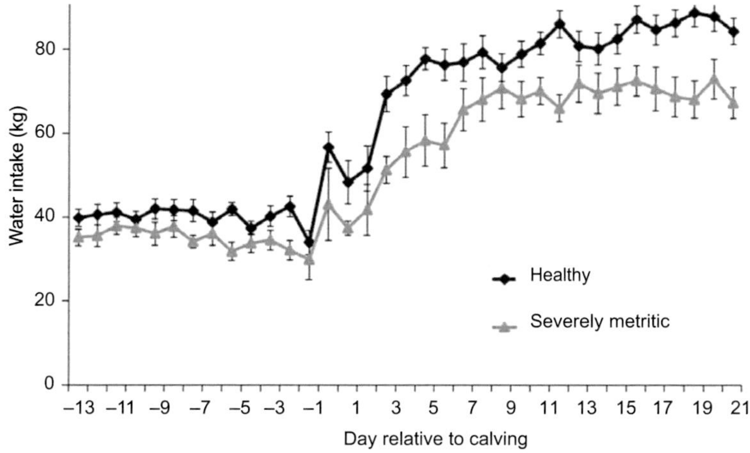

Fig. 9.1 Water intake of Holstein dairy cows that remain healthy and cows diagnosed with clinical metritis after calving (Healthy $n = 23$, Severely metritic $n = 12$) from day −14 to day +21 after calving. *Source*: adapted from Huzzey et al. (2007)

that are mounted at a height of ~60–75 cm (NFACC 2009). Young milk-fed calves must also be provided with access to fresh, clean water at all times (NFACC 2009), a requirement also forming part of European legislation (Council Directive 2008/119/EC). However, in many developing countries the access to clean water for cattle is limited.

Beef cattle may have to walk long distances to get water on rangelands, which may restrict their utilization of pasture around the watering point. In recent years, the availability of inexpensive polypipe has enabled farmers to extend water supplies to distant paddocks, thereby improving grassland utilization. Water supplies should be checked regularly to ensure that they have not become blocked or dirtied. Good access to the water source is important, preferably via a hard pad or concreted area which prevents mud build-up due to water spillage, but in rangeland situations water is usually provided from natural sources or dams, which may become bogged around the edges. This will lead to a dirty water supply and potential hazard to cattle if they cannot escape quickly from possible predators. The salinity of such water is crucial for the welfare of the cattle, although they will tolerate high levels for short periods. If the saline content is high, they may refrain from drinking and then rapidly consume large amounts with adverse effects on their health. Sulphates are particularly undesirable in groundwater, with detrimental effects on cattle (Grout et al. 2006).

Water is obtained not only from free water but also from water in the feed. Free water consumption and total water intake are positively correlated with milk production (Little and Shaw 1978; Huzzey et al. 2007), thus insufficient consumption will result in lowered milk production (NRC 2001). Water requirements can be separated into that needed for maintenance, approximately 0.09 L/kg body weight, and a requirement for milk production, 2–2.5 L per litre of milk output (Phillips

2010, p. 152). This includes some residual uptake, and an allowance of 10 % less than this will not affect production, but may still result in reduced welfare, due to thirst, competition and unequal allocation between cows. Cows voluntarily drink 2–5 times per day, thus providing water just whilst cows are waiting to be milked twice a day is not sufficient to ensure adequate water intake for all cows, as those milked soon after arrival have little opportunity to drink. As with feeding (DeVries et al. 2003), lactating cows appear to synchronize drinking with milking, probably because of the osmotic demand induced by milk removal. In cold conditions, buildings should be checked for freezing of pipes and any limitation to water supply of dairy cows. The recent trend towards large buildings with tall roofs reduces the heat output of cattle relative to airspace, increasing the likelihood of pipes freezing in cold weather. Farmers should consult with building engineers to determine whether water pipes require additional insulation to prevent freezing in cold weather that may result in interruptions of water supply to the animals.

Water requirements can vary considerably, depending on environmental factors and illness. Ambient temperature (Rouda et al. 1994) and relative humidity (Ali et al. 1994), in particular, influence the requirements for, and hence consumption of, water. In cool conditions (2.4–13.3 °C), there appears to be little effect of temperature on water consumption in cattle (Castle and Thomas 1975). Cattle experience heat stress above 25 °C (Hicks et al. 1990) and indeed, under ‘hot’ conditions (i.e. greater than 30 °C) water consumption increases rapidly as temperature rises (Rouda et al. 1994; Molina and Tuero 2000). This occurs in geographical regions that experience long periods of extremely hot weather, and failure to prevent heat stress can have detrimental impacts on animal welfare. There is a range of temperatures in which cattle can maintain homeostatic conditions without increasing their water consumption, and at temperatures above this range the animals must compensate through higher water consumption.

Beyond ambient temperature and relative humidity, climatic factors such as rainfall and wind affect water consumption. Stockdale and King (1983) found that daily water consumption was positively correlated with the amount of sunshine and evaporation. Conversely, rainfall and heavy dew reduce the DM content of herbage, sometimes to 110–130 g/kg dry matter (DM) in the case of young, leafy herbage. Feeding time, intakes, and production are reduced if DM content falls below approximately 180 g DM/kg (Butris and Phillips 1987). This is because the feed is less digestible, and it seems likely that this arises because the mastication process is less efficient (Phillips et al. 1991). In such circumstances, the voluntary intake of water from troughs will decline to zero and provision of a dry supplementary feed may be beneficial. Conversely, if cattle are only fed high-DM forage, in particular hay, they acquire additional capacity for voluntary water intake.

Day length may also influence water consumption, particularly for extensively housed cattle, as they rarely drink between sunset and sunrise (Sneva 1970). Heat-stressed cattle, however, are more likely to drink and perform other activities, at night than during the daytime. Even when heat stressed, they prefer water at 15–27 °C to cooler water (Albright and Arave 1997, p. 25).

Trough design features, such as physical dimensions (Pinheiro Machado Filho et al. 2004) and refill rate (Andersson et al. 1984), can affect water consumption rate, which can be up to 25 L/min. It has been suggested that 10 % of a herd must be able to drink concurrently (ARC 1980); therefore, trough availability must be able to accommodate several animals at once and refill at a rate sufficient to supply a continuous flow of water to the drinking animals. Competition and social interactions are known to affect feed intake (Proudfoot et al. 2009) and milk intake by calves (O'Driscoll et al. 2006) and are thus likely also to affect water intake of group-housed individuals when limited space is available at water troughs.

Water quality is a key factor affecting water consumption and is an issue that affects both the beef and dairy industries. Contamination of water with excreta, regardless of whether it is in a trough or surface groundwater, is highly undesirable. When providing water in a trough and competition for access is high, it may be necessary to fix a bar around the trough to stop cows getting too close, thus precluding their defecating in the trough. The provision of the bar may also aid in preventing wildlife, such as badgers, from drinking from the troughs and spreading disease (tuberculosis). Water drinking bowls with nose-operated levers do not allow cattle to drink as fast as from troughs and more need to be provided, and one per stall is needed in permanently stalled cattle.

Surface groundwater (ponds or streams) also serves as a drinking source, but is susceptible to contamination from excreta and mud (see Fig. 9.2). Willms et al. (2002) reported that when cattle were provided a choice of freshwater or water contaminated with 0.005 % fresh manure by weight they avoided the contaminated water. Moreover, growing yearling heifers provided with clean drinking water gained more weight than heifers provided with water from a pond (Willms et al. 2002). An additional concern with surface groundwater as the primary source of water for cattle is the potential impact that grazing cattle can have on riparian vegetation, fish habitats, and stream morphology, and, thus, every effort should be made to minimize potential damage when using natural water sites (Veira 2007).

If water contains compounds that diminish palatability, cattle will reduce their water consumption (Grout et al. 2006) or seek alternative water sources (Digesti and Weeth 1976). Biological pollutants, including algae, manure, and urine, lower the quality of drinking water, as do chemical pollutants such as minerals and the products of industrial development (Veenhuizen and Shurson 1992; Carson 2000). Water quality can also be impaired by the physical properties of the water, including temperature, colour, and turbidity (McKee and Wolf 1963; Price 1985). Temperature, for example, has the potential to alter the sensation of substances dissolved in water (Goatcher and Church 1970; Zoeteman 1980) and has been shown to affect water consumption by cattle (Ittner et al. 1951; Lofgreen et al. 1975). More recent work by Grout et al. (2006) reported that beef cattle reduce their consumption of water containing high concentrations of $MgSO_4$, even after being given time to adjust; such reductions may be accompanied by an increase in faecal dry matter.

picture courtesy of the UBC Animal Welfare program.

Fig. 9.2 Holstein heifers drinking from a groundwater site; an attempt is being made by one heifer on the right to gain access to the water. Note the algal growth in the water, which may be a result of excreta contamination and the mud around the perimeter of the watering site

9.3 Poultry (Broiler Chickens and Laying Hens)

It has long been known that the drinking behaviour of chickens is innate (Elliott 1896), with a distinctive scooping of water into the beak followed by raising of the head to allow swallowing. Drinking is influenced by the behaviour of other birds with evidence of synchronized drinking occurring (Hoppitt and Laland 2008). In layers, more drinking is seen immediately after egg laying (Shimada et al. 1987).

The water requirements of poultry are approximately twice the weight of the dry food consumed (Rose 1997). For fast-growing birds, such as the modern broiler chicken, this means an increase in daily water consumption of about 50 mL every week. At temperatures over 20 °C, this requirement of broilers increases by 6 % for every Celsius degree (Schlenker et al. 2010), and heat stress in layers has also been found to increase water intake (Xin et al. 2002; Dai et al. 2009). Under normal ambient temperatures, provision of very hot or very cold water reduces the amount drunk (Schlenker et al. 2010), but under extreme climate conditions, heated or chilled drinking water may provide a means for the birds to thermoregulate (GuoDong et al. 1999; Gutierrez et al. 2009), although this may not be economically viable and can give rise to wet litter.

The two most common types of drinkers for use in indoor poultry production systems are nipple drinkers and bell drinkers. However, it should be noted that the natural scooping motion for drinking by birds is only achieved when water is provided in the bell drinkers. Nipple drinkers do not take advantage of the birds' natural drinking behaviour as they require the birds to learn to peck at or push the nipple to allow water to be dispensed directly into the raised beak (SCAHAW

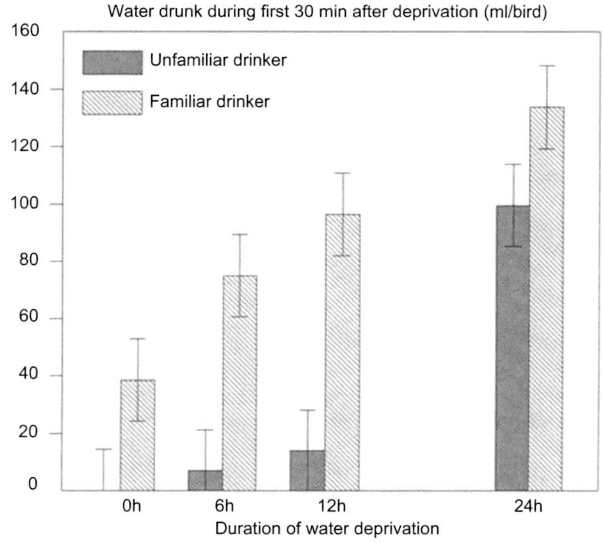

Fig. 9.3 Amount of water (mL/bird ± s.e.) drunk from a drinking trough either familiar or unfamiliar to the broiler chickens following different periods of water deprivation. *Source*: adapted from Sprenger et al. (2009)

2000). Bruno et al. (2011) found that bell drinkers were visited less often, but with higher water intakes per visit than for nipple drinkers. Houldcroft et al. (2008) also found a preference for bell drinkers over nipple drinkers, but showed that the height of the drinker was equally important as the design: broiler chickens preferred lower hanging nipples to higher ones, and when offered a choice between bowls and nipples of the same height they showed no preference.

Familiarity with a drinker influences the water intake of broilers. Sprenger et al. (2009) showed that water intake increases with duration of water deprivation and that prior experience with the drinker significantly influenced the amount of water drunk (Fig. 9.3). Clinical symptoms of water deprivation include restlessness, reduced feed intake, dry faeces, and discoloration of the comb. They occur after 48 h in newly hatched chicks and after 24 h in broilers and layers (Losing 1980).

There is often a fine balance between water accessibility and problems associated with wetness of the litter (Collett 2012). Rosete and Sarda (2006) found that nipple drinkers halved the humidity of the litter (from 47 to 24 %) compared with bell drinkers in broiler chickens, but following a heat wave on day 42, four times as many broilers with access to nipple drinkers died compared with those provided with bell drinkers.

The water supply of poultry is also used as a means to supply vaccines to large flocks of birds (Singh et al. 2010), as well as probiotics to improve production parameters (Han et al. 2010; Liu et al. 2012). Supplementation of probiotics in the water supply has been found to give better production results on farms where productivity rates are low (Timmerman et al. 2006). In some countries, pigmentation is added to the water to improve yolk colour (Perez-Vendrell and Hernandez 2006).

The source, and hence quality, of the available water for poultry production can vary from tap water to sewage effluents (Shoremi et al. 1990). Disinfection of drinking water can be achieved by UV radiation or by adding suitable substrates in appropriate quantities to the water source (Ono et al. 2007). It is important, however, to differentiate between the quality of the water entering the system and that of the water taken up by the animals. Zaman et al. (2012) found higher levels of *E. coli* (+57 %) and *Salmonella typhi* (+111 %) in the drinkers compared with the water tanks on commercial broiler farms. The presence of biofilms in drinker systems is a problem, especially in warmer climates. Biofilms consist of various microorganisms in a polysaccharide–protein layer, which is difficult to shift using common antimicrobial agents (Ahmad et al. 2008; Schlenker et al. 2010). When using disinfecting compounds, usually containing chloride, it is important not to adversely affect water uptake of the birds by using inappropriately high concentrations (Damron and Flunker 1993), even if this may reduce litter moisture.

The concentration of minerals and trace elements in water should be taken into account when calculating the amount to provide in the diet (Abbas et al. 2009). Tests of drinking water for poultry have also been suggested as a means to monitor avian flu virus, which has been found to be more detectable in drinking water samples than in faecal samples (Leung et al. 2007).

9.4 Swine

Drinking water for pigs is usually provided either from nipple drinkers or from water bowls, the latter enabling the motions synonymous with the natural drinking behaviour of swine, which is to obtain water from a surface with minimal movements of the jaw and extensive movement of the tongue, combining suckling and lapping (Thexton et al. 1998). Since 2003, the European Union legislation has specified that "*All pigs over two weeks of age must have permanent access to a sufficient quantity of fresh water*" (Commission Directive 2001/93/EC). Despite the simple wording, this has given rise to deliberations about what is meant by '*permanent access*' and even '*fresh*'. Is a pig whilst drinking from a drinking nipple preventing permanent access to water for other pigs in the group? And is water kept in a trough fresh, or does it have to be freshly drawn from the source? In reality, permanent access to water is determined by the *possibility* of the animal to access a source of water available at all times, i.e. 24 h per day. This permanent presence of water should, by default, also guarantee accessibility to '*sufficient quantities*' but unfortunately does not. Rather the wording, which complies with the current regulations, simply states that slowly dripping taps must not be present. Likewise, freshwater means drinkable (or potable) water, i.e. water that does not taste bad or make the animal sick.

In modern pig production, the first encounter with water to drink usually occurs shortly after birth when they are still nursed by the sow, and in Europe this is legislated in that access must be available from 2 weeks of age. The presence of drinkers in the farrowing pen can habituate the piglets to drinking water, but surprisingly little is known about water intake of piglets pre-weaning. It is assumed to correlate to intake of creep feed, which varies greatly between litters and among piglets within a litter (Pajor et al. 1991). At weaning, many piglets experience a growth check, which is associated with diarrhoea and the stress of environmental, nutritional, and social change (e.g. Bunger et al. 2000). As well as affecting intake of solid feed, drinking behaviour develops over the first days following weaning, with piglets having a large number of short drinking bouts. On average, according to Dybkjær et al. (2006), piglets have more than 60 drinking bouts of $\leq$10 s per piglet for each of the first 2 days following weaning, but with large inter-individual variation. Interestingly, these authors also found that the piglets which grew faster before weaning spent less time eating and more time drinking in the days following weaning than their smaller littermates.

A healthy growing pig kept in a thermoneutral environment with free access to water and a standard dry pelleted feed will usually drink between 2.1 and 2.7 L water per kg feed consumed (Li et al. 2005; Shaw et al. 2006), resulting in a daily water intake of about 3.4 L (Vermeer et al. 2009). This may double in situations with severe thermal stress (Huynh et al. 2005), with water intake increasing by approximately 0.1 L/day for every 1 °C increase in air temperature (Schiavon and Emmans 2000). Growing pigs spend about half an hour per day drinking water (Xin and deShazer 1991; Gonyou 1996), and the largest proportion (85 %) of water intake in pigs occurs in connection with feeding (Bigelow and Houpt 1988). However, water intake depends on the type of feeding. In liquid-fed pigs, where the daily feed ration is served in a water suspension, the need for additional water is obviously lower than for dry-fed pigs. Liquid feeding also imposes a certain water intake during feeding, and voluntary water intake decreases as the dilution ratio of the liquid feed goes up. However, Gill et al. (1987) found that additional water was still consumed, even when dilution ratios were above the normal water to feed intake of growing pigs (i.e. $\geq$3:1).

In many studies of water provision and consumption in pigs, the measurements of the latter (also called water use) are most often carried out by use of a water meter; this registers the quantity of water drawn from the water source and thus includes both water consumed and water not ingested by the pigs. Such water losses are normally a consequence of spillage or wastage and were thought to be a result of pigs playing with the water nipple. However, in a recent study using a standard drinking nipple for pigs, Andersen and Herskin (2012) found that none of the pigs were able to drink without spilling, and all pigs used 2–4 different ways of drinking from the nipple. Of the 5 L daily water use per pig, on average 35 % of the water was wasted, and all pigs wasted more than 15 % of the water (Andersen et al. 2014). This inadvertent wastage of water whilst drinking likely reflects the need for the pigs to alter their natural drinking behaviour, which is to drink from a water surface, in order to obtain water. Clearly, work is needed to identify solutions that reduce the extent of water wastage when pigs are forced to drink from nipples; our preference is that future designs take the natural drinking behaviour of pigs into consideration. Some recommendations regarding nipple design and placement have already been

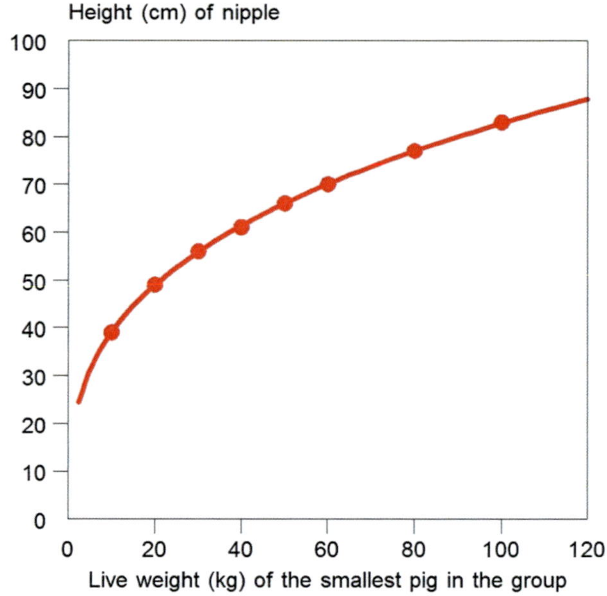

Fig. 9.4 Optimal nipple height (H) from the floor (in cm) in relation to the live weight (LW) of the smallest pig in a group. Curve equation is $H = 18.4 \times \mathrm{LW}^{0.33}$. *Source*: adapted from Gonyou (1996)

highlighted that should reduce, but not eliminate, water wastage when pigs are drinking. For instance, the height of the nipple should be adjustable and gradually raised as the pigs grow to prevent water spillage (Gonyou 1996; Fig. 9.4). Water flow rate also affects spillage, with more water being drunk but also a higher percentage wasted as flow rates increase (Li et al. 2005). As pigs cannot sweat, they may also use water to wet their skin in order to cool down in hot weather. Outdoors, pigs may use natural wetland or use water from the drinking trough to create a pond or a mud bath in which to roll (Bracke 2011; Bracke and Spoolder 2011). In a similar manner, water from drinkers may be used for thermoregulatory purposes in hot indoor conditions. It is therefore more appropriate to refer to water waste or spillage only when the water is not contributing to the health, welfare, and productivity of the pigs, and including only water leaking from the drinker (caused by malfunctioning or poor drinker design leading to spillage during drinking) and water being drawn by the pigs without a determinable purpose, such as drinking or thermoregulation.

In contrast to water wastage, water deprivation can occur in modern pig production, as when newly weaned pigs have not yet learned how to obtain water from the drinkers (McLeese et al. 1992; Torrey et al. 2008). Unless the deficit is extremely high, symptoms of water deprivation are difficult to identify in pigs because the methods conventionally used for other nutrients are not adequate in the case of water (Patience 2012). If insufficient access to water arises, pigs are able to conserve water by producing hypertonic urine, although there is an upper limit to the ability of the kidneys to concentrate urine (Andreoli 2000). Rainwater can also be an important water source for sows kept outdoors. In the days following rainfall above 5 mm, the water consumption from drinkers by sows is lowered by 5–7 L (Andersen and Pedersen 2014). In the same study, sows were found to consume 70–75 % of their water intake during the day, but that night drinking still persisted even during periods of frost.

Sows may not drink sufficient water during lactation (Fraser and Phillips 1989) but can have an excessive water intake during gestation, where feed restriction can lead to hunger-induced polydipsia (Yang et al. 1981). As in the case of piglets, excessive water usage may also arise without water ingestion, being caused by drinker manipulation performed in response to a state of frustration of either nutritional (food restriction) or environmental (insufficient rooting material) origin. The latter may also occur in growing and finishing pigs, and for sows the manipulation can take a stereotyped form (Rushen 1984).

9.5 Conclusions

Water forms the largest component of an animal's body and is an essential nutrient required for all biological functions, including temperature regulation, digestion, faecal development, and production. If water intake declines due to restricted access (for example, as a result of excessive competition) or inferior quality (for example, as a result of high salt content or manure contamination), feed consumption, productivity, and the welfare of the animal will be negatively impacted.

Acknowledgements The authors thank Amanda Grout for allowing some of her ideas from her MSc thesis on the importance of water for cattle (and their welfare) to be included in this chapter. Kathrin Schirmann is also thanked for her contributions to the Animal Welfare Program photograph library, specifically the photograph used in Fig. 9.2.

References

Abbas TEE, El-Zubeir EA, Arabbi OH (2009) The effect of saline drinking water on broilers and laying hens performance. Worlds Poult Sci J 65(3):511–516

Ahmad S, Ashraf M, Siddique M, Mehmood F, Arshad M, Khan HA (2008) Biofilm formation and drinking water quality in relation to *Escherichia coli* at commercial poultry farms. J Agric Soc Sci 4:77–80

Albright JL, Arave CW (1997) The behaviour of cattle. CABI, Wallingford

Ali S, Gooonewardene LA, Basarab JA (1994) Estimating water consumption and factors affecting intake in grazing cattle. Can J Anim Sci 74(3):551–554

Andersen HM-L, Herskin MS (2012) Why do pigs waste water? Description of drinking behaviour of growing pigs provided with water nipples. In: Proceedings of the 46th Congress of the International Society for Applied Ethology (ISAE), 31 July–4 August 2012. Vienna, Austria, p 214

Andersson M, Schaar J, Wiktorsson J (1984) Effects of drinking water flow rates and social rank on performance and drinking behaviour in tied-up dairy cows. Livest Prod Sci 11(6):559–610

Andersen HM-L, Pedersen LJ (2014) Drinking behaviour in sows kept outdoors during the winter months. Appl Anim Behav Sci 161:34–41

Andersen HM-L, Dybkjær L, Herskin MS (2014) Growing pigs' drinking behaviour: number of visits, duration, water intake and diurnal variation. Animal 8(11):1881–1888

Andreoli TE (2000) Water: normal balance, hyponatremia, and hypernatremia. Ren Fail 22:711–735

ARC [Agricultural Research Council] (1980) The nutrient requirements of ruminant livestock. Commonwealth Agricultural Bureaux, Slough

Balch CC, Balch DA, Johnson VW, Turner J (1953) Factors affecting the utilization of food by dairy cows. 7. The effect of limited water intake on the digestibility and rate of passage of hay. Br J Nutr 7(3):212–224

Bigelow JA, Houpt TR (1988) Feeding and drinking patterns in young pigs. Physiol Behav 43 (1):99–109

Bracke MBM (2011) Review of wallowing in pigs: description of the behavior and its motivational basis. Appl Anim Behav Sci 132:1–13

Bracke MBM, Spoolder HAM (2011) Review of wallowing in pigs: implications for animal welfare. Anim Welf 20:347–363

Bruno BLDG, Maiorka A, Macari M, Furlan RL, Givisiez PEN (2011) Water intake behavior of broiler chickens exposed to heat stress and drinking from bell or nipple drinkers. Braz J Poult Sci 13(2):147–152

Bunger B, Hillmann E, von Hollen F (2000) Effects of housing conditions during farrowing and nursing of sows on growth and behaviour of piglets before and after weaning. Archiv für Tierzucht 43:196–202

Butris G, Phillips CJC (1987) The effect of herbage surface water and the provision of supplementary forage on the intake and feeding behaviour of cattle. Grass Forage Sci 42(3):259–264

Carson TL (2000) Current knowledge of water quality and safety for livestock. Vet Clin N Am Food Anim Pract 16(3):455–464

Castle ME, Thomas TP (1975) The water intake of British Friesian cows on rations containing various forages. Anim Prod 20(2):181–189

Chiy PC, Phillips CJC, Omed HM (1993) Sodium fertilizer application to pasture. 3. Rumen dynamics. Grass Forage Sci 48(3):249–259

Collett SR (2012) Nutrition and wet litter problems in poultry. Anim Feed Sci Technol 173 (1–2):65–75

Commission Directive 2001/93/EC of 9 November 2001 amending Directive 91/630/EEC laying down minimum standards for the protection of pigs. Off J Eur Comm L 316:36–38, 1.12.2001

Council Directive 2008/119/EC of 18 December 2008 laying down minimum standards for the protection of calves. Off J Eur Union, L 10:7–13, 15. 1. 2009

Dai NV, Bessei W, Nasir Z (2009) The effect of sodium chloride supplementation in the drinking water on water and feed intake and egg quality of laying hens under cyclic heat stress. Arch Geflugelk 73(4):217–226

Damron BL, Flunker LK (1993) Broiler chick and laying hen tolerance to sodium-hypochlorite in drinking-water. Poult Sci 72(9):1650–1655

DeVries TJ, von Keyserlingk MAG, Beauchemin KA (2003) Diurnal feeding pattern of lactating dairy cows. J Dairy Sci 86(12):4079–4082

Digesti RD, Weeth HJ (1976) A defensible maximum for inorganic sulfate in drinking water of cattle. J Anim Sci 42(6):1498–1502

Dybkjær L, Jacobsen AP, Tøgersen FA, Poulsen HD (2006) Eating and drinking activity of newly weaned piglets: effects of individual characteristics, social mixing, and addition of extra zinc to the feed. J Anim Sci 84:702–711

Elliott HW (1896) Newly hatched chickens instinctively drink. Science 3(65):482

NFACC [National Farm Animal Care Council] (2009) Code of practice for the care and handling of dairy cattle. Dairy Farmers of Canada and the National Farm Animal Care Council, Ottawa

Fitzsimmons JT (1979a) Non-regulatory drinking and renal function. Behav Brain Sci 2 (1):105–106

Fitzsimmons JT (1979b) The physiology of thirst and sodium appetite. Monographs of the Physiological Society, no. 36. Cambridge University Press, Cambridge

Fraser D, Phillips PA (1989) Lethargy and low water intake by sows during early lactation: a cause of low piglet weight gains and survival? Appl Anim Behav Sci 24(1):13–22

French MH (1956) The importance of water in the management of cattle. East Afr Agric J 21:171–181

Gill BP, Brooks PH, Carpenter JL (1987) Voluntary water use by growing pigs offered liquid foods of differing water-to-meal ratios. In: Smith AT, Lawrence TLJ (eds) Pig housing and the environment. BSAP Occ Publ 11:131–133

Goatcher WD, Church DC (1970) Review of some nutritional aspects of sense of taste. J Anim Sci 31(5):973–981

Gonyou HW (1996) Water use and drinker management: a review. In: Annual report. Prairie Swine Centre, Saskatoon, pp 74–80

Grout AS, Veira DM, Weary DM, von Keyserlingk MAG, Fraser D (2006) Differential effects of sodium and magnesium sulphate on water consumption by beef cattle. J Anim Sci 84 (5):1252–1258

GuoDong D, YanKun H, LiGuo L (1999) The effect of drinking water temperature in winter on laying chickens. Poult Husb Dis Control 1999(11):41

Gutierrez WM, Min W, Chang HH (2009) Effects of chilled drinking water on performance of laying hens during constant high ambient temperature. Asian Aust J Anim Sci 22(5):694–699

Han SM, Lee KG, Yeo JH, Oh BY, Kim BS, Lee W, Baek HJ, Kim ST, Hwang SJ, Pak SC (2010) Effects of honeybee venom supplementation in drinking water on growth performance of broiler chickens. Poult Sci 89(11):2396–2400

Hicks RB, Owens FN, Gill DR, Oltjen JW, Lake RP (1990) Dry matter intake by feedlot beef steers: influence of initial weight, time on feed and season of year received in yard. J Anim Sci 68(1):254–265

Hoppitt W, Laland KN (2008) Social processes affecting feeding and drinking in the domestic fowl. Anim Behav 76(5):1529–1543

Houldcroft E, Smith C, Mrowicki R, Headland L, Grieveson S, Jones TA, Dawkins MS (2008) Welfare implications of nipple drinkers for broiler chickens. Anim Welf 17(1):1–10

Huynh TTT, Aarnink AJA, Verstegen MWA, Gerrits WJJ, Heetkamp MJW, Kemp B, Cahn TT (2005) Effects of increasing temperatures on physiological changes in pigs at different relative humidities. J Anim Sci 83(6):1385–1396

Huzzey JM, Veira DM, Weary DM, von Keyserlingk MAG (2007) Prepartum behavior and dry matter intake identify dairy cows at risk for metritis. J Dairy Sci 90(7):3220–3233

Ittner NR, Kelly CF, Hubert HR (1951) Water consumption of Hereford and Brahman cattle and the effect of cooled water in a hot climate. J Dairy Sci 84:225–232

Knowles HC Jr (1956) Hypertatrima. Metabolism 5:508–518

Leung YHC, Zhang L-J, Chow C-K, Tsang C-L, Ng C-F, Wong C-K, Guan Y, Peiris JSM (2007) Poultry drinking water used for avian influenza surveillance. Emerg Infect Dis 13 (9):1380–1382

Li YZ, Chénard L, Lemay SP, Gonyou HW (2005) Water intake and wastage at nipple drinkers by growing-finishing pigs. J Anim Sci 83:1413–1422

Little W, Shaw SR (1978) Individuality of intake of drinking water by dairy cows. Anim Prod 26 (April):329–339

Liu X, Yan H, Lv L, Xu Q, Yin C, Zhang K, Wang P, Hu J (2012) Growth performance and meat quality of broiler chickens supplemented with Bacillus licheniformis in drinking water. Asian Aust J Anim Sci 25(5):682–689

Lofgreen GP, Givens RL, Morrison SR, Bond TE (1975) Effect of drinking water temperature on beef cattle performance. J Anim Sci 40(2):223–229

Losing W (1980) Clinical symptoms and diagnosis of drinking water deficiency in chickens [Untersuchungen uber die klinischen Erscheinungen und zur Diagnose des Trinkwassermangels beim Huhn]. Tierarztliche Hochschule, Hannover, p 81

McKee JE, Wolf HW (eds) (1963) Water quality criteria, 2nd edn. Resources Agency of California, State Water Resources Control Board, Sacramento, CA

McLeese JM, Tremblay ML, Patience JF, Christison GI (1992) Water intake patterns in the weanling pig: effect of water quality, antibiotics and probiotics. Anim Prod 54(1):135–142
Molina A, Tuero O (2000) Effect of management method and access to the drinking trough on the water intake of grazing cattle. Cuban J Agric Sci 34(3):195–199
NRC (2001) Nutrient requirement of dairy cattle, 7th rev. edn. National Academies Press, Washington, DC
O'Driscoll K, von Keyserlingk MAG, Weary DM (2006) Effects of mixing on drinking and competitive behavior of dairy calves. J Dairy Sci 89(1):229–233
Ono T, Miyake M, Yashumoto R, Yamashita K, Fujii H, Nakayama E, Naito I, Sato K (2007) Disinfection of drinking water and evaluation of viability and egg production in the broiler hens using the weak acid hypochlorous solution. Jpn J Poult Sci 44:J148–J153
Padovan D (1980) Polioencephalomalacia associated with water intoxication in cattle. Cornell Vet 70:153–159
Pajor EA, Fraser D, Kramer DL (1991) Consumption of solid food by suckling pigs: individual variation and relation to weight gain. Appl Anim Behav Sci 32:139–155
Patience JF (2012) Water in swine nutrition. In: Chiba LI (ed) Sustainable swine nutrition. Wiley, Hoboken, NJ, pp 3–22
Perez-Vendrell AM, Hernandez JM (2006) Comparative effects of pigment inclusion in feed or drinking water on hen performance and egg yolk colour. Paper 339. In: Proceedings of the 12th European Poultry Conference, Verona, Italy, 10–14 September 2006, (p 5)
Phillips CJC (2010) Principles of cattle production, 2nd edn. CABI, Wallingford
Phillips CJC, Margerison JK, Azazi S, Chamberlain AG, Omed H (1991) The effect of adding surface water to herbage on its digestion by ruminants. Grass Forage Sci 46(3):333–338
Pinheiro Machado Filho LC, Teixeira DL, Weary DM, von Keyserlingk MAG, Hötzel MJ (2004) Designing better water troughs: dairy cows prefer and drink more from larger troughs. Appl Anim Behav Sci 89(3–4):185–193
Price M (1985) Introducing groundwater. George Allen & Unwin, London
Proudfoot KL, Veira DM, Weary DM, von Keyserlingk MAG (2009) Competition at the feed bunk during transition changes the feeding, standing and social behavior of Holstein dairy cows. J Dairy Sci 92(7):3116–3123
Riggs JK, Colby RW, Sells LV (1953) The effect of self-feeding salt cottonseed meal mixtures to beef cows. J Anim Sci 12:379–393
Rose SP (1997) Principles of poultry science. CABI, Wallingford
Rosete A, Sarda R (2006) Nipple drinkers versus bell drinkers for on-bed housed chickens [Bebederos de niples vs bebederos de campana para aves en piso]. Revista Cubana Ciencia Avicola 30(2):89–92
Rouda RR, Anderson DM, Wallace JD, Murray LW (1994) Free ranging cattle water consumption in south central New Mexico. Appl Anim Behav Sci 39(1):29–38
Rushen J (1984) Stereotyped behaviour, adjunctive drinking and the feeding periods of tethered sows. Anim Behav 32:1059–1067
SCAHAW [Scientific Committee on Animal Health and Animal Welfare] (2000) The welfare of chickens kept for meat production (broilers). European Commission, Scientific Committee on Animal Health and Welfare, Brussels, Belgium, SANCO.B.3/AH/R15/2000, p 150
Schiavon S, Emmans GC (2000) A model to predict water intake of a pig growing in a known environment on a known diet. Br J Nutr 84(6):873–883
Schlenker G, Bräunig I, Windhorst D (2010) Technological and health aspects of drinking water for broiler chickens. Lohmann Inf 45(1):48–56
Shaw MI, Beaulieu AD, Patience JF (2006) Effect of diet composition on water consumption in growing pigs. J Anim Sci 84(11):3123–3132
Shimada K, Tatematsu A, Sato K (1987) Effect of ovulation and oviposition on feeding and drinking activities in the laying hen. Jpn Poult Sci 24(5):288–294
Shoremi OIA, Sridhar MKC, Ekpenyong TE (1990) Effect of sewage and sewage effluents as drinking-water sources for point-of-lay chickens. Biol Wastes 32(1):1–7

Singh OP, Sharma A, Sharma N (2010) Comparative study of water stabilizer in drinking water vaccinations against Newcastle disease in chicken. Indian Vet J 87:650–651
Sneva FA (1970) Behaviour of yearling cattle on eastern Oregon range. J Range Manag 23 (3):155–157
Sprenger M, Vangestel C, Tuyttens FAM (2009) Measuring thirst in broiler chickens. Anim Welf 18(4):553–560
Stockdale CR, King KR (1983) A comparison of two techniques used to estimate the herbage intake of lactating dairy cows in a grazing experiment. J Agric Sci (Camb) 100(2):227–230
Sykes JF (1955) Animals and fowl and water. In: USDA yearbook of agriculture. USDA, Washington, DC, pp 14–18
Thexton AJ, Crompton AW, German AZ (1998) Transition from suckling to drinking at weaning: a kinematic and electromyographic study in miniature pigs. J Exp Zool 280:327–343
Timmerman HM, Veldman A, van den Elsen E, Rombouts FM, Beynen AC (2006) Mortality and growth performance of broilers given drinking water supplemented with chicken-specific probiotics. Poult Sci 85(8):1383–1388
Torrey S, Tamminga ELMT, Widowski TM (2008) Effect of drinker type on water intake and waste in newly weaned piglets. J Anim Sci 86(6):1439–1445
Veenhuizen MF, Shurson GC (1992) Effects of sulfate in drinking water for livestock. J Am Vet Med Assoc 201(3):487–492
Veira DM (2007) Meeting water requirements of cattle on the Canadian prairies. Rangel J 29 (1):76–86
Vermeer HM, Kuijken N, Spoolder HAM (2009) Motivation for additional water use of growing-finishing pigs. Livest Sci 124:112–118
Willms WD, Kenzie OR, McAllister TA, Colwell D, Veira D, Wilmshurst JF, Entz T, Olson ME (2002) Effects of water quality on cattle performance. J Range Manag 55(5):452–460
Winchester F, Morris FW (1956) Water intake rates in cattle. J Anim Sci 15(3):722–740
Xin H, DeShazer JA (1991) Swine responses to constant and modified diurnal cyclic temperatures. Trans ASAE 34(6):2533–2540
Xin H, Gates RS, Puma MC, Ahn DU (2002) Drinking water temperature effects on laying hens subjected to warm cyclic environments. Poult Sci 81(5):608–617
Yang TS, Howard B, McFarlane WV (1981) Effects of food on drinking behavior of growing pigs. Appl Anim Ethol 7(3):259–270
Zaman A, Amin A, Shah AH, Kalhoro DH, Muhammad K, Shakeebullah MH, Wazir H (2012) Prevalence and vulnerability characterization of poultry pathogens isolated from water tanks and drinkers in commercial broiler farms. Afr J Microbiol Res 6(3):691–695
Zoeteman BCJ (1980) Sensory assessment of water quality, vol 2, Pergamon series on environmental science. Pergamon Press, Oxford, See Chap. 6 (pp. 83–92)

Chapter 10
Providing Feed to Livestock During Emergencies

Larry W. Roeder Jr.

Abstract This chapter is intended to help animal nutritionists, food handlers, and other animal welfare experts intervene in disasters caused by natural phenomena and man-made events like conflicts. Concepts introduced include early warning, response, and recovery, as well as methods of collecting data and references to helpful institutional handbooks. Partnerships between humanitarian relief agencies are also proposed in order to take advantage of comparative strengths, especially for logistical support to transport feed and fodder. In addition, changes in international law and practice are proposed in order to make nutritional and animal health interventions easier to provide. It should also be seen as complementing, not replacing, the valuable Livestock Emergency Guidelines and Standards (LEGS livestock emergency guidelines and standards. Practical Action Publishing, 2009).

10.1 Introduction

Increasingly, as shown in Cyclone Nargis where the Burmese and American governments found common ground to protect livestock, relief agencies understand that by protecting livestock, a people's long-term sustainability is also protected. During Nargis in 2008, US planes flew in from Bangkok both traditional humanitarian supplies and livestock feed, medical supplies, and veterinarians from the NGO community. They also used remotely sensed data to assist in predicting rescue priorities for livestock. That was done with permission of the national authority. Animal welfare experts also responded to the conflicts in the Gaza with assistance to livestock. As far back as the Cyprus conflict in 1974, the United Nations facilitated entry of veterinarians to assist livestock. During the Bosnian emergency in 1994, vets brought in rabies vaccines to protect both people and livestock. The tradition, therefore, of protecting both livestock and people in the same crisis is a long one. Nonetheless, to do it right, during the early warning, response, and recovery phase of any emergency, there are many animal welfare and nutritional questions to answer if livestock are to be fed, grow stronger, and reproduce. More

L.W. Roeder Jr., MS (✉)
Catholic University of America (CUA), Washington, DC
e-mail: roederaway@gmail.com

C.J.C. Phillips (ed.), *Nutrition and the Welfare of Farm Animals*, Animal Welfare 16, DOI 10.1007/978-3-319-27356-3_10

difficult than those technical questions, however, may be how to surmount the physical and political obstacles to relief. As pointed out in Livestock Emergency Guidelines and Standards (LEGS),[1] before any intervention there should first be an assessment, often called a rapid needs assessment (see definitions at the end of this chapter); so one of the first questions is whether such is possible. Without this, relief workers will find it hard to develop any effective plan to deploy supplies or staff with strong feeding management skills. While caring for animals, what happens to the human victims, especially if they are also hungry and lacking in medical care? Will the social tension from trying to meet both needs fairly interfere with effectively providing either need? Are there dangerous environmental hazards adversely affecting relief such as irradiated ground or polluted water? In a conflict, how does an aid worker safely navigate mine fields to feed livestock? Obviously not: indeed, sometimes local populations even use livestock to seek out mines as was done in South Sudan during their civil war; but livestock experts do analyse the economic, environmental, and health consequence of minefields, often at their own risk. We saw this when WSPA entered the Kabul zoo in 1995 to feed wildlife, even though the institution was in the middle of a mines and unexploded ordnance. Will armies, bandits, or armed non-state actors (ANSA, see definitions) directly interfere? They might, if a proper balance of interests isn't found; but it's a delicate issue.

Will international political disputes prevent crossing a border to use a radio to coordinate activities? Assuming a plan has been developed, does it include modern treatments, and are they in harmony with the traditions, culture, and economic capacity of the owners? In other words, will they be acceptable and thus sustainable once the relief community departs?

If providing feed or other care in settings like Europe, Japan, or the United States of America, the answers to these questions are fairly easy; those regions have compatible standards of behaviour and local governments are responsible for emergency planning and response. Generally, feeds for animals and other resources, together with local experts, can also be found to help relief agencies prioritize the deployment of feeds. But is this true in the developing world? That is the geopolitical space to which this chapter is mostly oriented. The largest number of animals in need live there, as do most of the world's poorest, most vulnerable people, often heavily dependent on the animals for a living, for food, or for cultural identity. Experts are often less available, as is appropriate fodder. Roads can also be unsafe, requiring the transportation of feed pellets by air, which is very expensive.

A basic premise of relief ought to be that all animals (including wildlife and companion animals) have a right to nutrition and reasonable protection from the effects of natural and man-made disasters (see definitions), but interventions should also take into account local knowledge and social norms. In 2003 and 2004, the

[1] LEGS is a set of international guidelines and standards first established in 2009 by the Feinstein International Center, Tufts University and also used by the IFRC for the design, implementation, and assessment of livestock interventions to assist people affected by humanitarian crises. For full details, see http://www.sphereproject.org

Navajo Nation and various Pueblo tribes, in particular the Laguna and the Ohkay Owingeh, worked with the author and an international team of experts associated with the private sector, the Organisation for Economic Co-operation and Development (OECD) in Paris, NASA, and the US Departments of State and Defense to examine the potential for reducing intertribal conflict in Africa by sharing disaster early warning information between tribes with 'political differences'.[2] The experiment was to be replicated in East Africa once completed in New Mexico and Arizona. Part of the discussion, which was led by the Veterinary service of the Navajo, had to do with the potential impact of conflict and the environment on livestock. In addition, to the nutritional issues, a bark beetle infestation was killing entire forests, creating kindling that could explode into fires, as happened in Colorado in late 2012. In sustained droughts, because of the lack of roots, should rain then come, nothing will prevent the runoff from killing livestock (Davies and Roeder 2004). The Navajo then proceeded to show the author vulnerable pastures. Without that local information, his team's advice would have been far less effective.

Some policymakers with a primary focus on human relief in recipient countries, or even policymakers in international relief agencies, will not protect animals as a rights-based concept, but if they are practical, they should come to a similar protective philosophy that provides nutrition, because of the value of animals to society.

> 'About a billion of the world's poorest people depend on animals for food, income, social status, or cultural identification, as well as companionship and security. To them, losing animals ... is a catastrophe' (IWGAID 2008).

In Western, wealthier societies, livestock owners have the ultimate responsibility to provide proper nutrition to their animals. That is not the responsibility of those governments, which also do not provide individualized care, except in extraordinary circumstances the United States Department of Agriculture (USDA). Such support is even less likely in poor nations, especially in the droughts in Africa; yet, if pastoralists and farmers do not receive donations of food pellets and other assistance, livestock will die. Human beings of course are the first priority in an emergency. We don't let them starve; but if their livestock also die, the reality is that can lead to long-term poverty and an unsustainable local society. Contrary to common thinking, it's actually very hard to replace indigenous livestock with non-indigenous, and replacement is also very expensive. That is why it's very important that there be a partnership between traditional humanitarian and animal welfare relief agencies, not just intervention by one or the other. We saw that

[2] Between 2003 and 2005, the author conducted a series of expeditions pre-testing ideas to be used in Africa, in cooperation with local Native American governments in New Mexico and Arizona. A partner was the Organisation for Economic Cooperation and Development (OECD, Paris). The mission was to understand local risks to livestock and other assets, as well as coping mechanisms. Readers are recommended to read *OECD Studies in Risk Management: United States: Disaster Information Needs of Ethnic Minorities*, OECD: Paris, 2006.

especially clearly in NARGIS. Had the cattle not been rescued, the Country's economy would have been devastated, which is why the US Department of Agriculture and other agencies provided data to help veterinarians deploy resources effectively. The food and medicine were flown in on military aircraft, right along with needed humanitarian supplies.

International relief is therefore often a necessity, but animal nutrition and welfare specialists should not go it alone. For an intervention to be sustainable, it is better to build partnerships with humanitarians, taking into account local realities and the delicate balance between human and animal needs. Those are a lot of questions and issues to consider when thinking of whether or not to intervene, and then how to set and accomplish goals.

10.2 The Study Group and Assessment Concepts

10.2.1 Using a Study Group to Make a Plan

NGOs and national and international governmental organizations frequently assume that if an operation was successfully done in a certain way in the past, the current crisis should be done identically. Do not rush in without a technical evaluation. For example, a nutritional assessment on the ground or based on remotely sensed data will help identify where livestock are most in need, where useful plants and water may be located, and which routes may help in food distribution or an evacuation. That is exactly what was done during Nargis. No reason, it can't be done in other emergencies as well. In fact, crop acreage estimation and mapping projects based on remotely sensed data have been going on for decades. The science is dramatically improving, and as was shown by NASA's HELIOS and Peacewing projects could eventually be accomplished by solar powered UAVs.[3] This is often called a Rapid Needs Assessment (see definitions). However, even before deciding anything, a Study Group[4] should be formed, as stressed in a recent book for animal welfare NGOs (Roeder 2011) and a similar book for humanitarian NGOs (Roeder 2013). The two works focus on managing diplomatic initiatives, but Study Groups also develop operations to support nutritional and other needs in the field. The idea is to bring together the right experts from different areas of work to figure out the best course of action as quickly as possible, the resources needed, and a budget and funding stream.

[3] Peacewing was a NASA-Department of State experiment in the 1990s using the HELIOS solar-powered wing platform to prove that such tools could provide high-quality telephony and remotely sensed data less expensively and sometimes more effectively than traditional airplanes or satellites. See http://www.state.gov/1997-2001-NOPDFS/issues/relief/peacewing.html

[4] Study Group is a generic term to identify the team responsible for coordinating and implementing plans, as well as organizing the collection of critical data. The team can operate out of a single NGO, or be an interagency effort like a UN country team.

There are many possible formats for a Study Group; whichever is chosen, without the right information in a useable format, money and resources can be wasted, placing relief workers and animals in unnecessary danger. The team designs the assessment as well as the ongoing plan of activity, though as the operation advances, changes will be made. An emergency has phases and each needs to be planned for and properly resourced: (1) assessments (some from field observations, some from consultations); (2) creation of a proposal for field operations; (3) approval, with funding sources and potential partner agreement; (4) the actual intervention, including monitoring of progress; and (5) phase out and evaluation.

10.2.2 Field Assessment

Many handbooks exist on how to conduct field assessments, especially by the UN High Commissioner for Refugees (UNHCR), the UN Children's Fund (UNICEF), US Agency for International Development (USAID), the SPHERE Project, and its related effort, LEGS. Regardless of which model is used, the exercise should create an objective feel for the emergency, whether it is a rapid or sudden onset disaster like a war or flood, or a slow onset event like a drought or desertification. In the case of livestock, know the locations of animals (sheep, goats, camels, water buffalo, etc.), as well as the status of plants, since different species eat differently, and the location of clean water, shelter, and passable evacuation routes.

An assessment needs to also evaluate the resources required to provide help. Agencies conducting relief should also track assistance, and if more than one NGO is involved, understand what each is contributing, in order to avoid duplication of effort, as well as to take comparative advantage of each other's strengths. For example, if one NGO is providing feed and another medical care, there may well be a need to collaborate. During the crisis, it would be good to track by geographical sector what fodder and other resources have been requested and provided, as well as by whom. Are there critical gaps? Some vaccines require refrigeration and careful handling. Will the relief teams be able to transport appropriate cold storage units? How long will medicines be held in customs, hopefully not long enough to impair effectiveness. When considering nutrition, if feed is required for animals in a specific location, has it been provided, and for which species? How long can fodder remain in storage?

10.2.3 Preliminary Issues

There is no single set of correct questions, but the following is a sample of preliminary issues to be addressed, based on a review of the excellent work in LEGS and other emergency management manuals. Also, some answers may preclude or seriously limit assistance. In fact, making a decision on whether to intervene, not just how, is one of the most important functions of a Study Team. The United States of America did just this when it prevented certain groups from intervening in the Afghan conflict to protect animals. The relief workers were not properly trained for combat arenas and potentially could have been killed or endangered others, as well as the animals they wished to help. However, working with the American Zoological Association, the London Zoo, and other bodies, it did permit the World Society for the Protection of Animals (WSPA) and Brooke Animal Hospital to enter, as they were properly trained.

- What is the nature of the crisis, the stage of the event, and its consequences for livestock and society? A drought will have different consequences than a rapid onset flood. Some crises will also have a measureable impact in migrations.
- Who are the main players in the crisis? What other agencies are present and will they partner in some manner? If so, that can lower costs and increase efficiency and security, especially for logistical issues.
- Will the government be helpful or a hindrance? They can't be ignored; but if they will be a hindrance, that speaks to need for an effort to educate on the need for partnering and why such partnerships help bolster economies.
- Will the civilian population be helpful? It's important to show how helping livestock in particular is critical to preserving the income of families and small businesses, for example. As pointed out by Hayden Turner, a zookeeper in Sydney's Taronga Zoo in 2002, 'the Kabul Zoo was a strong symbol of hope' for the people of Kabul (Danielson 2002) but when entering a crisis, 'you go and talk to the people, ask them what they want, make sure they have ownership',
- Will ANSAs hinder the exercise, or can they be convinced to help, or at least permit assistance?
- Is the nation a signatory to the Tampere Convention on the Provision of Emergency Telecommunications, an international law intended to ease the movement and use of cell phones and other communications gear across international borders, even for NGOs? If not, it may be difficult to send situation reports or effectively manage security and transportation questions.
- What are the legal, political, religious, and societal hurdles? As an example of a political hurdle, during the Nargis crisis, Myanmar placed restrictions on which nationalities could participate. Religious or societal hurdles may hinder the use of women. Domestic legal hurdles could require specific licences for foreign veterinarians, visas, and work permits. International legal hurdles could include travel restrictions imposed by the UN Security Council.
- Are there veterinary services and a supportive governmental infrastructure? As an example, there is only a very weak veterinary infrastructure in Somaliland, though the government is supportive.
- What is the condition of shelters and medical facilities for livestock and workers? How will deficiencies be rectified and at what cost, following local construction regulations?

- What are the infection rates and the number of trapped, dead, or injured animals by species?
- Is there sufficient fodder or medicine appropriate to the species in order to maintain an adequate level of nutrition? If not, can these be obtained locally, or must they be imported, and at what cost, and by what means? If not, not only will nutritional levels decline, but vulnerability to disease will also increase. While most developing countries won't have the resources, understanding the extent of limitations in resources and infrastructure is essential information for planners.
- What is the extent of logistical support and infrastructure? Are trucks required (4×4 or not) and available? Is there fuel (petrol or diesel), what does it cost, and are the roads passable?
- Where are the animals and owners located and how will be movements be tracked? In emergencies, victims are frequently placed in tent camps. The United Nations is expert in designing such camps, but they are usually unable to house much livestock, for sanitary and safety reasons. However, if the livestock are just outside the camp and near the owners, perhaps the refugees or internally displaced persons (IDPs), as the case might be, can tend to their own animals. That supplements the need for relief staff and will also reduce social tension. It is also a good idea to place on each animal some form of identification, perhaps a brand or tag, in particular if they are going to be moved.
- What is the exit strategy, in other words, the transition from response to recovery or from recovery to sustainable development?

Note Logistics will be a critical area of discussion for the study team, especially regarding the distribution of animal feeds and other relief supplies. Will the right feed be available and how will it be delivered and at what cost? In some cases, this might mean trucking in hay to cattle herds stranded in snow. In the rescue of the zoo in Kabul during the invasion of Afghanistan by the United States of America and its allies, this meant that the team, which was supported by the Federation of British Zoos and other well-respected professional bodies like WSPA, had to negotiate an agreement with a local butcher to supply raw meat, highly sensitive and dangerous given the potential for Taliban attacks. Although major armed units of the Taliban were pushed out in advance, there was little doubt that members stayed behind and were a danger to local security. Despite that, the international team provided critical medical care, in some cases relocating animals. While organized resistance might not have been standing around, they were present in the population, still are. The citizens of Kabul were reportedly very grateful for this effort, which preserved an important social attraction in the capital.

During the Hurricane Nargis crisis, it meant negotiating a series of agreements with several governments. First, an NGO wanting to provide pellets had to negotiate a commercial contract with a feed pellet manufacturer, after making sure the company could provide the appropriate feed on time and at a specific quality. The NGO also had to load the bags onto trucks, to make sure that the right number were produced, test samples of pellets for quality, and then transport the bags to a distant military airbase, thanks to an agreement with the Thai government. The bags were then loaded on pallets and transported by the US Air Force to Myanmar under an arrangement with USAID. While the planes were allowed to land in Myanmar, the engines could not be turned off, so the pallets had to be quickly off-loaded and turned over to the UN and local Myanmar support staff.

10.2.4 Use of Maps

As many as possible of the questions posed should be answered in a quantifiable way and linked to specific coordinates on a map and then linked to other related information. As an example, we have already discussed the importance of water. Knowing its location is essential, but also its relationship to herds and roads, since maintaining a water supply will help stabilize the production of crops needed for nutrition. In addition, how is the water conserved and can additional water sources be developed for drought areas? This was one of the major issues discovered in Somaliland by the author on an expedition in 2010, when he found that instead of storing run-off where it might be used by livestock, water was often allowed to flow to the sea. Where is the water in relation to roads? This is especially important for moving livestock or water by truck; but all crisis information is temporal, in other words, just because a map says that a road is passable does not mean that conditions will not change. Also, while a map showing historical migration routes is useful, environmental factors can cause changes.

10.2.5 Data Analysis

One question always posed is to inquire as to the nature of the crisis. For example, was it caused by a high-wind event or earthquake of a certain magnitude? In that context, while an on-the-ground assessment will tell a lot, it can be possible to predict levels of damage by reviewing the impact of past phenomena on the same location. This kind of historical data, when coupled with fresh information arriving from the field, will help the team develop a triage to decide which area of the crisis zone is likely going to need more assistance and when.

Those wishing to support the future nutritional needs of animals should consider developing an understanding of the potential impact of the event in the long term. How hard will it be for the society to rebound? What will be needed that can affordably reduce risks from the next event and enable an efficient distribution of food? Another way to think about this is to remember that a hurricane is not a disaster. A disaster is nothing more than the measurement of an event in terms of the ability of the society to reduce damage or respond (see definitions). To predict the immediate impact of a current event or to predict the impact of an event on recovery and reconstruction, animal welfare bodies may wish to consult with the Center for

Research on the Epidemiology of Disasters (CRED), located in Brussels. Since 1988, CRED, in collaboration with the World Health Organization (WHO), has collected data on nearly 13,000 disasters that have occurred from 1900 to the present. Also consider La RED (La Red de Estudios Sociales en Prevención de Desastres en América Latina—the Social Studies Network for Disaster Prevention in Latin America). CRED's data goes back as far as 1900, but probably any trend analysis likely should begin its research from 1950, since the number of disasters recorded for the first half of the twentieth century were few in number or poorly documented.

10.2.6 An Assessment Form

One approach to gathering a good sampling of contemporary data for comparison with historical records is to have an Assessment Form in paper form or on an I-Pad or similar device. This should be used to quantify losses and injuries of livestock and identify levels of damage to structures as well as nutritional plants in a way that can also be understood by the general humanitarian community. In other words, many bits of information useful to animal welfare are identical to the humanitarian world. To the extent that both communities collect such information, the same symbols and terms should be used. That way, data collected by one body will be of value to the other.

As an example of a chart, consider the following, derived from one used by the UN; it is intended as a simple check list of structure damage in a format that could be loaded onto a computer and then placed on a map used by everyone in an emergency.

Images are used with text because not every worker will be fluent in the same language. Using a system like this, workers can collect data, enter the exact location and date, and then transfer the data into a geographical information system (GIS) that can be used for large-scale analysis. The same data can then be updated throughout the crisis and into the recovery phase as the situation evolves. The form should also be coordinated with other NGOs and relief agencies, using compatible text and images. As an example, data in the form of Fig. 10.1 track structural damage, clearly important when predicting whether shelters are likely to be available for workers, livestock, or medicine and fodder storage. If not, the team may need to haul shelters into the crisis zone, which may require transport, licences, and customs duties. Also keep in mind local cultural and social requirements. Initially, this might simply be to provide a tarpaulin roof to protect victims from the rain; over time more permanent structures will be needed.

Some symbols might not be normally used by humanitarians, especially those related to livestock, though they might use them if interested in sustainable development in an agricultural society. Here are a few examples (Fig. 10.2).

Shelter Damage Assessment

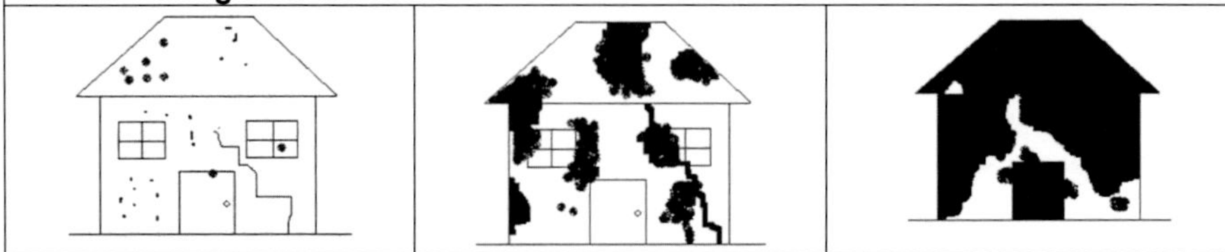

Mild	Severe	Destroyed
Broken windows, door locks, hinges, roof tiles. Cut-off electricity, water *Can be repaired*	Over 30% roof damage Fire Damage Doors, windows Damaged Can be repaired	Destroyed Needs reconstruction Can't be repaired

Fig. 10.1 Sample form showing levels of damage, thus reducing misinterpretation through language. Notice that only the *right side* is a disaster. The other two mark levels of recoverable damage. In any case, such information informs relief workers about the reliability of structures for storage or shelter

Try to collect both pre-disaster and post- disaster data, in order to understand the impact. Also, distinguish between alive, dead, in need and type of need as distinct data points. Humanitarians collect the same kind of data on humans.

Fig. 10.2 Clear images of livestock species to avoid confusion due to language

10.2.7 Other Principles

Other general principles should also be followed. Relief bodies should:

- be politically neutral in order to ensure that relief activities benefit all affected households, communities, countries, and regions;
- avoid being associated with an organization vested in the success of local interests;
- not be responsible for themselves only; they should be accountable for the beneficiaries as well, despite the fact that each relief body is expected to carry its own weight, provide its own supplies, and manage its own licences and permits. It is essential that intervening relief bodies help each other, especially in coordination committees or within clusters;
- act using humanitarian principles (see SPHERE standards in Definitions), which will probably ensure long-term acceptability of the relief effort. While bearing in mind that animal nutrition and health and safety are the first concerns of an

animal welfare body, it is critical that humanitarian bodies operate within local social norms and that standards retain family and village units as levels of intervention;

- be prepared to help livestock owners as well, at least until traditional humanitarian bodies can render assistance, or work in joint teams; and
- be practical. As an example, 4×4 vehicles are often required, but not in every situation. They are also more expensive. Regardless of the type of vehicle used, have on hand your own mechanic. Also, make sure the fuel used is easy to find locally.

10.2.8 Animal–Human Coordination

Protecting the welfare of livestock is one of their rights, but it also protects the human economy, and therefore offers an opportunity for partnerships wherein both the animal welfare and humanitarians communities can use the comparative strengths of the other to achieve success. Achieving such partnering is important since risks to animals may be less due to intentional neglect than the moral conflict of balancing human and animal needs. In a crisis, when travelling past hungry or injured human victims, if they are not adequately supplied with medicine and food, it will be natural for them to complain, perhaps even demand that aid be given to them, but as the Cyclone Nargis episode showed in Myanmar, a balanced, approach can work. When veterinary teams encountered human victims, they explained that their effort was needed to restore the original balance and prepare a foundation for a sustainable economy. They also mentioned that the UN was on the way with humanitarian assistance.

Feed is not enough to keep livestock healthy. They also need water, especially in a drought, so an essential assessment question will be asked if there is enough for both humans and animals. There are a variety of guidelines offered by different organizations, e.g. the well used UNCHR Water Manual for Refugee Situations (UNHCR Program and Technical Support Section 1992, Appendix B).

My advice is to remember that we often have to try to squeeze a humanitarian response to a drought condition into a typical non-hazard time feed and water recommendation. The reality is that sufficient water will always be a challenge and in short supply in disaster responses (let alone clean water!). This is why I strongly recommend an assessment of available water and a census of people and livestock in order to achieve a proper triage. Unfortunately, providing a fixed number here is fraught with the potential for controversy, given that every crisis is different. Therefore, the least confrontational recommendation in this context is to provide water at least once a day and more often, if available. Ideally, livestock should have access to water at all times. If we want lactating animals to produce milk, there is even greater need. Experiences in the drylands of the Sahel teach that some cattle breeds can go several days without water—up to 4 days in some cases; however, those animals will surely not grow, reproduce, or lactate; it is only survival. That (hopefully) is not the case in many instances. The bottom line is to stay away from fixed numbers and take into account the context to any recommendation, using all available data collection methods, such as remotely sensed data. Coordinate in advance on the latest thinking, taking into account the dryness of the air, the heat, as well as the distance to be travelled. The minimum amount of water for livestock is a function of what is needed for growth, lactation, and urine flow, as well as the nature of the food eaten. A diet high in protein and salt will mean more urine flow. Also consider that water lost through evaporation can exceed that lost by urine. Since, humanitarians often drill boreholes to provide water for remote villages, a useful partnership would be to ask them to develop separate holes for livestock, as well as irrigation systems. To sensitize humanitarians to the need for 'livestock water sources' and to learn which agencies are developing water systems in general, animal welfare interventionists should consider participation in the UN's cluster system, discussed later in the chapter, or in formal inter-NGO, inter-agency teams in the field (Wollen 2015).

Another form of partnering took place in April 2013, when the United States of America government, in coordination with the governments of the United Kingdom, Norway, and the European Union, hosted a forum of the South Sudan Economic Partners. They discussed challenges facing South Sudan and how the new government, together with partner nations, NGOs, and international financial institutions, could best find practical solutions (Stephens 2013). The April meeting would have been an excellent opportunity for livestock nutrition experts to explain their needs and offer proposals to integrate their solutions into the development process. After all, participants also included representatives from the International Monetary Fund (IMF) and the African Development Bank (AfBD). In similar fashion, the author attended a meeting in September 2005 of the Global Consortium on Tsunami Recovery chaired by former US President Bill Clinton. He recommended to the President and other national leaders that protecting livestock in the proposed recovery effort was essential (Roeder 2005). Understanding that the participants, some of whom were at Prime Minister level, primarily wanted to relieve human suffering, the author explained how their economic goals could be accomplished through integration. Many high-level meetings happen every year and should be a 'must-go' for experts wishing to improve animal nutrition in emergencies. They should also attend multilateral and industry-based conferences on disaster management. In the latter case, this will be to both learn the latest techniques and to educate professional disaster managers. An example of such meetings is the World Conference on Disaster Reduction held in Japan in 2005. This summit-level event, which was attended by thousands of NGOs, came up with the Hyogo Framework for Action, a new international policy on risk reduction. Although one panel did deal with animals, very few animal welfare NGOs attended, missing a major opportunity to plead their cause. On a much smaller scale, animal nutrition experts often must do the same thing during emergencies, with field-level discussions with local leaders.

To develop a joint humanitarian-animal welfare strategy for the Nargis crisis, animal welfare NGOs and veterinarians met with the Food and Agriculture

Organization of the United Nations (FAO), other UN agencies, and humanitarian NGOs around the table in Bangkok and Rangoon. Traditional humanitarian experts had a focus on public health, gender protection, etc., but many understood that protecting animals protected the society and its economy from collapse and so were willing to share operational information and suggestions. The United States of America government and FAO also provided some logistical support. Bangkok was the regional operational center for relief efforts, whereas Rangoon was the field operational centre. In both locations, representatives from both the humanitarian and animal welfare communities met and cooperated. This is a typical scene in many emergencies, so the recommendation is for anyone wanting to intervene to join such coordinating bodies. In some cases, these bodies will be managed through a Country Team approach, often led by the UN. In other instances, there will be cluster meetings (discussed later), which join together like-minded stakeholders. Often, both types of committees are involved.

10.2.9 The UN Cluster System as a Coordination Tool

The Cluster System emerged out of a UN reform effort in 2005–2006 to fill in gaps, strengthen response, ensure accountability, and build partnerships. These clusters are thematic groupings of NGOs, UN agencies, and government agencies, as well as partner corporations. They are led by a Lead Agency and pull together related efforts. That is supposed to avoid duplication. The concept offers significant networking opportunities for NGOs to build an infrastructure to identify and manage risks throughout the cycle of a crisis. For example, during Nargis, the World Society for the Protection of Animals (WSPA), which spearheaded much of the relief work for livestock, also joined the Agricultural Cluster of the UN, led by FAO. By participating in such clusters, those bodies wishing to help animals with nutritional needs will have an effective platform for articulating their requirements in a potential crisis, to respond to an ongoing emergency, or to act during the recovery phase. Another example is the 12 January 2010 earthquake that killed many people and flattened cities in Haiti, especially in the overcrowded capital. Humanitarian NGOs like *Groupe d'Appui Aux Rapatriés et Réfugiés* (GARR; The Support Group for Refugees and Repatriated Persons) ran a human rights monitoring programme along the border with the Dominican Republic and provided funds to over 200 farmers to replace or care for livestock (ACT Alliance 2010). Animal welfare NGOs also participated in the FAO Agriculture Cluster, as well as in the Health and Shelter clusters (Fig. 10.3).

Sitting above the cluster system, every UN disaster has a UN Coordinator who reports to the UN's Emergency Relief Coordinator (ERC) via the UN's Inter-Agency Standing Committee (IASC), which is the interagency body that coordinates how emergencies are managed. It is the UN Coordinator who often leads the coordinating committees discussed in the previous section. Members come from governments, UN agencies, the IFRC, and NGO networks. The FAO belongs to this

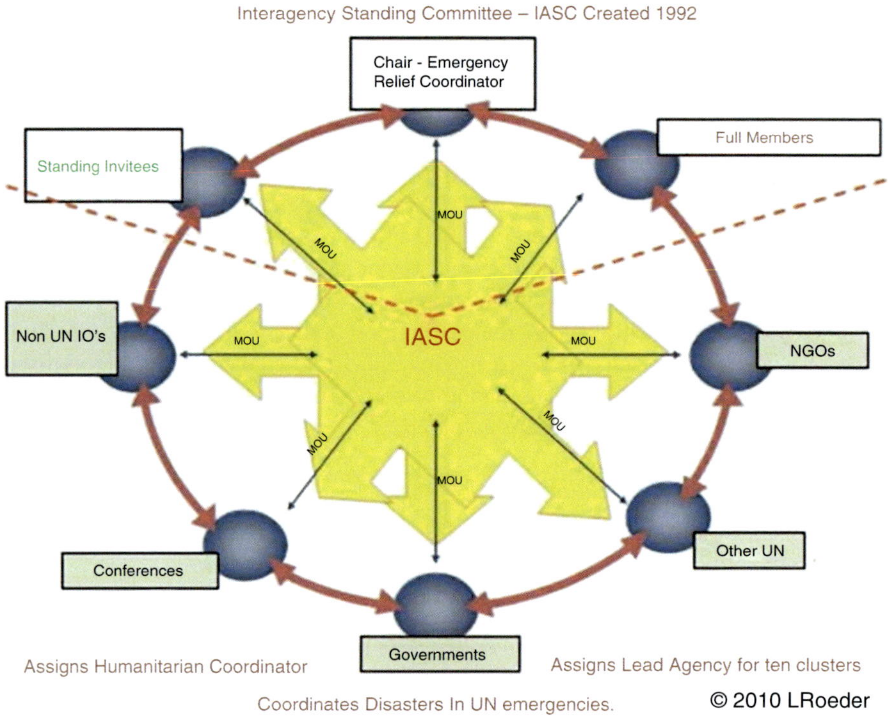

Fig. 10.3 The UN Inter-Agency Standby Committee (IASC)

constellation, and a few animal welfare NGOs also participate through an intermediary NGO umbrella body called InterAction. Because many of those organizations have overlapping mandates and skill sets, to avoid confusion in how assistance is applied as well as gain cost efficiencies, relief bodies operate as thematic clusters. For example, the Agriculture Cluster is led by FAO. The mix of clusters in an emergency will depend on the nature of the crisis, and the decision on which one will be activated is made by a UN Country Team, made up of the lead UN agencies in consultation with the host nations and donor nations. The whole idea is to provide predictability and accountability in humanitarian response, which is why organizations that intend to provide assistance to livestock should join the clusters. For example, in Somaliland, nine clusters have been active: Agriculture & Livelihood; Education; Food Aid; Health; Protection; Logistics; Nutrition: Shelter; and Water & Sanitation. The Food Security and Nutrition Analysis Unit (FSNAU) was created to provide situational analysis on food, nutrition, and livelihoods security. Managed by FAO, FSNAU collects relevant data, analyses them through a variety of techniques and information processing tools, and disseminates the analyses using various channels (reports, brochures, manuals, videos, etc.).

By being part of the cluster system, a veterinary group or animal welfare NGO can influence relief policy at ground level. One use of the system might be to set up a National Veterinary Service to improve nutritional standards. For example, during the Myanmar crisis, thanks to their participation in the cluster system, US government agencies provided free air support to fly fodder into Myanmar, and other agencies provided satellite-based imagery to show where livestock were likely to be located.

In 2010, WSPA deployed trucks with hay for cattle in Bolivia. To properly support efforts like WSPA's or that of the United States of America government in the deployment of nutritional supplies, a triage system is needed, based on legacy data and on-site information. The satellite-based maps used in Myanmar dramatically facilitated relief efforts.

10.2.10 To Evacuate or Not?

In some cases, the potential disaster is seen coming from a long way away, such as with a hurricane. Depending on circumstances then, if the advancing storm provides enough time, farmers should consider vaccinating and de-worming and then taking the livestock to high ground. As that is not always possible, coordination may have to take place far away from the crisis, as with protecting animals at the zoo in Kabul, Afghanistan, during the war in 1991. The Kabul Zoo had been reduced from 1000 animals to about 40, due to barbarism by the Taliban, like maiming a lion with a hand grenade (Talbot-Rosevear 1991). That report made it to international media, including CNN and the BBC, and caused many large-cat rescue organizations in the United States of America to approach the Department of State and various congressmen, asking to remove the lion. The government did some preliminary assessments, talking to sources in Kabul and experts in the Bronx zoo. Given the age and condition of Majul, an evacuation was considered counterproductive, so support was denied to the rescue bodies. What was needed was an intervention of professionals with real experience in the developing world who also would not get themselves killed in combat operations while providing proper nutrition and medical care. Allied forces were reluctant, so the Department of State gathered intelligence on the Zoo and in partnership with the British and American militaries, especially US Army Veterinarians in Afghanistan, a rescue operation by WSPA was supported, as well as an effort by the North Carolina Zoo. Nutrition was an essential part of the story, as the animals had not been properly fed for a long time. The lead for WSPA was John Walsh, the father of modern animal rescue, who developed some creative nutritional tactics. Conflicts are by their nature fluid situations, but there is good reason to believe that had the Department of State not coordinated with the British and American forces, assistance would not have been allowed.

Evacuation questions are among the first to come to the attention of a Study Team. In some cases, a medical evaluation will show that animals cannot be safely moved; in other instances there is no choice but to plan an evacuation, as happened in 1964 with the epic Operation Gwamba, often considered the first modern massive animal welfare rescue. John Walsh managed the evacuation of over 10,000 wild animals from certain death in a South American Rain Forest after the erection of a dam (Walsh and Gannon 1967).

In some situations, if livestock cannot be evacuated, the animals will be intentionally slaughtered (perhaps in horrific conditions due to the haste of the situation and need to gain as much income as possible before the owners depart). In fact, controlled slaughters were used to feed human victims in the famines in Ethiopia and Mali in 1984. There was not enough water for the livestock, due to the terrible drought. Such a procedure can mean an immediate flush of income, but the procedure is devastating to the economy in the long term because the natural cycle of replenishment has been disrupted. If, however, the livestock are evacuated during a conflict, soldiers, desperate civilians, or ANSAs may steal or kill animals, especially when human food is scarce. That can also be a security problem for organizations providing any form of assistance to animals.

If livestock cannot move with their owners, remaining relatives might maintain the animals in neighbouring 'safer regions', but in environmentally fragile environments the carrying capacity of the new land can be overwhelmed, leading to malnutrition for all animals, both indigenous and those just evacuated. This kind of unsuitable distribution is a common risk in many developing nations because reliable statistics often do not exist on livestock populations, which is why the Study Team should use remotely sensed data to examine high-risk areas in advance of emergencies. There was no such early warning data in Nargis, so livestock support was strictly in a response mode. Fortunately, the USDA provided satellite-based data to locate potential livestock victims and staging areas. Figure 10.4 shows a similar effort in response to the earthquake in Haiti, providing members of the Agriculture Cluster with maps based on satellite data that showed the location of cattle, measured in terms of cows per sq. km (USDA-FAS 2010).

In Nargis, if legacy data had been available, the relief effort would have been easier. It is always useful when considering nutrition, to know how many head of cattle or other species must be fed or medicated. That is not always easy. For example, if considering planning an evacuation of pastoral herds in East Africa, it has to be considered that the pastoralists probably have been less interested in the number of cattle owned than in the capacity to produce by-products like milk and blood. Therefore, animals are more probably sold only when fodder is scarce (Brown 2008). In that situation, determining actual stock distribution may be difficult without regularly collected remotely sensed data. That is why the time to collect data is before an emergency breaks out. Building a body of legacy data on livestock, nutritional plants, water sources, passable routes, etc., will become essential to veterinarians wishing to support livestock after any emergency hits, or to recommend preventative measures. In a rapid onset environmental hazard like a hurricane, it also provides the basis for planning a strategic retreat (Pielke 2001).

A way to determine where livestock might be is to use historical data on migrations. If that information is then connected with data on drought, roads, and conflict, that information can be very helpful determining where to pre-position

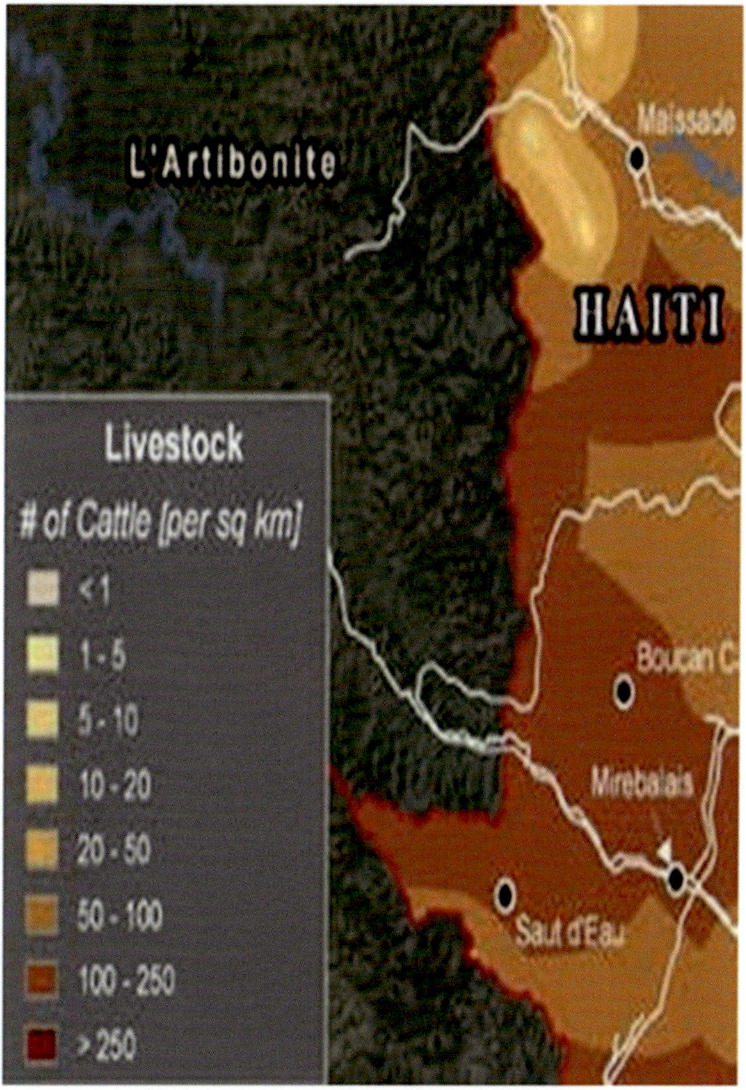

Fig. 10.4 Distribution of cattle in Haiti. *Note*: *White lines* are main roads

fodder, water, and other supplies. Figure 10.5 is an extract from a map developed by the US government to do just that for Darfur (US Department of State 2006).

Fodder at the destination point or along the evacuation route might be inappropriate if the species at the end point is different from that which is being moved. Even if appropriate, fodder may be in short supply for a variety of reasons. The existing herds may have exceeded the carrying capacity or stored supplies. Owners

Fig. 10.5 Traditional livestock migrations in West Darfur. *Notes*: *Green lines* are two-way traditional migration routes. Livestock tend to move from the north to the south in February and March and south to north in May. *Red lines* are paved roads. *Grey lines* are district boundaries

of existing livestock may limit access to supplies to their own herds. Bandits[5] can also be a problem. Any of these issues can result in serious livestock nutrition issues, and even unplanned mixing of genetic material that may weaken the stock. There can also be exposure to endemic pathogenic organisms (viruses, bacteria,

[5] These might simply be desperately poor people.

pests, etc.) against which the 'refugee' herd has no natural resistance. For example, during the Menelif famine of 1888 in Ethiopia, imported cattle from India and southern Russia carried Rinderpest, which killed off many local stock (Keneally 2011). If parasite control is weak along the route, that phenomenon will just make a desperate situation worse.

10.2.11 Complex Emergencies

When supporting livestock, natural and man-made disasters cannot always be separated into separate packages. A combined drought and war occurred in 2004 and 2005 in Darfur, Sudan. Following a series of technical assessments in 2004, it was determined that the supply of water for human and livestock consumption in North Darfur was critically low and declining. Rainfall in 2004 was approximately only half that of 2003, and the water tables were rapidly dropping. At the same time, approximately one million people were displaced in Darfur, driven from their homes by ANSA, resulting in people and livestock left without food or shelter (Muncy 2004). These drought conditions, along with continued insecurity and a long-neglected water infrastructure, resulted in large-scale population and livestock movements away from traditional grazing lands and watering points. The massive movements then advanced on camps and towns, increasing the existing high demand for water and animal feed in those drought areas. That led to more conflict. The Government of Sudan did report the presence of 1300 boreholes with hand pumps in North Darfur, but approximately half were not functional, due to poor maintenance. In addition, only 40 % of the approximately 166 water yards were operational. As a result, the UN had to work with NGOs and the private sector to rehabilitate neglected infrastructure and increase drilling capacity (USAID 2005).

Such terrible nutritional conditions can lead to massive livestock mortality, so one policy option can be to restock when the pastoralists or farmers return, or even while in a refugee camp once the environmental situation improves. However, such programmes are expensive, and there is a tendency to upgrade the genetic stock with exotic, non-native species. This urge should usually be suppressed. It is better for the indigenous animals and the owners if the livestock are allowed to move with the people, but only if protected along the route and if the end point has the proper carrying capacity.

10.3 Political, Legal, and Social Inhibitors

10.3.1 Risks to Society Due to Improper Emergency Care

The risks to any society from improperly caring for livestock are real. Consider if owners have to flee without their animals, due either to conflict or perhaps sudden onset natural phenomena like a hurricane or storm surge. In the West, companion animal owners will return, at great risk to their own lives to rescue their pets and livestock. That behaviour caused a significant change in post-Katrina North American laws, permitting joint evacuations in many instances, but the same is not always possible in the developing world. If not sheltered and left to wander, as often happened during Cyclone Nargis in Myanmar in 2008, those animals will be prone to serious injury. Cattle were exposed to winds reaching 215 km/h, eventually becoming bound up in loose barbed wire, suffering from wounds for extended periods before veterinarians could perform rescue. Despite the urgent need for care, moving veterinarians and food across the border was also delayed due to regional politics. For example, North American volunteers were not allowed in, whereas regional and British Commonwealth volunteers were permitted entry, but only after considerable direct diplomacy with Myanmar Department of Agriculture by the WSPA, the UN and the Government of the United States of America. In the interim, the livestock were also unable to eat properly. That led to lower levels of nutrition and increased vulnerability to disease. Unfortunately, for Myanmar, a cow also pulls the rice paddy plough. Herd damage threatened to turn the country from a rice exporter to an importer.

Not every animal welfare body will be able to conduct the direct diplomacy accomplished by WSPA in 2008. This again is why belonging to the Agriculture Cluster and participating in relief coordination committees will help. The committees will act for the entire group.

To reduce risks before a crisis or to properly manage it, experts should consider introducing compact and balanced feed supplements like urea-molasses blocks, as well as providing veterinary interventions such as vaccinations and prophylactic treatments such as de-worming. From a public health perspective, animals will need to be screened for zoonotic diseases and professional abattoirs developed, especially if considering breeding programmes or re-stocking, keeping in mind Manilif's famine. As noted in the Darfur crisis, separate water points will also be needed for livestock.

Those are some of the major interventions. But how will rural populations be convinced to change behaviour in order to manage the carrying capacity of rangelands to be sustainable through emergency situations. First, the herds should be mixed: sheep and goats, not just one species. Sheep graze and feed from ground vegetation. Goats prefer shrubs, bushes, and trees. Second, local knowledge must be integrated into risk-reduction strategies. Tensions over grazing and watering rights must be dealt with, especially in refugee situations when the host society is already under stress. Rather than create separate herds from the host population, it may be possible to combine them and hire local herders, with the payment being in-kind: rights to use milk or other by-products. This works best when the refugee and host populations are similar, but whether similar or not, the host herders will need to be convinced that the idea is beneficial, and the refugees can be convinced to trust the hosts. That will require a lot of diplomacy. Relief work is more than science, in other words. Finally, combatants, some of whom may derive from the local rural populations, need to be convinced to alter their behaviour, and not place land mines in pastoral fields.

10.3.2 Convincing Humanitarian Leaders to Support Animals

It usually is not hard to show that humane treatment of cattle or some other livestock could reduce poverty and disease and increase the GDP. This is the argument to make in clusters and coordination committees, with any humanitarian agency. The author faced this problem when talking with the Arab Red Cross/Red Crescent societies in Tunis. Their priority was humans, so were reluctant to deal with livestock, especially given limited resources. In response, the author suggested that NGOs could manage the livestock corrals and provide all the nutrition. He also pointed out that if the owners were in close proximity that would reduce their stress, since they would know where to find their livestock. The conference participants agreed that such a solution could provide the refugees some hope of taking their livestock back home after the crisis abated, and thus that livestock shelters could be next to refugee and IDP camps (Roeder 2008).

10.4 War and Conflict: Risk and Potential Opportunities

10.4.1 Special Risks and Tools

Destructive as mother-nature can be, she rarely harms animals as much as war, which can significantly reduce their nutrition. Many human-made disasters caused by industrial accidents, wars, and lower grade conflicts are similar in impact to those caused by natural phenomena such as earthquakes, floods, droughts, and high wind events like hurricanes, so all of the techniques discussed already will be essential. However, when the cause of an event is human violence, special coping skills are also required; otherwise the relief workers could be killed and the animals placed at higher risk. While touring the war zone in South Sudan, the author often came across the bodies of goats blown apart by land mines. In the Sinai, land mines were sown beneath the surface of the desert, which was just a sand ocean in many places. There were often markers delineating old mine fields, but goats and camels cannot read. And when no signs are available, humans are at risk, including the trucks delivering fodder. Better signage and training is needed, or in the case of land mines, implementation of the Ottawa Treaty.[6]

---

[6] The 'Ottawa Treaty', more formally known as the Convention on the Prohibition of the Use, Stockpiling, Production and Transfer of Anti-Personnel Mines and on their Destruction.

The 2009 conflict in the Gaza strip portion of the Palestine Authority raised serious animal welfare concerns; the people might move into bomb shelters, but the animals could not. In retaliation for missile strikes from the Gaza, Israeli Defense Forces attacked the region and reportedly dropped cluster bombs over farm lands. That created a safety and nutritional problem for livestock. In addition, livestock nutrition suffered once the food stock declined due to a naval and land blockade, since the Gazan government was going to feed the people first, and the animals second. Just like people, many animals were injured by shrapnel and flying debris, as well as stress. Unfortunately, the blockades inhibited an effective importation of fodder, though some inroads were made by animal welfare NGOs, due in part to direct appeals to the Israeli diplomatic mission to the UN in New York and to the IFRC. Diplomacy is an essential skill for animal welfare. In addition to direct lobbying or traditional diplomacy, NGOs could also consider using public diplomacy to convince society to pressure legislators and decision makers to change rules. The most effective tool, however, is likely to be direct diplomacy with the stakeholders to show that harming animals will undermine a stable peace. It is, therefore, useful to remember when talking to combatants that the international humanitarian community has already agreed that people have a 'right to food', which has been interpreted to mean that livestock should be protected in disasters caused either by natural phenomena or acts of man, because the animals are essential to the livelihood of people. This is reflected in the SPHERE standards (see definitions) (SPHERE Project 2011) and its companion handbook on Livestock (LEGS 2009).

10.4.2 Working with Armed Non-state Actors

In conflicts, there will probably be contact with the local government, but what about ANSAs? In some cases, in order to feed livestock, travel documents must be obtained from rebels.

> 'at least half the belligerents in the most widespread and most victimizing of armed conflicts around the world, i.e. non-international armed conflicts, are non-state armed actors' (Bellal and Casey-Maslen 2011).

GenevaCall is an NGO in Switzerland that focuses on armed entities that operate outside State control. Consider that prior to South Sudan gaining independence, ANSAs controlled the land and issued entry permits, like any government might. There will often be an operational need to develop a Memorandum of Understanding (MOU) with ANSAs in order to feed livestock. This may be possible, since GenevaCall has been working hard to make such combatants aware that if they want international recognition, they must comply with international norms, even though they cannot be signatories to a treaty. While formal agreements might convey a level of implied recognition beyond that which either the governments or the UN are capable of, NGOs like Geneva Call can negotiate 'Deeds of

Commitment', perhaps permitting nutritional support to livestock, and then publish them in Geneva much like the UN publishes state-to-state agreements.

10.5 Changing the Rules

This section takes a broader context than LEGS in that the author proposes that all animals are or should be seen as part of a larger sustainable development system that integrates the needs of the environment, people, and animals. In other words, the lessons of Operation Gwamba, Katrina, and recent conflicts have shown that consideration needs to be given to supporting companion animals, as well as wildlife and livestock. Field handbooks written by practitioners are essential, like SPHERE and LEGS, but to accomplish true integration or simply to efficiently enable nutritionists and animal welfare experts to enter emergencies will often require a cultural shift in domestic and international laws. Accomplishing that task will require diplomacy in all its forms, and people who provide feed or other animal welfare services need to be directly involved. The Katrina hurricane disaster brought to the North American public's eye that new laws were needed to enable livestock and especially companion animals to be part of the evacuation procedure for people, as well as to facilitate the entry of animal welfare relief workers during the response phase. As a result of public diplomacy, the attitude of legislatures was altered, and due to direct diplomacy between NGOs and local and state governments, as well as the USDA, and direct lobbying of legislators, new laws and rules have emerged. An important lesson here as well was the role of professional animal welfare experts in the NGO community.

Similarly, humanitarian relief workers (both government and private) realized that restrictions on telecommunications equipment were endangering the lives of victims and relief workers. That led to a partnership of practitioners, government, and UN diplomats who negotiated the Tampere Convention on the Provision of Telecommunications. Experts in the extraction of people from collapsed structures, called Search and Rescue (SAR), also realized that excessive travel restrictions often delayed assistance to that point that lives were placed at serious risk. This was a similar finding by many animal welfare NGOs after Katrina. That led to the creation of the International Search and Rescue Advisory Group (INSARAG), a network of experts from governments and private experts that live in either disaster-prone nations or which send SAR experts to emergencies. Thanks to their efforts, much has been done in the form of easing restrictions to make their work easier. Without proper nutrition and water, or medical care, animals will die in disasters, be those emergencies created by man like wars, or natural phenomena. Therefore, the author suggests a similar approach be conducted to ease the movement of animal welfare experts and their supplies and equipment. The more disaster management practices are standardized in a way that protects animals of all kinds, the easier it will be for veterinarians and animal nutritionists to make emergency interventions. This could be done by negotiating international conventions, which are very hard to achieve, but also through UN resolutions, especially those linked to sustainable development and the protection of the environment (Huertas and Murillo 2007). This does not mean something like the Universal Declaration on Animal Welfare (UDAW), proposed by some NGOs. Though UDAW is valuable, this proposal is for a specific set of rule changes related to disaster relief. To achieve, them, we recommend using diplomacy handbooks, such as in Roeder (2011, 2013).

Protection of relief workers and livestock in war will require special attention. Animal welfare NGOs have regularly assisted in wars. In 1974 during the Turkish invasion of Cyprus, understanding that livestock assistance was beyond the technical capacity of the stakeholders, UNHCR, UNDP and FAO asked the International Society for the Protection of Animals (ISPA) to intervene. This was the first known instance of a UN Agency requesting an animal welfare NGO to assist during an armed conflict. Assistance was provided to Nicosia, Kyrenia, and Limassol, with some staff flown to Cyprus by the Royal Air Force so that drugs could be provided by veterinary surgeons. WSPA and others also supported zoo animals in Afghanistan and Iraq during the recent conflicts. Those actions required negotiating agreements which helped animals, but what about changing the very rules of war? That concept could significantly improve nutrition and health of livestock, and should be considered.

First, consider that since 1945 the UN Charter has reserved conflict to self-defence or as authorized by the UN Security Council. Once war breaks out, *Jus in bello* takes over, which deals with the morality of the conflict's conduct, and may offer potential opportunities to protect the nutritional and other health needs of animals. Under the proportionality principle of *Jus in Bello*, regarding tactics against combatants, accidental damage to civil property is permissible, even inevitable. However, 'unnecessary collateral damage' is illegal unless the property is a tool of combat, like a horse carrying a machine gun. This could be interpreted as making illegal the laying of cluster bombs in farms. In other words, intentional or reckless acts against truly non-combatant animals may be illegal. There is a long history of armies moving through farms, and that practice will continue, but it is worth examining if the law of war could protect animals from direct attack as well as from suffering malnutrition due to blockades. If it is illegal to use food as a weapon of war against humans, that should also apply for livestock. Winning such litigation will not be easy, nor will the writing of new laws, but protecting livestock under *Jus in bello* has moral weight. Protecting farms can be logically connected to the principle that people have a right to food and a standard of care. While those rights are not well articulated in law, they must be considered relevant to successful ligation. Further, court action could provide a precedent for parliaments to enact national legislation aimed at protecting livestock and other animals from armed attacks, in other words, we should for authority utilize the philosophy of the 1899 Martens Clause and look beyond treaty law and custom by considering principles and distates of public opinion (Ticehurst 1997).

Consider the 1954 Hague Convention for the Protection of Cultural Property in the Event of Armed Conflict (UNESCO 2010), which could be amended to include animals. The Convention covers 'immovable and movable cultural heritage'. An amendment could include zoos, and farms, which are central to jobs and food security. This also relates directly to the Rome Statute of the International Criminal Court, used against Sudan's President in 2009 when he was indicted for crimes against humanity, in that the starvation of humans or destroying livelihoods (jobs) is illegal. Protecting livestock also specifically came up in 1982 in the case of the relocation of Nicaraguan Miskito Indians, when their livestock was destroyed by the Government, which was at that time fighting counter-revolutionary activities. The Inter-American Commission of Human Rights required compensation for lost livestock.

These precedents should lay a legal foundation to develop fresh animal welfare laws. The reason is simple. The inability of people to return to their lands and follow their previous agricultural lives fosters conflict and social upheaval, as discussed in Lautze (1997). Consider for example, the livestock-related Borana tribal massacres of 2005 in Uganda and Kenya. They resulted from ongoing rivalries worsened by drought-stricken livestock herds upon which both tribes depended. For five consecutive seasons prior to 2005, the winter rains all but vanished in much of East Africa, triggering violent competition for water, grazing land, and food. The extended drought resulted in a 50–80 % loss of livestock, mostly cattle. Consequently, violent livestock raiding became a huge regional problem that extended to Sudan and Somalia (Lal, pers. comm. 2006).

10.6 Conclusions

Before deciding to intervene in an emergency, whether to provide feed or any other aid, a Study Team should design the goals of the intervention and identify the scope of the crisis and the resources and procedures required to meet those goals. A Rapid Needs Assessment will help, drawing on data from the field, as well as remotely sensed data and legacy data, which is information gathered prior to the emergency and used for comparison. Very likely in a major crisis like an earthquake or a war, humanitarian NGOs and organizations will also be involved and perhaps operating under the UN's cluster system, so therefore the skills associated with negotiation and diplomacy should be learned. In addition, when trying to provide feed, the effort should often be done in cooperation with humanitarian entities in order to reduce the potential for tension from hungry human victims, as well as to potentially share resources like the construction of water boreholes or logistical support like aircraft and ships that can transport feed pellets. Lastly, the rules should be improved permitting intervention of those wishing to provide nutritional and medical aid to animals in disasters and conflict of disaster management and war.

Annex: Definitions

Armed Non-state Actors There is no single, universally accepted definition. Some describe them as 'terrorists' or 'bandits' while the groups often self-describe as ‘freedom fighters’, ‘liberation movements’, or tribal warriors trying to right a perceived wrong. The point is that the groups use armed means to achieve their goals and do not operate within formal State structures (Decrey-Warner, pers. comm. 2012). The importance to veterinarians and animal welfare NGOs is that while national military forces or even uniformed rebels might follow predictable rules of behaviour, this phenomena is less likely with ANSAs unless separate bilateral agreements are reached, such as are being developed by the Switzerland-based NGO Geneva Call.

One working definition for an ANSA could be

> ‘any armed group, distinct from and not operating under the control of, the state or states in which it carries out military operations and which has political, religious, and/or military objectives. See also: Committee on Armed Services, “Inquiry into the role and oversight of private security contractors in Afghanistan”, Report together with additional views, US Senate, 28 September 2010’.

Disaster A disaster is a hazard that overwhelms or nearly so the capacity of a social unit to respond, whether city, province, or nation. A natural disaster occurs when caused by natural phenomena (i.e. a hurricane or storm surge, earthquake, etc.). A man-made disaster occurs when caused by human being (i.e. wars, conflicts, bushfires, etc.). It is important to note that a hazard is not a disaster unless it causes overwhelming damage to humans. For example, a hurricane or cyclone might indeed bring nourishing water to animals during a drought (Roeder 2013). Another way to understand this is to think of a hurricane, storm surge, tornado, drought, flood, earthquake, a tsunami, etc., not as a disaster, but only as natural phenomena. In certain circumstances, those phenomena can cause great, overwhelming damage; then the event (not the hurricane for example) is a disaster. Most of the time, they cause little or no damage and thus should be thought of as simply ‘hazards’. Similarly, while a full-fledged war might cause a disaster, lower levels of conflict could be damaging events that can be coped with, not true disasters.

Consider if a hurricane marches up a coast that is suffering from drought or forest fires. Bands of moisture propelled by cyclonic winds can reduce the fire or lay down moisture useful to pastoralists and farmers. In that instance, while the hurricane’s force would have be a hazard, the rain event might only be beneficial, not a disaster.

If the same hurricane crossed over land and hit structures that are weak, such as the shanty towns around Johannesburg, then the event might be a disaster, certainly for the inhabitants if they didn’t get enough early warning.[7] The same hurricane when set against stronger structures with more disaster management infrastructure,

[7] Early warning is more than knowing when a hurricane might cross the Atlantic to the Americas and the point of impact. It is also about data pointing to long-term drought, heavy snow and rains, flooding and security issues, or the spread of disease or harmful insects, the latter of which can come to Latin America from increased ocean temperatures that affect prevailing winds.

as in Miami, Florida, might not be a disaster at all. The hurricane might cause damage, but damage easily taken care of by municipal authorities. Lots of those happen, whereas Katrina in 2005 caused massive damage and was a disaster in New Orleans.

The lesson for animals, as it is for people, is risk management. We need to provide pastoralist farmers with more early warning appropriate to the species they care for. If money allows, we might also want to see stronger protective structures, designated evacuation routes, and trucks to haul small animals like sheep to safe, feeding locations, as is done in Cuba, perhaps pre-positioning of supplies, if time and money allows.

Public Diplomacy The definition of Public Diplomacy in this book was developed by Michael W. McClellan, Diplomat in Residence, the Gerald R. Ford School of Public Policy, Michigan:

> 'The strategic planning and execution of informational programs by an NGO to create a public opinion environment in a target country or countries that will enable target country political leaders to be comfortable with changing their political paradigm and thus make decisions that are supportive of animal welfare objectives'. (McClellan, pers. com. 2004)

This is distinct from lobbying government officials, especially legislatures directly to change laws, or traditional diplomatic initiatives with governments to gain permission to cross a border, or to negotiate with governmental representatives at the UN to change international law.

Rapid Needs Assessment This is an examination of an emergency in order to:

- craft a rapid assessment of damages, nutritional needs, and other issues, using on-the-ground resources and remotely sensed data;
- identify life safety issues for both animal and people;
- identify imminent hazards to animals;
- identify sources of food and water;
- identify potential evacuation routes and shelters;
- identify communication sources and problems;
- identify potential sources of veterinary assistance;
- determine the scope of damage;
- identify needs for external logistical support, such as airlifting feed pellets;
- identify legal and political hurdles to assistance; and
- prioritize response activities

The Sphere Handbook: Humanitarian Charter and Minimum Standards in Humanitarian Response The Sphere Project is a voluntary initiative of humanitarian agencies including the International Federation of Red Cross and Red Crescent Societies and major international NGOs which wish to improve the quality of humanitarian assistance and its accountability. Often used with institutional field handbooks, the Sphere Handbook is a standard text for all relief agencies (Sphere Project 2011).

Livestock Emergency Guidelines and Standards (LEGS) This is a crucial piece of literature for any organization wishing to engage in emergency work with livestock. The book contains international guidelines and standards for the design, implementation, and assessment of livestock interventions and mirrors SPHERE's effort for humanitarian intervention. Indeed, some SPHERE experts participated in the development of LEGS, including the Feinstein International Center of Tufts University, FAO, the International Committee of the Red Cross, the African Union, and Veterinaires sans Frontieres Belgium (LEGS 2009). The book does not address either wildlife or companion animals, but many of its useful standards would be valuable.

References

ACT Alliance (2010) SitRep Haiti—17/2010. ACT Alliance, Geneva

Bellal A, Casey-Maslen S (2011) Rules of engagement. Geneva Academy of International Humanitarian Law and Human Rights, Geneva

Brown ME (2008) Famine early warning systems and remote sensing data. Springer, Berlin

Danielson S (2002) Conditions improving at Kabul Zoo. National Geographic News. June 10, 2002. http://news.nationalgeographic.com/news/2002/06/0610_020610_kabulzoo.html. Accessed 30 June 2015

Davies G, Roeder L (2004) Report by standing committee on Native American disaster information affairs. Global Disaster Information Network (GDIN), Washington, DC

Decrey-Warner E (2012) Personal communication. Discussion on GenevaCall, interviewed by L. Roeder

Huertas G, Murillo JC (2007) Surviving the flood: a risk reduction process for the cattle industry in Sante Fe, Argentina. WSPA, Costa Rica

IWGAID [International Working Group on Animals in Disasters] (2008) Protecting animals from disasters. IWGAID, Paris

Keneally T (2011) Three famines. Public Affairs, New York, NY

Lal P (2006) Personal communication in discussion with L. Roeder on the livestock situation in Kenya and Uganda in 2005. Prakash Lal was Development Co-operation Coordinator, Fida International

Lautze S (1997) Saving lives and livelihoods: the fundamentals of a livelihoods strategy. Feinstein International Famine Center, Tufts University, USA

LEGS [Livestock Emergency Guidelines and Standards Project] (2009) LEGS livestock emergency guidelines and standards. Practical Action Publishing, Bourton on Dunsmore

Muncy L (2004) Darfur paragraph. US Department of State, Washington, DC

Pielke JR (2001) The role of models in prediction for decision. In: Cary conference IX: Understanding ecosystems: the role of quantitative models in observations, synthesis, and prediction. Carey Center for Global Good, Rensselaerville, NY

Roeder L (2005) Global consortium on tsunami recovery, Washington, DC. US Department of State, Washington, DC

Roeder L (2008) Protecting animals from disasters—a new perspective on humanitarian relief. In: Annual Conference of the Association of Arab Red Cross and Red Crescent Societies, Tunis

Roeder L (2011) Diplomacy, funding and animal welfare. Springer, Hamburg

Roeder L (2013) Diplomacy and negotiation for humanitarian NGOs. Springer, New York, NY

Sphere Project (2011) The sphere handbook 2011: humanitarian charter and minimum standards in humanitarian response. Hobbs, Southampton. http://www.sphereproject.org/. Accessed 19 Nov 2013

Stephens A (2013) United States to host South Sudan economic partners forum. USAID, Washington, DC

Talbot-Rosevear MG (1991) Email report: history of Kabul zoo. British Federation of Zoos, London

Ticehurst R (1997) The martens clause and the laws of armed conflict. International Review of the Red Cross, No. 317. Available at http://www.icrc.org/eng/resources/documents/misc/57jnhy.htm. Accessed 6 Nov 2014

UNESCO (2010) Convention for the protection of cultural property in the event of armed conflict—1954. http://portal.unesco.org/culture/en/. Accessed 28 July 2010

US Department of State (2006) Darfur livestock migration route, map 8. US Department of State, Khartoum

USAID [United States Agency for International Development] (2005) Darfur humanitarian emergency—March 4, 2005. Internal report. USAID, Washington, DC

USDA-FAS [United States Department of Agriculture, Foreign Agriculture Service] (2010) Haiti, Center Province, January 21. Livestock analysis, earthquake impact 2010. US Government, Washington, DC

Walsh J, Gannon R (1967) Time is short and the water rises. Dutton, New York, NY

Wollen, T (2015) Former Director of Livestock Advocacy at Heifer International, interview by Larry Roeder, 8/5/2015

Chapter 11
Animal Welfare, Nutrition, and Economics: Past, Present, and Future

C.J.C. Phillips

Abstract The nutrition of food animals has a major influence on their welfare. A direct link between feed provision and welfare is evident for farm animals, with undernourished animals demonstrated to have poor welfare. Currently, animal production systems are poised at a crossroads. Increased demand for animal products is encouraging the establishment of more intensive production units, such as battery cages for chickens or feedlots for beef cattle, which have adverse effects on animal welfare. However, increasing concern about intensive livestock production, including for their welfare, may reduce demand for products from these systems and encourage production of human food from plant material for an expanding human population. A key to this transition is the greater resource use efficiency of plant-based food for humans, compared with livestock production. The outcome will depend on governments' willingness to recognize adverse human and animal welfare implications of concentrating animal production into intensive units. Regulatory control will require more stringent systems of monitoring and controlling animal welfare than those currently available. A market-led trade in animal products will only sometimes produce the good welfare outcomes required by the public, and intervention may be required to regulate animal welfare and human food supply, as basic rights for those concerned.

11.1 Introduction

In the past, a close connection between standards of nutrition and animal welfare has been largely driven by the economics of food production. For the most part, underfed animals were less profitable than those given adequate nutrition. For example, beef cattle intensively finished with a mixed silage and concentrate diet grow more rapidly and reach a suitable weight for marketing faster if they are fed a high level of concentrates (Table 11.1). There is a substitution of concentrates for silage and the finishing period is kept to within a year for cattle fed the extra

C.J.C. Phillips (✉)
Centre for Animal Welfare and Ethics, School of Veterinary Science, University of Queensland, Gatton, Australia
e-mail: c.phillips@uq.edu.au

C.J.C. Phillips (ed.), *Nutrition and the Welfare of Farm Animals*, Animal Welfare 16, DOI 10.1007/978-3-319-27356-3_11

Table 11.1 Production characteristics of beef cattle fed high and low concentrate levels (Phillips 2010)

	High concentrate	Low concentrate
Concentrate intake (kg/day)	2.3	1.9
Silage DM intake (kg/day)	2.5	2.8
Liveweight gain (kg/day)	1.06	0.96
Slaughter weight (kg)	454	444
Finishing period (months)	12	14
Effective stocking rate (cattle/ha)	10.4	8.6
Relative gross margin (high concentrate = 100)	100	76

Table 11.2 Economics of growth in calves fed restricted milk replacer in buckets or ad libitum acidified milk replacer via teats, from birth to 3 months of age (Phillips 2010)

	Restricted milk replacer	Ad libitum milk replacer
Milk substitute intake (kg)	11–16	20–30
Concentrate intake (kg)	115–130	105–120
Hay intake (kg)	5–9	5–9
Liveweight gain (kg/day)	0.5	0.8
Relative cost per kg gain (restricted system = 100)	100	125

concentrates, offering the opportunity to finish a group of cattle annually. In this example, the gross margin is increased by offering more concentrate feed; the cattle are well fed and are more fully able to achieve their growth potential. The relative gross margins will depend on relevant prices at the time, especially of cereal grains and fertilizer. In other situations, increasing the level of nutrition can increase the costs of the growth of cattle. For example, this is likely if there is substitution of milk powder, which is expensive per unit of energy, for concentrates, which are cheaper (Table 11.2). However, from a welfare perspective, ad libitum feeding of milk replacer is likely to avoid the stereotyped sucking behaviour that is associated with restricted feeding (Vieira et al. 2008).

Although we cannot be certain of the cognitive responses to a nutrient-rich diet, it seems likely that there are welfare benefits from a diet that contributes to faster growth, increased energy for performing normal behaviour, less disease risk, and an absence of the abnormal behaviours that develop if nutrients are deficient (Phillips and Kitwood 2003). However, if the diets offered are too concentrated, based mainly on cereals for example, ruminant livestock are at risk of developing acidosis and other associated problems because of excess acid production in the rumen (see Chap. 8).

11.2 Livestock Welfare in Low Output Systems

Research in animal welfare economics is sparse. A concept that has been widely presented is the proposal that animal welfare will be optimal at medium levels of production, with reductions at low levels because of risks posed by undernutrition, predation, parasitism, and exposure to climatic variables, and also at high levels by highly intensive farming methods that constrain space allowance per animal (McInerney 2004). McInerney argues that a moderate rather than a low level of production leads to maximal welfare because of improved availability of food, shelter, physical comfort, health, safety, and social interaction. No empirical data are provided to support this contention, and it is likely that the outcome of any comparison would be context specific, i.e. in some situations welfare might be improved by increasing production, and in others it would be reduced. If increased production requires housing of animals and more intensive management, it must be remembered that, by analogy with humans at least, removal of the ability of animals to move where and when they want is a fundamental restriction on liberty. McInerney cites reduction in predation risk as a benefit of increased production. Although the threat of predation to semi-wild livestock is real, particularly to the neonate, it is unlikely to be perceived with persistent fear or anxiety except for very short periods of a high level of threat. Constant alertness would be maladaptive. The impact on welfare is therefore likely to be small. For example, rangeland cattle are at risk of predation by wild dogs, dingoes, and crocodiles in Australia. They may feel fear briefly when they go to waterholes as they are most likely to be attacked there. This functions to heighten their alertness and readiness to escape. Such escape is a normal behaviour and fear could be seen as contributing to their ability to satisfy the 'Freedom to perform normal behaviour', even though it creates temporary negative emotions.

McInerney controversially labels his low production level animals as in a state of 'natural welfare', which has the implication that animals' welfare is reduced in the wild compared with in systems managed by people. Charles Darwin recognized the benefits of life in the wild when he wrote

> 'We may console ourselves with the full belief, that the war of nature is not incessant, that no fear is felt, that death is generally prompt, and that the vigorous, the healthy, and the happy survive and multiply' (Darwin 1859).

Significant negative effects on welfare in low production systems are likely to be confined to situations in which animals are subjected to prolonged stress and distress, such as when individual animals are diseased. If these systems are situated in locations where there are major disease risks, such as ectoparasites in extensive tropical rangeland systems, it is likely that welfare can be improved with more intensive management, with regular use of ectoparasiticides for example. In northern Australia, buffalo fly (*Haematobia irritans exigua*) is an important welfare problem for cattle, and in the south, sheep blowfly (*Lucilia cuprina*) and the sheep body louse (*Bovicola ovis*) feed on the skin of sheep, the blowfly creating large oedematous lesions (Phillips 2005). Ectoparasites also affect welfare by disrupting grazing behaviour and creating itchy skin, which encourages animals to rub and bite themselves. The economic argument for controlling parasites rarely takes into account the welfare benefit (Whan 2005). Whereas a simple cost–benefit analysis argument is often used to advocate use of parasiticides, taking into account the cost of the chemicals, application, and the benefit in terms of increased weight gain, the on-farm situation may produce substantially less benefit than predicted from trials run by pharmaceutical companies. In addition, there are other options to control parasites than just chemical application, and an opportunity cost to investing in expensive treatment options (Whan 2005). There are complex relationships between the various diseases to which livestock are susceptible in rangeland conditions, mostly synergistic, and an integrated pest management strategy is advocated to address all potential disease risks.

Parasitism is a greater concern in extensive than intensive systems of livestock production, and an animal's susceptibility depends greatly on its level of nutrition (Valderrabano and Uriarte 2003). If it is satisfactory, particularly around parturition when susceptibility is increased, colonization by gastrointestinal parasites is much reduced compared with underfed animals. Thus, stocking densities must be set to provide for adequate nutrition at critical times, especially in low rainfall and low herbage growth periods. A key nutrient in this respect is the protein content of the diet. During gastrointestinal parasitization, the amino acid requirements of the gastrointestinal tissue are increased by the damage to and sloughing of the epithelial cell layers, leakage of plasma and extracellular fluids, and an increase in mucus production. This potentially leads to competition between the immune system and tissue requirements. At the same time, feed intake may be reduced and animals will preferentially select high-protein diets. Minerals are also important nutrients during parasite challenge to sheep, with phosphorus, cobalt, molybdenum, and copper all believed to have anthelmintic properties. Sheep reduce the depth of biting during parasite load, potentially reducing ingestion of larvae (Hutchings et al. 2000).

Indirect relationships between ectoparasites and the nutrition of livestock exist, such as the need to minimize diarrhoea in grazing sheep, which attracts flies (Phillips 2009). Ensuring adequate fibre in the diet, through supplementation with stored forages and restriction of consumption of young, leafy herbage, will help to achieve this. Some supplements are also likely to be beneficial, particularly those with condensed tannins. Plant poisoning will be more likely if the pasture is deficient in quality or quantity, and the consumption of bones of dead stock remaining in the paddock can lead to botulism or osteomalacia.

Drought management is central to ensuring that extensively managed 'range' livestock do not experience undernutrition. A strategy should be planned in advance of drought conditions, to include feeding supplements, agistment of stock, or sale of stock. Early action is vital, as transport and sale of stock that are undernourished is in itself a welfare problem. The effects of drought are prolonged and usually not felt by livestock until subsequent seasons. If the livestock are of low productivity, such as Merino sheep, several years of low rainfall are required to affect the animals, whereas high producing stock, such as dairy cows, can be affected by as little as

1 month without rain. Drought susceptibility can be predicted to a certain extent from rainfall records, although climate change is making this more difficult.

11.3 Livestock Welfare in Intensive Systems

Welfare problems are most associated with highly intensive production systems, as recognized by McInerney (2004). In these systems, animals may be subjected to unnaturally high stocking densities and group size, disease challenges, excessive demands on production and reproduction, and unnatural feed types. These problems are more common in, but are not confined to, pig and poultry production systems because their monogastric digestion system makes them suitable for intensive feeding systems, based usually on cereal grain. Cattle and sheep, being inherently less efficient at digesting high-energy feeds because of the involvement of micro-organisms and rumination to assist them in digesting low energy feeds, are less likely to be used for intensive production. However, if they are, such as when individual stalled cows are fed concentrated diets, they develop the same stereotyped behaviour problems seen in monogastric animals (Redbo 1992).

11.4 Regulatory Frameworks

The widely recognized 'Five Freedoms' are used to assess compliance with basic standards of welfare in many jurisdictions. For example, in the 2001 Queensland *Animal Care and Protection Act* (ACPA 2001), people have a duty of care towards animals, which obliges them to provide 'appropriate care' by providing for their needs in a reasonable way. This includes:

- providing food and water;
- providing accommodation or living conditions;
- understanding the animal's normal behavioural patterns;
- treating disease and injury; and
- handling the animal appropriately.

Notably, freedom from hunger and thirst is replaced by provision of food and water, but neither the quantity nor quality is prescribed.

At an international level, the World Organisation for Animal Health (OIE) has as one of its guiding principles that

> 'the internationally recognized "five freedoms" (freedom from hunger, thirst and malnutrition; freedom from fear and distress; freedom from physical and thermal discomfort; freedom from pain, injury, and disease; and freedom to express normal patterns of behaviour) provide valuable guidance in animal welfare' (OIE 2012).

The scientific basis for their recommendations is centred on welfare being assessed from 'the many elements that contribute to an animal's quality of life, including those referred to in the "five freedoms"' (OIE 2012).

In this context, it is relevant to ask whether 'Freedom from Hunger and Thirst' is the best way to describe welfare problems arising from nutritional management. Tolkamp and D'Eath (Chap. 2) and D'Eath et al. (2009) suggest that chronic hunger is experienced by monogastric animals that are restricted qualitatively or quantitatively. Hunger is defined as 'a negative subjective state experienced by an animal that is chronically undernourished'. The evidence that they provide includes increased feeding rate and motivation, as determined from operant tasks, abnormal behaviour, and increased activity in pigs and poultry that are underfed in quantitative or qualitative terms. However, Hogan and Phillips (Chap. 3) find limited evidence of chronic hunger in ruminant livestock. High-yielding dairy cows probably do not work harder than low-yielding cows to obtain high-quality concentrate feed (Cooper et al. 2010), suggesting that the notion of 'metabolic hunger' in dairy cows offered low-quality diets proposed by Webster (2005) requires evidence. There is, however, evidence of increased cortisol concentrations in cattle spending time off feed and water when transported, but it is likely that this is caused by the transport process (Hogan et al. 2007). Irregular feeding times do not adversely affect the growth or health of calves, although there is some evidence of frustration behaviours (Johannesson and Ladewig 2000). I conclude that there is only limited evidence of chronic hunger in malnourished animals, and there may be differences between ruminants and monogastric animals that require further study. Freedom from malnutrition would more effectively describe the possible welfare challenges associated with nutrition, than would freedom from hunger and thirst, or a requirement to provide food and water.

There are no international standards yet available that control nutrition for 'optimal' farm animal welfare, other than those prepared as guidelines by the OIE themselves. Acknowledged to be in their infancy, they have been established to provide international guidance that will be sufficiently broad to cover all systems in operation in the 178 member countries. For example, their guidelines on beef production state that 'Attributes of physical appearance that may indicate compromised welfare include ... dehydration and emaciation' (OIE 2013). In relation to nutrition of beef cattle, the OIE welfare standards state:

> The nutrient requirements of beef cattle have been well defined. Energy, protein, mineral, and vitamin contents of the diet are major factors determining the growth, feed efficiency, reproductive efficiency, and body composition.
>
> Cattle should be provided with access to an appropriate quantity and quality of balanced nutrition that meets their physiological needs. Where cattle are maintained in extensive conditions, short-term exposure to climatic extremes may prevent access to nutrition that meets their daily physiological needs. In such circumstances, the animal handler should ensure that the period of reduced nutrition is not prolonged and that mitigation strategies are implemented if welfare is at risk of being compromised.
>
> Animal handlers should have adequate knowledge of appropriate body condition scores for their cattle and should not allow body condition to fall outside an acceptable range. If

supplementary feed is not available, steps should be taken to avoid starvation, including slaughter, sale or relocation of the cattle, or humane killing.

Feedstuffs and feed ingredients should be of satisfactory quality to meet nutritional needs. If appropriate, feed and feed ingredients should be tested for the presence of substances that would adversely impact on animal health.

Cattle in intensive production systems typically consume diets that contain a high proportion of grain(s) (corn, milo, barley, grain by-products) and a smaller proportion of roughages (hay, straw, silage, hulls, etc.). Diets with insufficient roughage can contribute to abnormal oral behaviour in finishing cattle, such as tongue rolling. As the proportion of grain increases in the diet, the relative risk of digestive upset in cattle increases. Animal handlers should understand the impact of cattle size and age, weather patterns, diet composition, and sudden dietary changes in respect to digestive upsets and their negative consequences (acidosis, bloat, liver abscess, laminitis). Where appropriate, beef producers should consult a cattle nutritionist for advice on ration formulation and feeding programmes.

Beef producers should become familiar with potential micronutrient deficiencies or excesses for intensive and extensive production systems in their respective geographical areas and use appropriately formulated supplements where necessary.

All cattle need an adequate supply and access to palatable water that meets their physiological requirements and is free from contaminants hazardous to cattle health.

[There is a need to record] outcome-based measurables: mortality rates, morbidity rates, behaviour, changes in weight and body condition score, reproductive efficiency.

In the face of this information, it is important to consider what are the measurable components, and how can they be measured on farms, and by whom. The advice in the OIE standards will assist in providing a summary of basic knowledge about cattle nutrition and its relation to welfare, but it does not (yet) provide measurable standards, against which production systems and units can be evaluated.

11.5 Future Trends, Feed Availability, and Animal Welfare

By 2050, the world will need to feed an additional two billion people and will require approximately 60–70 % more milk and meat (2000 as the base year) if current levels of consumption are maintained (FAO 2009). This demand is driven by a growing world population, urbanization, and income growth. The current per capita consumption of animal products is much higher in developed countries than in developing countries. However, it is increasing in the latter at a much faster rate than in the former. As an example, in the last two decades the meat consumption per person has increased by 30 kg/year in East Asia. In some parts of the developed world, meat consumption has shown a decreasing trend. Both production and consumption of animal products are shifting from the North to the South—the contribution of Europe and North America to total meat and milk production has decreased in recent years. Currently, the contributions of South and East Asia to world milk and meat production are growing (FAOSTAT 2012). This is happening amidst increasing land, soil, and water scarcities, ongoing global warming and associated frequent and drastic climatic vagaries, and an anticipated increased competition for arable land and nonrenewable resources, such as fossil carbon sources, water, and minerals (especially phosphorus).

Good feed availability is a prerequisite to an efficient output of animal products. One consequence of the acceleration in this dietary and demand transition in the human diet has been an increase in the need for high quality animal feed. IAASTD (2009) pointed out that of the over one billion additional tonnes (relative to 2000) of grains projected to be required by 2050, ca. 40 % will be used for animal feed. There has been high and increasing dependence in the developing regions on feed concentrates. For example, the proportion of cereal grain used for animal feed in China increased from 7 % in 1960 to 22 % in 2007 (FAOSTAT 2012). This increase in the use of concentrates has seen the total quantity of compound feed being traded globally increasing to 959 million tonnes (Alltech 2013). This does not include forage or concentrate feed produced 'on-farm'. Pig and poultry systems are the dominant users of the compound feed. Grains are major components of monogastric diets and grain production is resource hungry. The resources are under severe pressure due to biophysical constraints noted earlier. Both availability and safety of feed supplies are at risk due to climatic extremes, such as floods and droughts, which can adversely affect animal welfare and productivity. Examples are the drought-related increase in the levels of aflatoxins in maize in North America in 2012–2013, and the severe fodder shortage in Ireland that caused the death of a large number of animals in January and February 2013, due to prolonged winter weather, in combination with a poor harvest the previous summer. With the continuing increase in environmental temperatures, a shift can be expected of forage plant types from C_3 to C_4 (Morgan et al. 2011), the latter having lower nutritive value than the former, while in the rangelands, more browse and plants that contain defensive compounds with anti-nutritional and toxic action are anticipated. It is expected that welfare of free-ranging animals will be affected unless carefully managed. Poisonous plants can impose serious health and productivity issues, not only by increasing the incidence of deaths but also through productivity losses, emaciation, and negative and chronic impacts on cells, tissues, and organs. In grazing regions, planting of drought-resistant and high temperature-tolerant crops could be a way forward to maintain or enhance productivity and to maintain animal welfare under the expected future conditions of global warming. At the same time, the use of browse plants and forbs containing high levels of defensive compounds, such as tannins and saponins, has been shown to be effective in the control of parasitism. Plants with anti-parasitic properties, indigenous to many parts of the world, have been reported by several authors (Makkar et al. 2007; Githiori et al. 2006; Hoste et al. 2006). Strategic use of such plants can have improved animal welfare and productivity benefits.

During extremes in weather, ruminant livestock can also experience particularly cold conditions as well as hot conditions. Comfort and survivability of animals under very cold conditions can be improved by achieving adequate body condition (or body energy reserves) by feeding nutritious diets before the cold weather arrives. It is recognized that some animals subjected to extreme cold conditions can to some extent compensate by lowering their basal metabolic rate.

The increased spread of weed species in many pastures around the world is leading to their colonizing novel regions, thus posing new threats to grazing livestock (Hogan and Phillips 2011). Frequent overgrazing sometimes accompanied by drought allow increased weed colonization of existing grazing lands. Animals may consume significant quantities of soil. Secondary compounds in plants have specific functions in the plants and do not necessarily cause acute toxicity, but can lead to chronic, subclinical effects in livestock. In turn, these may affect nutrient utilization, or liver and kidney function, without overt signs of morbidity. Use of appropriate supplements that block or attenuate the effects of toxins is possible, which could avoid animal health or welfare issues. An appropriate plane of nutrition will also improve the animal's ability to detoxify certain secondary compounds, thus reducing their potential negative impacts on the animal. Some soils have high salinity to the point that might challenge both the vegetation and ruminants reared on them. The welfare and productivity of animals reared on such soil types can be improved with the use of sodium-tolerant animal breeds and plants, the removal of all sodium provision in the mineral premixes, and the use of alley cropping of trees and pasture, to lower the water table and prevent salts accumulating in the soil surface strata.

The livestock sector has gone through a number of structural changes, the main being a movement from small-scale farmer systems to large and intensively managed livestock production systems, with increased incidences of some contagious animal diseases (FAO 2009). This trend is likely to continue in the absence of sound policies to promote small-scale and mixed-production systems, which will potentially adversely affect both animal welfare and human well-being. Another trend has been the increased global sourcing of feeds and their transport around the world, potentially leading to nutrient (including minerals) depletion at the place of origin and to nutrient loading at the place of their feeding. Long-term depletion of nutrients could lead to conversion of crop lands to rangelands, in which the quantity and quality of forages will be poor. Thus, an effective and efficient cropping system may, without adequate management and control of inputs and outputs, end up as a low output animal production system. This will affect not only the availability of grain but also animal productivity at the source and may cause reproductive problems in the animals due to mineral and other nutrient deficiencies. At the destination of grain exports, there may be loading of nutrients, especially at and around the site of feed use. This may lower animal welfare due to increased environmental pollutants around the farms, especially in developing countries because of lack of regulations for proper management of manure or of willingness and means to strictly enforce them. So the disruption of the nutrient cycle, if it continues to increase and in conjunction with a lack of public policies, may have negative impacts on animal welfare.

In some parts of the world increased pressure on grazing lands due to crop intensification will cause pastoralists to either change to crop production or be driven to poorer lands. Overgrazing of rangeland will cause degradation and as a result reduced availability of feed resources. This may lead to hunger in animals, giving rise to animal welfare concerns. This could be further aggravated by increased energy spent by animals foraging and in high ambient temperatures.

Reductions in stocking density could alleviate these stresses but are rarely achieved. Rotational grazing of land could improve animal well-being and productivity by preventing overgrazing of specific, favoured plants. Silvipastoral systems, with plantations of tree cover for shade, offer some protection for the animals, and they also increase plant biomass in the degraded land for animal consumption, thereby improving both animal welfare and productivity.

During the last three decades, inexpensive grain, energy, and protein enabled the intensification of livestock production systems, mainly poultry and pig production systems, including a 34 % increase in meat production from 2000 to 2011. It has resulted in the pig and poultry monogastric sectors contributing 75–80 % of current global meat production (FAOSTAT 2011; USDA 2013). The intensively managed monogastric sector relies heavily on feed ingredients such as cereals and soybean that compete with human food, and some of them are sourced from far-off places. The systems have become highly capital intensive, raising many environmental and animal and human disease challenges. It is unlikely that the growth rates of the past can be sustained in the future due to increasing costs and decreasing availability of energy, grains, and other inputs. The present-day animal-based human food production system needs redefining and restructuring if it is to sustainably contribute to future human population needs. Increased competition between food, feed and fuel, increased energy costs, and anthropogenic climate change will most likely lead to decreasing amounts of grains in animal diets. A number of animal welfare issues have been highlighted in the literature due to excessive consumption of grains by both ruminants and monogastrics (FAO 2012), and decreases in the level of grain use will help mitigate these problems and in many situations might increase productivity and profitability. Any feed- or feeding-related intervention that is supposed to enhance animal welfare will be difficult to apply on-farm if the farmers are not convinced that the intervention enhances profitability of their system. There are many examples from the field where animal nutrition-based strategies increased both profitability and animal welfare (FAO 2013).

11.6 Conclusions

Feed quality and supply and the ongoing structural changes in the livestock sector will have critical implications for livestock welfare over the course of this century. Changes anticipated to occur in response to increased requirements for human food will also have a bearing on animal welfare, which may deteriorate considerably if there is greater reliance on highly intensive production in battery cages, feedlots, and highly stocked housing systems. However, if governments can be persuaded to recognize the adverse human and animal welfare implications of using expensive resources, in particular grain, for expansion of livestock production at a time when the human population is expanding rapidly, there is hope that high welfare, sustainable systems of production can be found. Regulation and control of the impact of nutrition on animal welfare will require better systems than currently in use.

Acknowledgement The author is grateful to Harinder Makar, FAO, for contributions to this chapter.

References

ACPA [Queensland Animal Care and Protection Act] (2001) https://www.legislation.qld.gov.au/LEGISLTN/CURRENT/A/AnimalCaPrA01.pdf. Accessed 19 Nov 2013

Alltech (2013) Feed survey. http://www.alltech.com/news/news-articles/2013/01/29/alltech-feed-survey-findings-world-increases-production-959-million. Accessed 19 Nov 2013

Cooper MA, Arney DR, Phillips CJC (2010) The motivation of high and low yielding dairy cows to obtain supplementary concentrate feed. J Vet Behav Clin Appl Res 5(2):75–81

D'Eath RB, Tolkamp BJ, Kyriazakis I, Lawrence AB (2009) 'Freedom from hunger' and preventing obesity: the animal welfare implications of reducing food quantity or quality. Anim Behav 77(2):275–288

Darwin C (1859) On the origin of species by means of natural selection, or the preservation of favoured races in the struggle for life, 1st edn. John Murray, London

FAO [Food and Agriculture Organization of the United Nations] (2009) The state of food and agriculture—livestock in the balance. FAO, Rome

FAO (2012) Impact of animal nutrition on animal welfare. Report of an Expert Consultation, 26–30 September 2011, FAO Headquarters, Rome, Italy. In: Animal Production and Health Report, No. 1. 44 pp. http://www.fao.org/docrep/017/i3148e/i3148e00.pdf. Accessed 19 Nov 2013

FAO (2013) Enhancing animal welfare and farmer income through strategic animal feeding—some case studies. In: Makkar HPS (ed) FAO Animal Production and Health Paper, No. 175. FAO, Rome, Italy. 100 pp. http://www.fao.org/docrep/017/i3164e/i3164e00.pdf. Accessed 19 Nov 2013

FAOSTAT (2011) and (2012) From FAO statistical database, Rome. http://faostat.fao.org/. Accessed 18 Apr 2013

Githiori JB, Athanasiadou S, Thamsborg SM (2006) Use of plants in novel approaches for control of gastrointestinal helminths in livestock with emphasis on small ruminants. Vet Parasitol 139 (4, Special Issue):308–320

Hogan J, Phillips CJC (2011) Transmission of weed seed by livestock. Anim Prod Sci 51 (5):391–398

Hogan JP, Petherick JC, Phillips CJC (2007) The physiological and metabolic impacts on sheep and cattle of feed and water deprivation before and during transport. Nutr Res Rev 20(1):17–28

Hoste H, Jackson F, Athanasiadou S, Thamsborg SM, Hoskin SO (2006) The effects of tannin-rich plants on parasitic nematodes in ruminants. Trends Parasitol 22(6):253–261

Hutchings MR, Kyriazakis I, Papachristou TG, Gordon IJ, Jackson F (2000) The herbivores' dilemma: trade-offs between nutrition and parasitism in foraging decisions. Oecologia 124 (2):242–251

IAASTD [International Assessment of Agricultural Knowledge, Science and Technology for Development] (2009) In: McIntyre BD, Herren HR, Wakhungu J, Watson RT (eds) Agriculture at a crossroads global report. Island Press, Washington, DC

Johannesson T, Ladewig J (2000) The effect of irregular feeding times on the behaviour and growth of dairy calves. Appl Anim Behav Sci 69(2):103–111

Makkar HPS, Francis G, Becker K (2007) Bio-activity of phytochemicals in some lesser-known plants and their effects and potential applications in livestock and aquaculture production systems. Animal 1(9):1371–1391

McInerney J (2004) Animal welfare, economics and policy. Report on a study undertaken for the Farm & Animal Health Economics Division of Defra. vi + 68 pp. http://archive.defra.gov.uk/evidence/economics/foodfarm/reports/documents/animalwelfare.pdf. Accessed 19 Nov 2013
Morgan JA, LeCain DR, Pendall E, Blumenthal DM, Kimball BA, Carrillo Y, Williams DG, Heisler-White J, Dijkstra FA, West M (2011) C4 grasses prosper as carbon dioxide eliminates desiccation in warmed semi-arid grassland. Nature 476(7359):202–205
OIE [World Organisation for Animal Health] (2012) Terrestrial animal health code, vol 1. World Organisation for Animal Health (formerly Office International des Epizooties), Paris
OIE (2013) Animal welfare and beef cattle production systems. Chapter 7.9. In: Terrestrial animal health code. OIE, Paris. http://www.oie.int/en/international-standard-setting/terrestrial-code/access-online/?htmfile=chapitre_1.7.9.htm. Accessed 19 Nov 2013
Phillips CJC (2005) The effects of ectoparasites and their control on the welfare of livestock. In: Holdsworth P (ed) Ectoparasiticide use on sheep and cattle in Australia today. AVCare, Canberra
Phillips CJC (2009) A review of mulesing and other methods to control flystrike (cutaneous myiasis) in sheep. Anim Welf 18(2):113–121
Phillips CJC (2010) Principles of cattle production, 2nd edn. CABI, Wallingford, p 233
Phillips CJC, Kitwood SE (2003) The effects of feed supplements on the production, metabolism and behavior of dairy cows fed low levels of concentrates during lactation. J Appl Res Vet Med 1(3):187–195
Redbo I (1992) The influence of restraint on the occurrence of oral stereotypies in dairy cows. Appl Anim Behav Sci 35(2):115–123
USDA [United States Department of Agriculture] (2013) Livestock and poultry: world markets and trade. Foreign Agricultural Services, Office of Global Analysis, USA. 29 pp. http://www.fas.usda.gov/psdonline/circulars/livestock_poultry.pdf. Accessed 19 Nov 2013
Valderrabano J, Uriarte J (2003) Effect of nutrition in early pregnancy on the periparturient relaxation of immunity to gastro-intestinal parasitism in prolific ewes. Anim Sci 76 (3):481–489
Vieira ADP, Guesdon V, De Passille AM, von Keyserlingk MAG, Weary DM (2008) Behavioural indicators of hunger in dairy calves. Appl Anim Behav Sci 109(2–4):180–189
Webster J (2005) Animal welfare: limping towards Eden. Blackwell Scientific, Oxford
Whan I (2005) Economics of ectoparasites in Australia. In: Holdsworth P (ed) Ectoparasiticide use on sheep and cattle in Australia today. AVCare, Canberra, pp 58–62

Index

C.J.C. Phillips (ed.), *Nutrition and the Welfare of Farm Animals*, Animal Welfare 16, DOI 10.1007/978-3-319-27356-3

FSC
www.fsc.org
MIX
Papier | Fördert gute Waldnutzung
FSC® C083411

Kolibri 360 GmbH
Ferdinand-Jühlke-Straße 7
99095 Erfurt, Deutschland
produktsicherheit@kolibri360.de